FUTURE FIRMS

FUTURE FIRMS

How America's High Technology Companies Work

ERIC J. BOLLAND &
CHARLES W. HOFER

New York   Oxford   ▪   Oxford University Press   1998

Oxford University Press

Oxford   New York
Athens   Auckland   Bangkok   Bogota   Bombay
Buenos Aires   Calcutta   Cape Town   Dar es Salaam
Delhi   Florence   Hong Kong   Istanbul   Karachi
Kuala Lumpur   Madras   Madrid   Melbourne
Mexico City   Nairobi   Paris   Singapore
Taipei   Tokyo   Toronto   Warsaw

and associated companies in
Berlin   Ibadan

Published by Oxford University Press, Inc.
198 Madison Avenue, New York, New York 10016

Oxford is a registered trademark of Oxford University Press

Library of Congress Cataloging-in-Publication Data
Bolland, Eric J.
Future firms : how America's high technology
companies work / Eric J. Bolland, Charles W. Hofer.
p. cm.
Includes bibliographical references and index.
ISBN 0-19-510436-6
1. High technology industries—United States.
I. Hofer, Charles W.   II. Title.
HC110.H53B65 1998   97-49370
338.7'62'000973—dc21

1   3   5   7   9   8   6   4   2

Printed in the United States of America
on acid-free paper

For Whit, Anneliese, and Polly
—E.B.

For Judy
—C.H.

PREFACE

The subject of this book is high technology businesses. The aim is to provide an understanding of what these businesses do and what factors, if any, make a difference in their business performance. Also, the intent was to discover if high technology is really a grouping of similar kinds of businesses or if high technology companies differ among themselves. It is quite common to refer to the "high tech" business sector as a cohesive and distinctive collection of science and engineering-based businesses. Part of our aim is to explore that grouping and see if this is the right net in which to collect all the specimens.

Our look at high technology includes both relatively new ventures with less than five years of operations and the more mature, established businesses. This way, we have been able to evaluate changes as companies grow. Our interest is in privately and publicly held technology businesses and not governmental research and development laboratories or not-for-profit organizations.

This topic is worth exploring for three reasons. The first is that the subject itself is important. High technology workers make up 8 percent of the American workforce. High technology companies, especially software and computer companies, are our fastest growing firms.

The second reason is that a better understanding of high technology business is important for high technology business executives and managers. There are lessons in practical management to be drawn from this exploration. Identifying factors that may foster success and avoid failure are addressed in this book.

The third reason is that the subject of high technology business has not been comprehensively studied before. This book is an initial effort at broadly describing and analyzing the business field. Other authors and researchers have looked at more narrowly defined high technology businesses such as computers or have investigated certain phases of business development such as the venture phase. This book covers more completely the whole spectrum of high technology enterprises.

We have explored the subject through the survey research of high technology companies. We questioned both venture founders and established company managers, over three hundred in all, and collected information on business structure, operations, and performance. Other parties to high technology, the venture capitalists, economic development directors, and dozens of suppliers and customers, were also questioned. The addition of these answers provided a more complete depiction of high technology businesses and their environments. More complete data about how this information was gathered and analyzed are found in the appendixes. On-site visits to technology companies and our own work experiences also formed a basis for study.

The book is divided into four parts. Part one introduces the subject and defines terms. The history of high technology is presented and we visit the daily operations of four technology firms.

Part two is devoted to new technology ventures. The role of venture capitalists and their views on funding ventures is described. We explore the founders and their impact, where ventures come from, their business strategies, and their environment.

Part three compares ventures with established companies. We consider human resource issues and impacts, CEOs and their impacts, strategies, and results along with various issues in organization and the environment. Marketing operations are described and evaluated and we discover if company size makes a difference.

In part four, we look at where high technology businesses are located in the United States and describe state and federal support for high technology businesses. Conclusions about this business are also reported in this section.

The appendixes contain descriptions of the methodology and a guide to high technology resources for incipient entrepreneurs and managers of existing high technology companies.

E. B.
C. H.

ACKNOWLEDGMENTS

The authors would like to thank Corporate Technology Information Services, Inc., and Steve Parker in particular for providing the database of established companies. The database for technology ventures was provided by Venture Economics, Inc., through Jess Reyes.

This book owes much to the diligent research efforts of two young market researchers, Beth Hooper and Ali Muharremoglu. Renaldo Inamato collected information for the state survey.

Without the responses of over three hundred ventures and established companies, there would be no book. There were a number of people from these companies who made a special contribution. They are Robert Horsch of Agracetus, Gary Dalgaard and John Kessler of Isthmus Engineering, Laurie Mittelstadt and Kevin O'Connor of Hewlett-Packard, and Bill Reining of Reining International.

When the book was in its venture research phase, two people, Jim Johannes and James B. Wood, provided valued advice. G. Dale Meyer also contributed ideas.

There were many organizations that aided this effort: the state economic development offices as well as chambers of commerce. They provided us with information that would have been extremely difficult to obtain otherwise. The U.S. Small Business Administration also assisted.

Thanks also go to the University of Wisconsin School of Business and Associate Dean R. D. Nair, who helped facilitate the research on established companies. The library at Cardinal Stritch University provided updated literature searches. Nova Southeastern University faculty and Ron Needleman especially aided the initial research on ventures.

We also thank the technical entrepreneurs who reviewed drafts: David Fisher and Phil Elleray. Sharon Stark kindly assisted by providing incoming survey processing.

Our executive editor at Oxford University Press, Herb Addison, gracefully guided this book along, making substantive comments on many of the chapters. We were delighted to work with Herb.

CONTENTS

I

THE NATURE AND
BACKGROUND OF
HIGH TECHNOLOGY
BUSINESS

1

UNDERSTANDING
HIGH TECHNOLOGY
BUSINESS

What we have come to know as high technology is a very modern happening, at least in comparison to the conception and usage of the term *technology*. Yet it is readily identifiable as its own activity and distinctive in its business form. This chapter evaluates and constructs a definition of high technology business and thus represents our effort at reaching common ground on the subject.

A Pervasive but Obscure Form

The term *high technology* quickly summons mental images of computers, microscopes, and advanced aircraft, among other forms of technology. These images are peopled with figures hunched over computer keyboards or adjusting electron microscopes. We might also envision technicians at control boards remotely operating robots as part of this vision.

High technology's most obvious examples, like those mentioned, are easiest to conjure. They are the forms we constantly see in the media. They are also deceptive. They serve only as the most visible examples of a pervasive but obscure enterprise. What are some of the less visible examples? How about coal-burning electric power generation stations? Or garbage bins in recycling plants? Few would place these activities in the classification of high technology. Yet high technology is incorporated into these activities. Its manifestations may be more obscure, but high technologies are nonetheless present. Electric power plants are now controlled by distant computerized command centers and recycling plants have dozens of microprocessors controlling the separation processes. But these high technology control functions cannot be detected during a casual stroll through a power plant or a recycling center. They are largely hidden from view. So what high technology powerfully and rapidly conveys in immediate images it also shadows in obfuscation. Getting a fix on this activity is no easy prospect. It introduces the problem of going beyond the readily apparent to find what is at the core of high technology. To a certain extent, we need to reshape our thinking to understand this term. As we'll see, high technology is multifaceted and, in many other cases, a diamond in the rough, taking the form of a seemingly obsolete technology that has at its heart very advanced machines or processes. This is exactly the case in the examples of power generation and recycling just cited. Generating electric power or running a recycling plant are technology intensive, but the technology seems very basic. However, embedded within are state-of-the-art circuitry and complex electric-demand modeling programs that belie its mundane technological wrapping.

Pursuit of the immediate and apparent in high technology, as was done here by naming computers, electron microscopes, and aircraft as examples, does the injustice of displaying the iceberg's tip, not its base. In a cursory review of the subject, we may be able to list a few examples of well-publicized high technology firms, but this business classification contains dozens of other business activities that qualify as high technology endeavors. There are also subclassifications of high technology that peel away, onionlike, to show how technology bases of manufacturing processes support each other in increasingly complex layers. One of our main objectives is to show just how rich and diversified this field is, to tip the iceberg rather than simply take its tip as the whole. Doing this will help us get its true mass and dimension.

Another difficulty in approaching the subject of high technology is that the organizations that do high technology are obscure themselves, even in a physical sense. They don't occupy the terminus of rail yards nor do they herald themselves with boiling smokestacks. They often lie past grassy berms, around curved drives leading to single-story campuslike buildings, the very models of urbane corporate citizenship. You don't wear a corporate uniform or show rank as you do in traditional firms. You don't arrive and depart in a limousine. There is utter commonness in the way its practitioners dress, a commonness that adds to the difficulty of discovering which are high technology businesses and which are ordinary occupations.

These businesses are also concentrated in the well-known clusters of California's Silicon Valley and Massachusetts's Route 128. Other less-known concentrations are located near major research universities and are not as noticeable as the vast manufacturing swaths across the East and Midwest, the mining expanses of Appalachia, the agricultural empire that is the middle of America, or the timber baronages of the West. The concentration of high technology firms, a well-recognized characteristic that gives them notice, serves also to make them invisible to most of us because most of us do not live where high technology is concentrated. Geographically, high technology business occupy only a few thousand square miles, a mere duchy among the industrial and manufacturing empires.

Even when they are concentrated, they do not give us a monolithic impression of a uniform industry that is in the same stage of overall development, a stage that may give a singular impression of newness or of industrial power if it existed. Silicon Valley has its genealogy stretched out from Stanford University's research park in Palo Alto to San Jose and beyond. The early technology firms are closer to Stanford and the later firms are farther away. The high technology equivalent of being born in a log cabin and becoming president is being born in a garage and becoming a major corporate power, as Hewlett-Packard did. This company actually did start in a garage. That structure is now a local landmark and a nativity scene that other high technology firms proudly replicate by having their own garage gestations.

The steps along the way to corporate success are the occupation of larger and larger buildings in Silicon Valley. So there is a mix of structures there as companies either make it to prestige locations or falter and lose their leases. Different software and engineering firms show up along the valley with different architectures and in varied landscapes. But they do not give the impression of corporate evolution into a giant center of industrial concentration. They are nothing like the striking megacomplexes of industrial America that appear elsewhere. High technology businesses are discrete, not ostentatious, further adding to their mystique. The people they have made wealthy, the nouveau techno-rich, are not, for the most part, conspicuous consumers. They are a well-behaved second generation of postwar industrialists, both mannered and poised, still waiting for the torch to be passed. For the most part, we have learned their names only recently, more than a decade after high technology became its own business class.

The concentration of high technology is also deceptive. While it's true there are concentrations of these businesses in certain geographical areas, the fact remains that most high technology firms are not located in Silicon Valley or along Route 128. Most are in smaller pockets around other university cities, and many are freestanding, very small, independent firms all across the United States. This is really where the bulk of the "grunt work" of high technology is done. One might even characterize this version of high technology as having a garage genesis and evolution into a cottage industry. As our topic develops, just how decentralized high technology is will become evident.

The work setting of high technology clearly distinguishes it from other types of businesses. This demarcation line is on one side of methods and processes. And that side is methodology and process dominant. High technology work is plainly not the same as manufacturing work. It is what Thomas Hardy said of the past: it being like a foreign country, they do things different there in high technology land. You'll see more people following laboratory-like procedures. Big binders with testing procedures and source codes will be lined up on the top of the cubicles of software engineers. The work of high technology is marked by precision and procedure. If there is any distinctive incongruity about the business it is the regimen of technique practiced by legions of professional workers who are pridefully informal in appearance and personal demeanor.

High technology abounds in most business operations yet it is often difficult to locate exactly where it is happening. For all its impact on the economy as a whole, its physical dimension is often quite small. The work of high technology is done with electrons and DNA. Even when the scale is large, it is often distant, as in the case of space exploration. In this business, the products of high technology are flung across the universe to be replaced by more sophisticated spacecraft. We are lucky to see their faint glow in orbit, but we'll never see them up close. This is a business of the minuscule, not the mighty. Consider some of its recent advances— micromachines and gene splicing—and you'll see how scale figures to differentiate this business.

Peopling High Technology

There is also the human energy of high technology. This is an ideas business, not a physical labor business. In fact, high technology is aimed at the obliteration of useless physical labor, its metaphoric and actual antipode. The men and women who are engaged in high technology are cut from a different cloth, almost literally, as they move about in semi-conductor clean rooms, cool and sweatless in antiseptic "bunny suits" that make them seem the nemesis of good, honest laborers. The research and development laboratories of high technology are inhabited by our most educated workers. Production workers who do the repetitious job of cutting layers and layers of silicon are engaged in delicate and specialized work. Almost all are women who have been trained to nimbly operate precision machinery. The professional employees are also highly specialized in particular branches of engineering or science, adding a panache to their business we don't see in more commonplace occupations. Go into what most people would identify as a high technology business and you will not see a cross section of the total labor force. You'll see something quite different. Chances are that the demographic skew will be toward the younger side. Chances are, too, that you will not be able to typecast workers based on appearance. The salespeople will not be obvious nor will the usual showing of authority be based on dress. Stories abound about the informality of Edson deCasto at Data General or the barefooted Steven Jobs at Apple, and these stories are not exaggerations. They are accurate depictions of the milieu of high technology work. The informality of its practitioners belies the intensity of the work itself. As we explore the day-to-day work life of these firms, we'll show how long hours and commitment are part and parcel of this activity. We remarked about the commonness of high technology workers. It is a quality that hides high technology professionals outside their work environment yet identifies them within the workplace.

Another difference between high technology businesses and other businesses is the absence of a corporate bourgeoisie. In high technology firms that have both a research and development and a production facility, there is a two-class structure with an upper echelon of engineers and a low rank of clean-room production workers who are often minority, semi-skilled laborers. The machines these workers use may be state of the art and their tasks may require high dexterity, but the educational requirements for the jobs are minimal.

In some of the larger, more diversified high technology businesses, there are more mid-level positions. On the engineering side these are electronic technician positions and on the administrative side they are a variety of support positions such as purchasing agent, financial analyst, and technical writer. High technology may be hard to find in companies that produce goods and services, but it is especially hard to find in continuous-process companies. We have to look hard in continuous-process businesses to find high technology. Natural gas and water companies use advanced microprocessor flow control systems, but these companies do not register with us as high technology firms. Their technology is tucked away, invisible to all but trained technicians. For companies that produce goods, we have an easier time finding evidence of high technology in the product itself. We may be able to find the chip in our calculator, but water is water and gasoline is usually nothing more than a ringing pump at the station. You simply don't see the computerized control systems back at the refinery that blend your brand to its combustible reliability.

The Intrigue of It

Much of our fascination with high technology has to do with its attractiveness. Businesses that grow are attractive and high technology has grown. Its most visible member, the computer industry, had a growth streak of thirty years.[1] You can't argue with success and most admire it.

There is also the fact of the hidden high technology operation nestled within a larger corporate structure. Corporate Technology Information Services, which publishes the CorpTech Directory of Technology Companies, estimates there are 8,914 such operating units within larger corporations and independent private companies. This amounts to nearly a third of the organization's total directory listing of technology firms, the leading such directory in the United States. Some of these operating units may be research and development units, but others are essentially independent "skunk works" type operations that could be spun off by the parent firm as an independent company. In a way, these units can be thought of as incipient high technology firms that enter the commercial market with products or services that have been proven internally after a well-funded gestation period within a well-established larger company. Naturally, these hidden high technology units are found in larger firms, not small independent companies, so they can be discovered by the intrepid. They do, however, further demonstrate that we are dealing with an often obscure activity.

If we were to consider high technology as a god-given miracle for making all kinds of labor easier in many different fields, we would need to encounter both its pantheism and its unique character. It is itself seemingly unique and ubiquitous, an inexplicable contradiction. It is an activity of a select group of different professional workers, yet also increasingly intrusive in even the most mundane of businesses. It is hard to imagine how any business could survive without computer technology, a technology that is the very essence of high technology. These are observations that will let us better frame the whole portrait, a depiction of this dynamic and important activity that has some peculiar definition problems.

Definition Stakeholders

Since the ab initio point for serious study is defining the subject, we do so now. That search begins where it would for anyone curious about the term.

Who should be curious about the term *high technology*? In addition to researchers whose careers are based on the discovery of truth, practitioners of high technology business have something to gain by having a common definition. These practitioners are better served if their art is defined, if it has what most people might call an occupation, or sets of tasks and outcomes. It would be difficult to see high technology as a viable activity without such organization connected to some type of business definition. Those close to the practitioners—their capital and raw materials suppliers, as well as their customers—need to know what the term means.

The capital suppliers most often take the form of venture capitalists. There is a subgroup of the larger group that specializes in investments in the field of high technology. It is plainly important that the high technology investment area have some kind of definition for these venture capitalists to use. We'll see later how they operate, but for now we simply want to stress the value of defining their art.

The suppliers of high technology are many and varied. Some may literally be raw material suppliers such as the silicon substrate manufacturers for the microprocessor business. Or they may be the providers of very expensive electron microscopes for the biotechnology business. In either case and the many others that lay in between, actual and potential high technology suppliers can benefit by a definition of what the field encompasses. Can suppliers accommodate this business sector by providing like goods to all high technology businesses or are some types of business different from one another? Establishing a definition helps get answers to questions such as this.

Because of the explosive growth of media interest in the field, it is equally important that journalists have a definition of the subject. Most of us get our information about high technology from journalists, so it is especially important that they come to terms with a definition of high technology. High technology subjects such as new software releases, cellular communications, and the Internet are frequently cover stories for the business press, and the subject is constantly being covered. So accurate reporting of high technology based on a general understanding of its characteristics is needed. Indeed, coverage of high technology subjects is unrelenting and will remain so in the future. As an example, *Scientific American* forecasted in September 1995 that key technologies would continue to emerge in such diverse areas as optical networks, intelligent software, high-speed rail, gene therapy, self-assembling materials, and solar energy. These developments will certainly attract considerable public interest that, in turn, should be a function of a technologically adept media.

The practice of high technology involves complex social and organizational interdependencies, and in that setting, a common language is not only desirable but also necessary. Some of these interdependent organizations are the educational institutions that produce high technology's scientists and engineers, the business organizations that employ them, the governments that tax and regulate their market activities, the capital and raw material suppliers, and the consumers and buying organizations that purchase their goods and services. These entities cannot communicate and interact with one another in a broad field of high technology if there is no commonly accepted definition of the term or if the accepted definitions contradict one another. So the quest for a definition is well justified.

Organizational theorists look at business systems as being "closed" or "open," with open systems having many more and more influential external forces that impact operations. High technology organizations are truly open business systems that are dependent on many factors, both internal and external to the organization, some of which we have described earlier. We would, in fact, be hard pressed to find any other business organization with such a high number and level of interdependencies. This will be demonstrated later, but it is worth remarking that having a definition of high technology would help facilitate the transfer of capital, labor, and knowledge between the high technology organization and its environment.

Defining High Technology

A modest immediate goal of this chapter is to arrive at a usable definition of high technology business. That will be done in steps, by deconstruction of the entire expression of *high technology business* into its parts, starting with *high technology*. Then the definition will be populated with example firms.

The search for a definition of high technology begins and ends with failure in the pages of most dictionaries. The search is short because dictionary definitions are very few. High technology simply cannot be found easily. Since the dictionary is the first reference for most people curious about the term, it seems evident that people do not have access to a convenient and accepted definition of the term. Thus, the definition problem is understandable. However, the *Random House Dictionary of the English Language* in its second edition offers the illumination of high technology as being, "any technology requiring the most sophisticated scientific equipment and advanced engineering techniques as microelectronics, data processing, genetic engineering or telecommunications." The etymology of the definition is traced to 1965–1970, a notation we will explore later in uncovering the history of high technology.

The definition contrasts high technology with low technology, which is said to be "any technology utilizing equipment and production techniques that are relatively unsophisticated." This definition originated between 1970 and 1975. With high and low technology we have the complements of yin and yang, but we do not have the contrast, the symmetrical curve that makes the boundary between the two.

Neither of the definitions truly pave the way for an understanding of the high technology business, although the *high technology* definition offers examples of true high technology businesses. Both definitions focus on *technology* without elaboration of that term, thus necessitating a further definition.

This definition of *high technology* positions it at the apex of lesser technologies. It implies that only technologies requiring use of the most sophisticated tools qualify as high technology. Putting the definition into operation poses difficulties. It would require setting both a scale for technologies from lowest to highest and selecting a point near the top where high technology resides and where lower technologies are absent. Neither of these requirements are met in the practical world of high technology operations. There is no standards manual with a ranking of technologies that engineers refer to when considering how their particular technology might rank. The issue simply has no bearing on them because it is inconsequential for engineers to come up with a word that describes how their business is linked to other technologies. They are more concerned with how the program works or if the machine functions. There is also no commonly agreed upon demarcation point between high and low technology anywhere inside or outside science and engineering, in spite of its frequent usage in the business and popular media. The theme of media stories is the new forms of high technology, not contrasts with low technology. And since the coverage is frequently about leapfrogging technologies and new advancements, the viewer, listener, or reader is left without a way to sort out the higher and lower forms of technology. High technology can then only appear as an incremental improvement over yesterday's progress and thus a nuance rather than a fundamental leap. The comparative base, low technology, is almost always missing from media accounts.

To further compound the matter, the application of a technology rather than the inherent technology may make the development seem to be high technology in one case and low technology in another. The microprocessor in a greeting card playing a simple tune is not on a par with the same microprocessor used in a cockpit avionics alert system. However, the microprocessor in the aircraft appeals to us as high technology while we are inclined to dismiss the greeting card example as diversionary entertainment.

The definition we've cited amounts to saying that technology uses technology. That is not rigorous enough for our purposes and far too close to being a circular tautology. There is too much dependence on the degree of technology to convince us it is a satisfactory definition. We need to differentiate the degrees of technology to make use of this definition. That is a difficult task because such differentiation is subjective and transitory, most often a function of one researcher's impressions at one point in time. Instead, we need to look more closely at the elements of high technology to arrive at a usable definition.

A starting point is discovering what technology is. Then we can tackle the difficult matter of assigning it a degree, as we must to distinguish high technology from its variants. So we continue to deconstruct the term in eventual pursuit of a practical and research-based definition of high technology business.

Our same reference for high technology defines *technology* as "the branch of knowledge that deals with the creation and use of technical means and their interrelation with life, society and the environment drawing upon such subjects as industrial arts, engineering, applied science and pure science." This definition at least offers a base of identified interactions. It gives us something to build on, and that is where we take our cue for building a definition.

Unlike the case of defining high technology, there are many definitions of technology available. These are found in dictionaries and textbooks alike. They echo the theme of using science-based knowledge to improve life. Technology historians have also developed working definitions, one of which has simple elegance belying prolonged study of the subject: technology is "the varied body of knowledge and devices by which man progressively masters his natural environment."[2] It is a definition that leads to something tangible because the definition mentions devices as an element of technology. This adds to the more intangible aspect of technology—the body of knowledge it produces. Technology produces tools but also the knowledge needed to create tools. Thus the definition does not attempt the difficult job of distinguishing science from technology. Derry and Williams pose technology as the servant of humankind. It is not science, which is aimed at discovering the physical laws of the universe, irrespective of humanity's harnessing those laws to better the world.

What is still absent from the depictions of technology is a system for clearly indicating what particular activities are high technology pursuits. At this point, we need to depart from the dictionaries and look for ways we can use definitions in the real world of high technology. Here, we must look for a classification scheme.

Classification of High Technology Businesses

Although not a definition of high technology, there is a classification scheme for high technology. This classification method tells us what types of business operations are high technology and what the operations do rather than what they are intrinsically. It does move us closer to what is needed, however.

Business organizations being complex, ambiguous, and mutating, they cannot be forced into precise alignment such as in the Periodic Table of Elements. Yet there is a logical arrangement of the many types of business activities that exist. A taxonomy of business organizations is found in the Standard Industrial Classification (SIC) system of the Office of Management and Budget. Within the major divisions of Agriculture; Forestry and Fishing; Mining; Construction; Manufacturing; Transportation, Communications, Electric, Gas, and Sanitary Services; Wholesale Trade; Retail Trade; Finance, Insurance and Real Estate; Services; Public Administration; and nonclassifiable establishments, as well as 89 subdivisions, virtually all business enterprises can be found. The subdivisions take the form of more narrowly defined types of businesses. As an example, in the Services division, Legal Services is the more narrowly defined business subdivision.

The SIC uses numbers to notate business types. The leftmost two digits indicate the division and the rightmost two digits indicate the subdivision. Thus, the four-digit code is as precise as the SIC gets about business types. Private companies such as Dun & Bradstreet have added digits for further specification of business types, but the private variations on the theme are not as widely used as the basic SIC system.

It is within the SIC system that we can try to ferret out what businesses are high technology. And what we find is elusive. Virtually every subdivision (three-digit industry group number and simple-digit industry number) can exhibit some element of high technology. Even the lowly 3295, Minerals and Earths, makes use of computer-based proportional, integrative, and derivative (PID) weighing systems as a crucial part of its operations. Technology, even high technology, is invasive in all industries, a point that was made earlier but bears particular relevance here.

You must also consider that assignment to one SIC over another depends on which activity the establishments in the industry are "primarily engaged in," according to *The Standard Industrial Classification Manual*. There is evident difficulty in classifying large, well-diversified, strategic business unit–structured firms such as General Electric. Firms such as General Electric are engaged in diverse businesses. The obvious solution of combining SIC codes in such cases obscures the true activities of a diversified firm and thus defeats the purposes of industrial classification. These firms may not be primarily engaged in high technology when business divisions are aggregated into a SIC identified entity, but a number of their business units may be totally dedicated to high technology, a feature that becomes diluted through upward aggregation.

The problems with SIC classifications of high technology companies have been recognized in a study of high technology entrepreneurship sponsored by Marquette University. Some categories such as Analytical Instruments or Electromedical Equipment may be correct designators of high technology, but other categories are more obscure. The study observed that using SIC codes alone to pick high technology firms runs the risk of denigrating analysis because there are a small number of true high technology firms that are detected by the SIC method.[3]

Why then use the SIC as a basis for extracting a definition of high technology? One answer is in the universality of SIC. It is known and used by business people and academics alike in the United States, though it is not a worldwide system. It is also a system with sufficient scope to classify all types of business organizations. An examination of businesses captured with SIC codes include the arcane (within 3291, Tripoli), the minute (within 3295, Fuller's Earth), and the thankfully obsolete (3292, Asbestos Products).

The SIC system is a logical system as well, having groupings of somewhat similar establishments located within more generally described product headings. For the benefit of researchers and librarians, SIC publications have cross-references of alphabetical headings and numerical listings of short titles.

An additional useful feature of the SIC system is that it is the sole parent of a variety of proprietary industrial classification systems. As will be shown, there are private organizations with their own business classification systems that can be traced back to certain derivative SIC classifications.

Still another endorsement of the value of SIC is its application in the field and study of marketing. Almost invariably, marketers use the SIC as a way of segmenting their markets. It is quite often a first step in marketing segmentation.

Credit must also be accorded the SIC on the basis of its sheer ambition. The system is "intended to cover the entire field of economic activities," according to the SIC manual. Any classification method with such a broad scope must at least be recognized for its aim of avoiding errors of omission. If the field is economic activity, chances are good the system will collect a large population of economic production units and that high technology organizations will not be missed.

Without the SIC, the field of economics—particularly industrial organization theory—would be stranded on ambiguous conceptual shoals. Though not having the depth of the Linnaean biological classification system, the SIC has the breadth to give most of the social sciences a tested reference point for myriad studies.

The SIC has endured over time and has been modified over time. As new types of businesses emerge, the federal Office of Management and Budget relies on a host of subcommittees to suggest additions and subtractions to the system. Many of these suggestions are adopted periodically. Considering that our subject matter, high technology, is constantly changing as new technologies replace old forms, this regular revision of the SIC system is a welcome accommodation of reality.

There is a challenge in extracting high technology organizations from the vast skeletal protoplasm that is SIC. On one hand, the brief descriptions of what businesses do in each SIC code do not contain enough clues to indicate if the representative firms actually do high technology. On the other hand, once you dig into the enormous base of SIC-based industrial data that reside in dozens of government and private agencies, there is so much descriptive material about the activities of these firms that it is difficult to decide exactly what may be crucially defining criteria for high technology.

Another task faced by those wanting to find high technology within the SIC system is that the organizational unit needs to be specified. It is not enough to define *high technology*. A high technology business unit needs definition, too. The SIC sets up enterprises as establishments with more than 50 percent common direct or indirect ownership. This is parenthetically described as a company in the SIC manual. The individual establishment does not have this requirement and can be thought of as more like a family-owned small business. Thus the classifier of high technology organizations has to meld *high technology* and *organization* amid the definition and classification nuances of the vast SIC system.

The solution devised by Riche, Hecker, and Burgan (1983) was to look at the composition of the workforce and the amount of resources spent on research and development. In a series of three successively constrictive then more relaxed definitions, these authors succeeded in distinguishing high technology operations from less technologically intensive businesses. Instead of taking an output approach, which would involve dissecting the high technology part of the product or service offered by each organization, the authors instead focused on the business process itself and examined the transformations from inputs to outputs. This is a systems approach to understanding organizations that has its basis in organizational theory. There are two elements used to characterize high technology organizations: the percentage of technical workers and the percentage of research and development expenditures to sales revenues.

The three definitions are collected into these groups:

Group 1. Industries in this group employ a proportion of technology-oriented workers greater than 1.5 times the average for all industries or 5.1 percent of total employment. Industries with fewer than 25,000 employees were excluded and this resulted in a list of 48 three-digit SIC industries. Three out of four industries on the list are manufacturing industries.

Group 2. Industries in this group display ratios of research and development expenditures to sales greater than twice the average for all industries, or a minimum of 6.2 percent. Nonmanufacturing industries were excluded from the analytical cut owing to data limitations. This screening resulted in only six three-digit industries.

Group 3. Industries placed in this group had to satisfy criteria concerning both the relative research and development expenditure and the proportion of technology-oriented workers. The proportion of technology-oriented workers had to be greater than the average for all manufacturing industries (6.3 percent), and the research and development to sales ratio had to be close to or above the average for all industries (3.1 percent). The authors also excluded some industries on the basis of their products. The result of this criteria was 29 three-digit industries.

The Group 3 definition has come closest to becoming a universal definition of high technology businesses. The Brookings Institution constructed a similar definition using a Dun & Bradstreet database. This definition includes 96 four-digit SIC industries. There is another definition by Glasmeier, Hall, and Markusen (1984) at the University of California, Berkeley. This effort used data from the Bureau of the Census and resulted in 99 four-digit industries that qualified as high technology organizations.

A compilation of various definitions of high technology was done in 1985 by the Commerce Department (Hatter, 1985). The report uses a definition developed by the department's Lester Davis that is somewhat different from the others we have reviewed. Davis estimated the technology intensity of industry production based on research and development expenditures needed to make the product. He also included research and development expenditures for unfinished products, not just final products. Davis further required that research and development expenditures on products be significantly greater than the average for all industries, not simply above average, another departure from the other definitions. However, Davis did use the SIC code as the basis for groupings.

The National Academy of Science has also contributed a definition in which industry research and development expenditures, as well as employment of scientists and engineers, defines high technology industries.

A definition of high technology by the National Science Foundation is based on the number of scientists and engineers employed in research and development, along with company research and development expenditures, as a percent of total sales. There is still another Commerce Department definition based on research and development expenditures relative to shipments.

Some writers have added variations on the theme as a means of describing high technology. Rogers and Larsen (1984) have added two conditions to the Office of Technology Assessment definition. They assert that high technology has a "fast rate of growth" and "a worldwide market" (p. 29) for its products. While the first of these additions is evident in high technology businesses, the second is not. True worldwide markets are more characteristic of commodity goods than of the products of high technology firms.

There are those, too, who have struck their own course and gone the route of simplification in trying for a definition of high technology.[4] A definition by the Jacob France Center of the University of Baltimore divides high technology into classifications of very high technology, moderately high technology, and somewhat high technology. Definitions of high technology are now varied by the number of industries included and the intensity of high technology activities.

Within the financial community, *high technology* has come to mean something as well. The term refers to the group of technology-oriented companies trading on the exchanges. A now extinct magazine *High Technology* once set and reported on high technology stock in an index within the magazine. The index was composed of 35 business groupings, such as large computers, robotics, laser and infrared equipment, process and industrial controls, test equipment, genetic engineering, home computers, and telecommunications equipment. The magazine would track performance of these stocks in comparison to the Standard & Poor's index.

In the myriad definitions and classifications, we find no precise, free-standing definition, but instead see a series of relativistic embracements of the term that suggest high technology operations have relatively more technical workers and relatively more research and development expenditures. We can find problems in this. For example, biotechnology businesses frequently have zero sales. We'll show how the pursuit of patents for biotechnology firms often obliterates the search for product sales and thus renders research and development to sales a meaningless definer of high technology for these companies. Some high technology firms such as Chiron spend more on research and development than sales bring in.

The research and development to sales criterion also has the problem of linking the qualification of high technology company to what American industry as a whole is spending on research and development and sales. As overall sales decline and research and development are constant, more industries can become high technology without any additional, or possibly even a reduced, research and development expenditure. The same consideration is applicable to the technical employment factor in the definition.

The Office of Technology Assessment has noted that the commonly used definitions are not satisfactory in three respects. First, because they are based on SIC codes, individual firms are not considered. The code does not differentiate among firms on the important characteristics of size and structure. The code only reflects similar firms that produce similar products. Second, because the codes are product oriented, high-tech process industries can be excluded or misplaced. Third, service firms tend to be buried in an inappropriate code. For example, software engineering services are contained in the larger computer programming code even though they may be highly specialized software engineering companies that serve only industrial engineering needs.

The Department of Commerce report previously noted (Hatter, 1985, p. 36) raises two issues regarding high technology definitions. One concerns the data used to make the classifications of high technology and the second has to do with the now familiar core problem of any definition of high technology—its transitory nature. In the first instance, the data used to identify products and employment "are available for only very broad product groups. These product groups are often so broad they include products with significant differences in technological intensity." In the second instance, a fixed product list will become outdated as products go through their life cycles. The abbreviated product life cycle for high technology is one of the defining criteria for the activity, as we'll soon see. So rapid change both clarifies high technology and confuses it.

What is also quite apparent is that the problems of defining high technology have not been confined to academe or government. Some who have corresponded with *High Technology* magazine have complained that the SIC-based high technology definitions are founded on the primary business activity that is taken to be the output of the business. The input side is ignored even though it may be the locus of advanced technology. Many other business press articles have taken this output view of high technology, and so there is little wonder that we do not have a widely held definition.

All the definitions we have reviewed gravitate toward the end result of high technology activity: the goods, the services. The definitions largely rest on research and development expenditures and on technical workforce composition. These criteria are applied to a sliding scale that continuously redefines high technology. These considerations pose serious reservations about attaining the most basic goal of an operational definition.

A Definition at Hand

Are we now in a morass of the indefinable or is there a way we can find a coherent subject to examine? The importance of the subject pushes us to a usable definition, and there are two choices here. One is to formulate an original definition of high technology and the second is to adopt or modify an existing definition. The first option requires both the creation of a definition and the rejection of existing definitions. Because many aspects of existing definitions are highly usable, it would be ill advised to reject former definitions. As a consequence, the approach will be to select the best of what is available and employ it with modification if necessary.

For our purposes, we will build from the Group 3 Bureau of Labor Statistics definition. It represents the most systematic refinement of employment and research and development parameters, a refinement that is needed because other definitions are far too broad. This definition is also consistent with the Brookings and University of California definitions. In short, we come closer in this definition to having the beginnings of a commonly accepted definition. We note as an item of curiosity the fact that the definition emerged from the academic and government side rather than the practitioner side. As an explanation, it may be that no businesspeople have felt comfortable with the daunting challenge of defining not only their own business but hundreds of others as well. It's understandable, too, because the task of businesspeople is to do business, not tell us how they would classify their occupations amid thousands of other close and distant businesses.

Who then can call themself a high technology business according to this definition? The industries that meet the research and development and employment thresholds of Group 3 are:

Natural gas liquids
Industrial inorganic chemicals
Plastic materials and synthetics
Drugs
Soaps, cleaners, and toilet preparations
Paints and allied products
Industrial organic chemicals
Agricultural chemicals
Miscellaneous chemical products
Petroleum refining
Ordnance and accessories
Engines and turbines
Specialized industry machinery except metalworking
Office, computing, and accounting machines
Electrical transmission and distribution equipment
Electrical industrial apparatus
Radio and TV receiving equipment
Communication equipment
Electronic components and accessories
Miscellaneous electrical machinery
Aircraft and parts

Guided missiles and space vehicles
Engineering, laboratory, scientific, and research instruments
Measuring and controlling instruments
Optical instruments and lenses
Surgical, medical, and dental instruments
Photographic equipment and supplies
Computer and data processing services
Research and development laboratories

These 29 industries fall into the Group 3 definition yet we can easily pick out the discordant within the harmony. Guided missiles and space vehicles, computer and data processing, and research and development laboratories certainly fit, but do soaps, cleaners, and toilet preparations? Or paints and allied products? As the definition is pushed to more narrow limits, the problem of having seemingly low technology industries in a high technology grouping remains. Even the most constrictive Group 2 definition includes soaps, cleaners, and toilet preparations (along with office, computing, and accounting machines; communications equip ment; electronic components and accessories; aircraft and parts; guided missiles and space vehicles as the only other Group 2 members).

Based on a review of all the industries in the Group 3 category and what products they produce, we chose to exclude both soaps, cleaners, and toilet preparations and paints and allied products. The product groups for these industries have little intrinsic technological sophistication and, just as important, are not used principally by high technology firms.

As a needed supplement to these industries, we'll elect to include the biotechnology industry since this is a serious omission in the SIC system. Businesses in the biotechnology field are research dominated and clearly exceed the minimum technical workforce and research and development spending percentages. The result of this addition and the two subtractions is a list of 28 industries that qualify as high technology industries.

Natural gas liquids
Industrial inorganic chemicals
Plastic materials and synthetics
Drugs
Industrial organic chemicals
Agricultural chemicals
Miscellaneous chemical products
Petroleum refining
Ordnance and accessories
Engines and turbines
Specialized industry machinery except metalworking
Office, computing, and accounting machines
Electrical transmission and distribution equipment
Electrical industrial apparatus
Radio and TV receiving equipment

Communication equipment
Electronic components and accessories
Miscellaneous electrical machinery
Aircraft and parts
Guided missiles and space vehicles
Engineering, laboratory, scientific, and research instruments
Measuring and controlling instruments
Optical instruments and lenses
Surgical, medical, and dental instruments
Photographic equipment and supplies
Computer and data processing services
Research and development laboratories
Biotechnology

We make no claim that this is a definitive list of high technology industries. New industries may join the list as new technologies are discovered and used as a way of organizing a business. Such was the case with biotechnology, which was not in the SIC system only a few years ago. The point of having any listing at all is to have at least an anchor group of technologically intensive industries to use in exploring the unique character, if any, of this economic sector.

The types of enterprises we find with this application of the definition offer real face validity. They appear to be true high technology classifications, and this simple but necessary review furthers our confidence in using this definition as an operational bridge to our subject.

In all, this definition is suitable albeit imperfect. We take our cue from the history technology itself and not reinvent the overworked wheel but instead refine it.

Locating High Technology Businesses

The issue now is how to operationalize the definition, how to get names of high technology firms from the digits of SIC. Fortunately, others have tackled this particular challenge, and we are their beneficiaries.

We have noted that there are various commercial databases that have been developed as variants of the SIC system. These have been instigated as improvements over SIC, usually to simplify SIC classes or better capture service industries. One of these databases will be used to explore new high technology firms in their early years of operation. This source is the Venture Economics, Inc., database of venture organizations. There will be more detail about Venture Economics later.

Venture Economics gives us industry codes that correspond closely to two three-digit SIC classifications. Venture Economics has a Computer-related firm code of 2000, which is similar to the SIC 737 (computer and data processing services), the Other Electronics Related firm code of 3000, and a Biotechnology code of 4000, which has no direct

SIC analog. The lack of a broad biotechnology classification is a distinct disadvantage of the SIC system and reason to search for an improvement of the SIC system.

Another source used for finding high technology companies is the *CorpTech Directory of Technology Companies*, published by Corporate Technology Information Services. There are approximately 35,000 United States–based manufacturers or developers of high technology products listed in the directory. SIC codes are cross referenced in the directory, but Corporate Technology Information Services claims a superiority over SIC by asserting that for every SIC code there are about 20 codes that it uses, thus providing a richer classification system. According to the directory's overview, the technologies included are advanced materials, factory automation, biotech, chemicals, computer hardware, defense, energy, environmental, manufacturing, medical, pharmaceuticals, photonics, computer software, subassemblies and components, telecommunications, test and measurement, and transportation.

One thing to notice about this commercial high technology database is that, for the most part, it is a product-oriented classification system. Almost all the technologies listed are ones we can associate with a tangible good. Biotech, computer software, and factory automation have strong service elements, but overall we get the sense of material goods in this directory. That is a feature that makes it easier to confirm that these are truly high technology businesses, not purely service-oriented occupations. More details will be provided later about Corporate Technology Information Services and its connections with high technology firms.

The *CorpTech Directory of Technology Companies* is our primary source for established high technology companies and the Venture Economics database is our primary source for high technology ventures. Together, they enable us to follow the maturation of high technology enterprises.

In a sense, we are only half as far as we need to go by arriving at a definition of high technology. The other half of the task is to successfully graft business onto high technology to produce a definition of high technology business.

A broad definition of business typically designates it as an activity or an organization that aims to create profits by providing products and services. Fortunately, defining business is far easier and the result far better understood than is the job of defining high technology. It is practically teasing the superfluous to even pose a definition. Nonetheless, there are a few definition variants worth mentioning. We relax the usual definition of a business as essentially a profit-seeking venture and allow organizations that pursue high technology processes without a singular orientation to profit. This enables us to accommodate firms in the business of innovation, not necessarily profits. This broader definition also reflects the fact that profit is as alien as a distant universe for high technology companies in their early venture periods. In this respect, they are not too different from most other ventures where seeking profit is playing a waiting game.

As we relax this part of the definition, we constrict another. We require that the business be an actual commercial enterprise rather than a concept of a business or even a legal entity with articles of incorporation. Our high technology business needs to have a product or service in the market. There needs to be cash inflows and outflows, some financial performance and actual management of the enterprise. In addition, our expectation is that the high technology operation will have a location of operation. There are no minimum size requirements, although the operation must have at least one person associated with the operation. Other than that, there are no minimal requirements. We do, in short, expect our high technology operations to be in business. They must have economic meaning, not theoretical musing.

The result of our explication of high technology and business having been done, we can now move to a generalized definition of high technology business. High technology business is defined as *an activity and/or an organization that seeks to provide products and services that incorporate advanced technologies or are created through technologically intensive processes.* Because there is no commonly accepted definition, it is appropriate to introduce this as means of coherently exploring our topic. The generalized definition provides us with a plain-language definition that is not entwined in the problems of SIC and SIC surrogate-based depictions of high technology enterprise. While the Bureau of Labor Statistics Group 3 variant definition we have developed will serve as the more complete definition for this book, the generalized definition can be transferred outside this book and used for related purposes.

The search for a definition of high technology has led to the discovery that our subject is somewhat elusive. This happened in an atmosphere charged with vague ideas about what high technology is. There are general notions about what high technology is, but they are inadequate for our purposes. Most of us could sort out high technology firms from traditional firms based on simple information about what the firms produce and what kind of workers inhabit it. The devil, however, is in the detail and the parting of water molecules, not oceans. Our general sorting of high technology firms from more conventional firms doesn't help us when a new high technology organization emerges, and this is a business that reinvents itself as it itself invents. With only general ideas about high technology, we can but classify a particular new example based on its similarity to established forms. The Group 3 definition variation helps us in some respects, however. We can look at the employment and research and development substructure of even a new form of a high technology business and call it a high technology firm. We have adopted an operational SIC code–based definition as well as a generalized definition. These let us proceed with the study. Considering the difficulty inherent in the subject, that is progress. As we look at high technology businesses, we'll occasionally return to the definitions to see how well they apply and how well they serve to help us understand the business of high technology.

Now that we are done with our noun, high technology business, we ought to give some consideration to the associated verb. What do we call the process of doing high technology? We have here an activity that, unlike most others, combines science, engineering, and the social arts. We'll confront this by saying that the practice of high technology may emphasize one or more of these activities at different points, but our overall consideration is with the business of high technology, which can be best described as an art since there is no scientific law that guides how people organize themselves to do the work of high technology. It's appropriate, then, to consider high technology as the output of the system and engineering and science as the processes by which the output is created.

With our definition of high technology business in place, we can turn to the history of the field. We will then enter the daily lives of the businesses and see if the life of high technology is what it has been conceived to be or if it something quite different. Or, quite possibly, we will not cast the dichotomy so distinctly and will need to fabricate another understanding of this form of business that has so captivated public attention with its promise of world competitive leadership, better living, and mastery of the physical world. In any case, our view of high technology business is a glimpse into the future. The companies that do high technology are truly future firms. What they do and what they invent will be passed on to mainstream businesses, an occurrence we alluded to when we brought up the examples of power generation and recycling firms. So it is well worth exploring this special form of business in all its quite varied dimensions.

With both an outline and a definition, we're prepared to look at the origins of high technology. This is an essential part of this exposition because, as dynamic as these businesses are, they all owe something to their roots. These roots are the basis for our understanding of contemporary high technology.

Notes

1. Causes for this are technological advances and burgeoning demand, as stated by J. R. Norsworthy and S. L. Jang, *Empirical Measurement and Analysis of Productivity and Technological Change: Applications in High Technology and Service Industries* (North-Holland, Amsterdam: 1992). They add that dramatic price decreases and better computer performance originated in the technological advances within the semiconductor industry.

2. This definition by T. K Derry and T. Williams, in their *A Short History of Technology* (London: Oxford University Press, 1961) reminds us that technology is both the knowledge and the results of knowledge in stating that technology is a body of knowledge and the devices born from that knowledge.

3. Based on discussions with Paul Reynolds and Tim Stearns of Marquette University, August 1994.

4. One such simplification by M. Maidique and R. Hayes is in the Summer 1985 *McKinsey Quarterly*. It proposes that companies that spend more than 3 percent of sales on research and development are high technology firms. However, these authors note that only five industries would qualify as being high technology under the definition. They are chemicals and pharmaceuticals, machinery, electrical equipment and communications, professional and scientific equipment, and aircraft and missiles. As we'll see from the diversity of firms that engage in high technology, this is a far too restrictive definition.

References

Glasmeier, A., P. Hall, and A. Markusen (1984). "Recent Evidence on High Technology Industries Spatial Tendencies: A Preliminary Investigation, University of California Institute for Urban and Regional studies as reported in the Office of Technology Assessment." *Technology, Innovation and Regional Economic Development OTA-STI-238*. Washington, DC: U.S. Government Printing Office.

Hatter, V. (1985). "U.S. High Technology Trade and Competitiveness Staff Report." *Office of Trade and Investment Analysis*. Washington, DC: Department of Commerce.

Riche, R. W., D. Hecker, and J. Burgan (1983). "High Technology Today and Tomorrow: A Small Slice of the Employment Pie." *Monthly Labor Review*, November, pp. 50–58.

Rogers, E., and J. Larsen (1984). *Silicon Valley Fever*. New York: Basic Books, 1984.

2

HISTORY OF HIGH TECHNOLOGY

Can there really be a history of high technology when the term itself applies to the most advanced and probably short-lived forms of technology? Can you have a history of something so seemingly transitory? Or when the definitions of the term are so varied? We answer affirmatively by viewing high technology as a series of successive improvements in technology, with no known end point. The history is then the look at past technologies, with the value of seeing if there are common lineages and possibly direction in those lineages. We consider both the words and the deeds of technology—how the words came about and what activities characterize high technology work.

Emergence of High Technology through Organizations

When did high technology begin? When did technology join hands with the business organization? Who first uttered the words *high technology* and how did they become ingrained in our language? These questions drive our present exploration, and we provide answers by tracing events and forces that culminate in our modern understanding of the term. An expedition into the evolution of the term will explain why we have come to the point of having this business classification. It will also lead to consideration as to whether the term remains meaningful. If the special circumstances that created high technology no longer exist, then perhaps the entire classification has no value. A look at the history is a look at the circumstances of high technology creation, so it is instructive to do so.

To make a point about the complex and interwoven set of concepts and activities that became high technology business, we need to do a brief review of history and extract those defining incidents and advancements that created the high technology business organization. History is where technology and business have joined.

For all of our history as humankind, technology has been with us. But *high* technology is nothing more than the last tick of our clock. Technology is as old as modern humanity and perhaps older. Meave Leakey and Alan Walker reported in 1995 that walking hominids strode eastern Africa at least 4 million years ago. These were not Cro-Magnons and perhaps not truly human. Their technologies are unknown, but their close descendants certainly employed some devices to move, kill, eat, and find shelter. Even primitive weapons and tools were needed to survive. Much later the Neolithic, Bronze, and Early Iron Ages were times when small iron ploughs and axes were the greatest manifestations of technology. These tools were fashioned by individual craftsmen. What we see in early history is that technology is the province of individuals rather than organizations. The tools are individual tools used by their creators for the benefit of family.

Later, technology served community purposes. There is some evidence of an early connection between technology and organizations in 3500 B.C., when the walls of Jericho were built. One of the western world's first examples of a civil engineering project, these walls were the result of structural technologies and public edict. People and materials were organized to produce a walled city under direction of Canaanite kings. That the product of technology was destroyed by the arts—in this case music—marks an early and poetic rendering of the dialectic between science and the arts.

Later and greater civil engineering projects occurred with the building of the Pyramids and the Great Wall. The great Greek and Roman cities, roads, aqueducts, and irrigation projects demonstrate how visible and enduring technical accomplishments can be when state resources support technological innovation.

These large-scale civil engineering tasks were developed by empires and kingdoms. But magnificent as they were, they are still a long way from the modern high technology firm. Nevertheless there is a heritage, a theme that continues in high technology today. The great public projects demonstrated how human labor could be combined with technology to produce devices and structures that endure and are used to this day. What should not be lost sight of is this: as monumental as these projects were, they were not isolated tributes to gods or humans but the progressive development of new ways of improving life—exactly the definition of technology we cited earlier. That observation is important, especially as we marvel at the artistry of their execution. In ancient Mesopotamia, the shapes and ways of laying bricks were different from now, with the tops of the bricks curved. Better clays and kilns improved this simple technology until it made possible the great buildings. Together with newer stone cutting tools, the stunning European cathedrals rose. But all the time the brickwork was the technology and all the time it evolved. This is not meant to be a tangent on ancient brick making but a reminder that technologies continuously improve. However, they improve only to a certain point. Then they have substitutes, just as stone cutting replaced bricks for monumental buildings. The history of technology is one of both technology refinement and technology substitution. This is a key concept in Richard Foster's *Innovation: The Attacker's Advantage* (1986). He uses examples such as the substitution of steam power for sail power and the transition from cotton to rayon to polyester tires to illustrate his contentions about technology substitutions. In essence, technologies develop slowly over time, then accelerate rapidly and finally evolve slowly again, an S-curve. Near the most mature peak of a technology, a discontinuity occurs and a new technology emerges to follow another S-curve. We look at this idea later in the book to see if it applies to modern high technology.

When we try to answer the question of when technology became a function of the business organization, we need to look at when the first business organization was formed. Technology, as we have defined it, predates the business organization, so we are grafting the business organization on to technology in this brief review. This is important because for most of our history, technology was a separate entity, something that could be put aside after labor. The harness of business organization necessarily grappled with a dynamic activity that could not be well contained. Things changed. Technology spread quite rapidly during the industrial age. It shed the shackles of social class and nation, creating the bourgeoisie and setting up a worldwide search for resources. Technology was increasingly tied to daily life as people adjusted themselves to the machine, not vice versa.

Rail transportation developed quite rapidly in industrialized countries. It was not easily bound by organizations, either governmental or business, and this is key to appreciating the evolution of high technology. The rails became community goods, needed for business and community development. When businesses died out, the railroads were still needed. It may even be true that advanced forms of technology still transcend the business organization, because there are now those who forecast the extinction of the corporation but there are none we know of who forecast the end of high technology. So high technology preceded and may outlive the business organization.

When did technology join with business? Although the business firm existed in functional form, if not actual legal organization, from the charters of kings in the Middle Ages, it is not until the industrial revolution that the particular dynamic of capital—textile automation, steel production, and steam transportation—combined to stimulate the creation of large, privately owned companies. Whatever might be the future of these business organizations, they were certainly in their element as sophisticated organizations using sophisticated technologies in the late 1800s.

The early industries were substitutes for mercantilist occupations rather than fully developed industries. There were no large, multibusiness firms until the latter part of the industrial revolution. But the successful companies turned into capital machines that could fund research and development activities, as did the Krupp Firm and the duPont Company in the nineteenth and twentieth centuries. These big firms incubated and closely held what we might consider high technology activities today. The Krupp factory laboratories experimented with steel alloys. Chemical firms also had major laboratories. In short, technology improvements came from industry.

One particularly memorable reaction to the industrial revolution and the growth of technology was not against the social and economic transformation itself but the technologies that accompanied the industrial revolution. The Luddites destroyed the mechanized loom and not the forces that destroyed their work lives. Their hate was directed at the machine alone, though they understood its implications. It was not until Marx that the rampage of the industrial revolution was put into category of historical determinism and worker anger was explained under the labor theory of value. Until then, it was the machine, the high technology robot of its time, that took the brunt of worker hatred.

The other edge of the technology sword, the side that cuts the social fabric, is a drama nearly three hundred years old. And this is the edge that has become even sharper toward the end of this century, leading to a reevaluation of the benefits of technology and its social and environmental costs. That's a major factor that started in the 1960s, however. For most of our nation's history, technology has been seen as liberating.

High technology within business organizations—or, relatively speaking, the high technology businesses of their time—came into being when big business was big enough to sponsor internal research and development functions, as suggested earlier. Or it happened when small research companies were founded around certain breakthrough inventions, usually by single inventors. When AT&T became a truly national and prosperous company, it set up its internal research and development operation. The same applied to Standard Oil. The consumer and durable goods companies such as the automobile manufacturers internalized research later.

In more recent times, technologies have become enmeshed with human organizations to the point where individuals have been lost in the process. We can rarely identify the sole technologist—a genius such as Edison, Firestone, or Bell. Even when we can, they are almost as notable for their solitary selves as they are for their achievements. Some of the names— Edison, Westinghouse, Marconi, DeForest, the Wrights, Goddard—are not contemporary. They are the inhabitants of history. Yet companies, industries, and other huge enterprises were built around the innovations of these individuals. We now have joint discoveries and multiple patents and can barely identify the innovative giants. The American space program is a perfect example of the merger of organizational systems and technological advancement, now in the hands of hundreds of unheralded technical specialists. Aside from Wernher von Braun, there are no individual inventors associated with this adventure. And von Braun himself never published a scientific paper, a point his Peenemunde boss Walter Dornberger delighted in mentioning as von Braun eclipsed him in America. Von Braun succeeded only in the context of his engineering team. Even he did not build the V-2 alone. He and fellow members of Dornberger's team did.

Technology and Culture

The evolution of what we now call high technology happened on many different fronts for a longer period of time than may be immediately apparent. It also had a cultural dimension about which we find critiques in the time between the World Wars.

A kind of proto-history of high technology can be found in the popular culture beginning as early as 1925. The popular culture treatment starts with articles about technology rather than high technology. *The Reader's Guide to Periodic Literature* first lists Technology as a subject in that year. What follows in the next few years are a handful of articles about industrial production and related subjects. The first purely intellectual slant on technology was taken by radical historian Charles Beard, who wrote about a government by technologists. This effort resulted in the word *technocracy*, reflecting Beard's theme, and this word so seized public discussion that the *Reader's Guide* created a category of Technocracy just to accommodate the articles that followed. By 1932, articles for and against technocracy reached a peak. The technocracy movement advanced the idea that technology was wholly beneficial and that it was the job of people to adjust to it rather than vice-versa. In the Soviet Union, it was actually practiced by Gosplan planners, though the exigencies of world conflict ended all that.

Some of the most troubling aspects of technology took hold very early. In 1936, an article on how services can absorb employees lost to factory technology clearly described the down side of automation (Independent Woman, 1936). In film, Fritz Lang portrayed the dark vision of a future *Metropolis* where men serve machines. In *City Lights*, Charlie Chaplin starred in a factory were workers have mechanical lunch feeders. The seamy side of industrialism was hammered at by the Progressives, Lincoln Steffens, and Sinclair Lewis, and was mocked by H. L. Mencken. This period does appear to be when the human issues regarding technology were raised most intensely—more so than now, at least in terms of literary passion. However, for the most part, articles of this era were positive and welcoming of technology. People were told about how it would end manual labor, about the prosperity it would bring, and about how it would create a world community. It's only the intellectually intrepid who questioned technology at all.

We turn now to some of the more immediate roots of high technology, the growth of which accelerated quite quickly in the late 1930s.

Stages to Modernity

Many of the technologies we think of as contemporary actually emerged during World War II. Such is the case with radar, atomic energy, television, large-scale use of penicillin, and jet aircraft. The postwar era accelerated development of these technologies, but it did not initiate them. Even what we might regard as the ultimate technology—space travel—actually emanated from the Germans of Peenemunde, whose V-2s tore through the troposphere into space. The United States then brought in von Braun and the key engineers. The moon-bound *Saturn* of 1970 was a variation, albeit a dramatic one, of the theme in which "our" Germans beat the "Soviet" Germans.

The emergence of high technology in the United States can be best understood as happening in three periods. The first was from 1940 to 1950, a period marked by warfare technology; the second was from 1950 to 1975, marked by atomic and aerospace technology; and the third was from 1975 on, marked by the computer. Of course, leave room here for any further phases that may occur as long as high technology is a separable and identifiable endeavor.

The division into these periods is based on major shifts in the scope and focus of high technology activity within the postwar time frame. It is not simply convenience that makes these cuts, but several distinct groupings of high technology and an evolution from one stage to the next.

The initial phase is marked by the war effort's dominance of the field. High technology was war technology. The results of this technology were intense and horrible, but they were confined to small areas of battle— the V bombs over London, the Me-262 jet over Germany, and radar on the coasts and in only a few military vehicles. Most civilians were not touched by these technological advances except for the atomic bomb, which created its own culture.

It is true that World War II interrupted technological development. Some high technology firms were up and running before the war.[1] But the fact is that the war pushed technology ahead on many fronts, with parallel progress in each. It did so even if most were not exposed to technology products. This first phase set the foundation for what we now recognize as high technology. Indeed, World War II was the tragic epic that gave rise to a whole spectrum of enabling technologies. These technologies were so advanced that they spawned rapidly accelerating technologies that produced long-term, even permanent changes. The most evident example is nuclear energy, which harnessed fission. That enabling technology having proven feasible, the path was beaten to the fusion bomb just five years later. Another example is the transition from reciprocating aircraft engines to jet engines. Still another is the development of radar or the use of liquid fuel rockets. Proximity fuses, rubber substitutes, and the emergence of synthetic fibers all came in this period.

One of the clearest indications of the aims of technology during the war years was the Kilgore-Patman Act of 1943, which was passed by Congress and had the explicit intent of mobilizing science for the war effort. The conscription of science meant that its discoveries were directed only toward destruction or defense during the war years.

Another way that the war produced an acceleration of technology was that it internationalized the whole matter of weaponry. The Allies and Axis governments were in global competition for better weapons, clashing with each other every day. When a new weapon of war fell on the field of battle, it was quickly dissected by the enemy and replicated, the Norden bombsight being the example here. Allied bombardiers were sworn to protect it with their lives, but it was retrieved by the Germans in a downed B-17 and they raced to duplicate it themselves. So the war became a showcase for technology as the finest and best weapons were sent as quickly as possible to the field and then reverse engineered as quickly as possible when they were captured. Unlike the technology of the old age, weapons technology was on the world stage.

As C. P. Snow (1962, p. 116) wrote, "Technological history, even in the confusion of war, is more precise and ruthless than any other kind of history." His particular warning was that the blindness of technology advocates, in his mind F. A. Linderman, allowed for secret and unethical employment of destruction during war. The hubris of these advocates is that they thought that superior technological weapons would solve problems.

In the United States, the dark side of technological progress was blindingly displayed first in the New Mexico desert. It was a grim technology that defined an age, the Atomic Age, that will remain as long as we do, giving endorsement to Robert Oppenheimer's awed recitation of the *Bhagavad Gita* at the Trinity site, "I am become death, the destroyer of life."

It was during World War II that scientists such as Linderman in England and Oppenheimer in the United States were given considerable power, and in their pursuits they demonstrated the very destructive side of technology—a side so dark that no innovation can pull us away from it.

It's been said that World War II's acceleration of technology is its distinction. That is not exclusively true, though. Edison's invention of the "magic lantern" accelerated the formation of an entire industry. And the industrial revolution, spurred by technology, accelerated virtually everything at that time. What this war did do was make human extinction possible through technology by launching a quarter-century of defense-oriented weapons development.

High Technology Emerges

By sheer inertia and postwar fixation on the Communist menace, much defense-sponsored technology research continued. Still, there was a pronounced shift toward private business sponsorship of technology projects, as we now describe. The second phase largely occurs in the domain of aerospace, nuclear energy, electronics, and later, computers. This focus sets the tone for this phase, between 1950 and 1975.

The business of defending the United States certainly predates the genesis of high technology, but there is no other activity that so dramatically accelerated the pace of high technology. To appreciate the push of defense on high technology, we need to look at the characteristics and growth of defense and then its interactions with high technology.

The business of war making has gone on for as long as there has been a United States. In the immediate postwar period, war making developed its own characteristics, which one researcher has divided into eight categories (Gansler, 1981): the cyclic nature of procurements; a lack of structural planning; inadequacy of industrial-preparedness planning; lack of industrial readiness; the importance of technology and research; the differences among industries that serve as military providers; high concentration within industries; and a heavy dependence on foreign sales. These characteristics mark the 1950–1975 period. Much was invested in military technologies and much of it was invisible in the consumer market. Even the greatest advances, in aircraft design, were for fast military fighter aircraft, a different breed from commercial aircraft.

Concentrating on the most significant characteristic for our purpose—that of technology and research importance—we will echo Gansler's observation that maintaining technological superiority has been a justification for large defense research and development spending. Such spending has often been seen as a compensatory factor for low numerical forces according to Gansler, who noted "history shows that the U.S. Defense industry has been driven by technology (rather than by costs). This has resulted in an extremely heavy emphasis on R&D funding, often at the expense of production funding" (p. 11).

What we got in the second period was advanced construction materials, aircraft and rocket propulsion systems, and microelectronics—the space age. Of the three periods, it was the most exploratory, most purely research based, quite possibly made possible by the emphasis on R&D funding and fear of the Soviets.

But the period is not totally dominated by defense and space. Technology also creeps into the workplace. One view by the noted industrial organization economist F. M. Scherer (1970) is that it was not until the 1950s that technological change was a significant factor of production. Before then, optimal combinations of labor and capital were seen as twin harnesses of economic efficiency. Since then, technological and scientific knowledge has been recognized as contributing more to worker productivity and less to capital intensity, as summarized by Scherer.

There are two types of technology-driven productivity improvement: process innovation and consumer product innovation. In describing these, Scherer adds that the latter is the creation of better things for better living, a deliberate repetition of an advertising tagline.

Though factories were slow to automate, their floors changed during this period as production and operations managers realized (or confronted) the idea that technology improved productivity. Mechanical robots did simple tasks in some machine shops. The auto industry finally found a place for robotics.

To get to the position of having a history of high technology, we need to include the ideas of research and development, as well as innovation and technology, because high technology is really a mix of these. If they were explored as rivers, we would call technology and innovation our Tigris and Euphrates and research and development our Yellowstone River as it cascades down the young Rocky Mountains. Innovation we can take to be any improvement over present products or processes—that has certainly existed throughout humanity, as has technology. The relatively newer river is research and development, which has become the formalization of scientific and engineering practices.

The other feature of the modern era of technology was how multidimensional it had become. Prior to the 1940s, areas of rapid technological improvement were more limited. Fields such as electronics, chemistry, and transportation advanced rapidly, but fields such as printing and electric power production made slow advances. That is nothing like the broadscale advancement today, where virtually every area of human activity is being revolutionized through high technology. The speed with which things are done, and the accuracy of how they are accomplished have been affected by one form of high technology or another. Many activities have been affected by the compounded effect of technology, as in the case of biomedical engineering and gene therapy.

The second period gives us the transistor at the start, then the integrated circuit and microprocessor. Vacuum tube computers give way to the PC. As noted by Cross and Szostak (1995), growth happened in other areas as well. The use of plastics increased by 50 percent between 1945 and 1955. The Salk vaccine, psychotropic drugs, the birth control pill, ultrasonics, nuclear medicine, and MRI joined the repertoire of technology triumphs.

A Term Enters the Language

There was still no verbal net to capture all this; it was all "science" or "engineering" or "invention." The language caught up only in the late 1960s, as we now describe.

If we look first at the emergence of the term *high technology* in the language, we can see the forces that stimulated the creation of the word. A first step is to probe the dictionaries that trace the use of the word. Here, there is disagreement. *The Oxford Dictionary of New Words* introduces *high tech* as the "phrase which started to be used as an adjective in the early seventies when electronics began to affect consumer goods and the design of homes." The dictionary adds that the term became so popular in the early eighties that some people considered it as meaningless. A different genesis is offered in the *Third Barnhart Dictionary of New English*. This source suggests the term was used since 1967. Returning to our definition offered by the *Random House Dictionary*, we obtain yet another first-use estimate—that being in the mid-sixties.

None of these sources provides sufficient information to pinpoint the first use of the term, so it is necessary to take a broad sweep of popular literature over several decades to determine that. The tools for doing this are the various reader's guides to technical and popular literature.

A broad sweep leads us to a curious discovery. The first article using the term *high technology* in its title is a 1972 piece in *Business Week* entitled "Where High Technology can Help the Cities." It is curious because it is a business press article that embraces this new technology description, not a technology or science magazine.

During the 1960s, a variety of terms were used to describe and place this activity. Prominent among these was the phrase *new technology*, but there were other—*sci-tech*, for example. There were also terms that described the times: *the atomic age, the electronics age,* and later, *the computer age.* All these are terms for high technology, but it appears that throughout the 1960s there was no single term that lashed together what we now call high technology.

There is a somber tone in much of the popular literature about technology. It is at this time that the environmental movement entered the cultural milieu, and we see the debate emerge between technological progress and environmental preservation. This is exemplified in the SST debate, the Clean Air and Waters Acts, and other flashpoints.

It is not until 1981 that *High Technology* was added as a category to the *Reader's Guide to Periodic Literature*. Thus, its acknowledgment as a subject in and of itself comes late—54 years after the category of Technology was created in the *Reader's Guide*.

Textile Mills and Almond Groves: The Geographic Bases

High technology has more to it than the game of words. Its physical manifestations are in the geography it occupies. In this section, we look at the continental poles of geographic identity—the East and West Coasts of the United States. These are the places where high technology has the most meaning for most of us. The geographic identification of high technology on the East and West Coasts represents its third phase, a phase dominated by the computer.

One of the geographical regions once well known as a hotbed of high technology activity is the area surrounding Boston, with Route 128 its catalytic collector. High technology firms were once strung like pearls along the road and their collective value to the region became famous as the "Massachusetts Miracle." That miracle had to do with the economic revitalization of the area, owing to the number and success of high technology startups.

In 1973, Massachusetts experienced very high unemployment, at times as much as 11 percent, plus corporate relocations outside of Massachusetts and major state budget deficits. By 1988, however, unemployment dropped to below 3 percent and the economy rebounded from the miasma of the 1970s. Taxes even went down. Most observers attribute the turnaround to the growth of the high technology sector, a growth led in many ways by industry leaders, not small shops. The flagship business was Digital Equipment Corporation, the first to push its prow through the old seas of textile mills and industrial manufacturing. Digital Equipment's suppliers followed and the high technology fleet emerged.

The computer businesses managed to reduce the size of computers while increasing their computing power and lowering their costs. This itself created a whole host of new technology needs. These needs included computer networking and new applications software for the PDP series of computers that DEC was churning out for a rapidly growing minicomputer market.

As reported by Lampe, Lester Thurow theorized that the reasons for the revitalization were the low need for energy and raw materials, as well as low transportation costs. Furthermore, Massachusetts was a rich source of high skill, low-cost labor that sprung from its educational and training institutions (Lampe, 1988).

The Massachusetts Miracle had numerous antecedents, which contributed to both its rise and its fall. High technology in its earlier form consisted of massive federal investments in World War II weapons programs. MIT received these grants to conduct war-related research. However, after the war two things occurred. There was a shift in the focus of research toward high-energy physics and astronautics, and there was increasing resistance from MIT to defense-sponsored research. This reluctance opened the way for private-sector defense research that flourished, at least for a time, until it also succumbed to the erosion in defense spending during the 1970s. In a sense, then, the Massachusetts Miracle was high technology's rebirth.

But now the miracle is no more. The cost of skilled labor has risen and the Route 128 pearls have chipped. There is still, however, a concentration of high technology businesses. If nothing else, the Massachusetts experience stands as an example of the limitations of any expectations of a high technology economic transformation. Its lesson of sobriety amid revelry remains appropriate today as we learn about domestic high technology companies setting up their production and sometimes their research and technology operations outside the United States.

More recently, there has been a restoration of high technology companies in the medical and biotechnology areas, as described in Chapter 16, so Massachusetts has prevailed in its high technology heritage. An agency dedicated the high technology, the Massachusetts High Technology Council, fosters development of high technology businesses in that state.

West Coast High Technology

To an extent, there was a cross-continental umbilical cord that explains the birth of West Coast high technology. Once again, academe was involved in the western maintestation. Fred Terman, a Stanford University vice president, and President Wallace Sterling invented an original and highly successful way of building permanent university-industry ties: give them land.

The Terman-Sterling land arrangement was really a solution to the problem of attracting top researchers rather than a way to build a breeding ground for technical entrepreneurship. To attract these researchers, money was needed. Stanford University rested on nearly 9,000 acres of increasingly valuable real estate, but sale of the land was forbidden by the Stanford family. Much of it remained vacant—at least until Terman made a forbidden technical city out of it. While it could not be sold, it could be leased. So first came Varian Associates in 1951, and later came Hewlett-Packard in 1954. Now there are some 90 high technology firms on the property, all in close proximity to campus classrooms and laboratories.

The beginnings of West Coast high technology were again related to the defense industry. Technology had shifted northward from the southern aircraft plants. Lockheed Aircraft relocated to northern California in 1956. The Defense Department became an instant and substantial customer of transistors, a team creation of Dr. William Shockley, John Bardeen, and Walter Brattain.

The transistor was invented in AT&T's Bell Laboratories in Murray Hill, New Jersey, and it probably would have stayed there if Shockley did not have fond memories of his hometown, Palo Alto. He organized his firm, Shockley Semiconductor, in 1956, but by 1957 the semiconductor business had its first schism. Several engineers bolted to form Fairchild Semiconductors, whose engineering progeny populate nearly all semiconductor companies in the area. The Fairchild genealogy is about as legendary as it was damaging. So concerned was Fairchild President Wilf Corrigan that he abolished the company's research and development division. Why run a laboratory for your competition? That was a question that certainly occurred to him.

But Fairchild's internal hemorrhaging wasn't its only problem. It also stuck with geranium transistors rather than move toward the more reliable silicon transistors. He who hesitates in high technology is lost, and Fairchild hesitated. If the atomic structure of geranium was only slightly more conducive for low DC voltage and had more elemental integrity, Fairchild would have taken the place of Intel and Silicon Valley would be Geranium Valley. The transistor, and later the integrated circuit and still later the large-scale integration, formed the core development stream around which Silicon Valley was created. There are the host of computer firms and software companies that we are quite familiar with, but it was the semiconductor that formed the core.

Route 128 and Silicon Valley Comparisons

Although they are certainly geographically distinct and popularly separate in the public mind, we can see a large number of connections between these two high technology areas. The first is in the men and women who made high technology a truly significant economic activity. Consider Fred Terman, the MIT professor who transplanted himself to Stanford. With him, we can actually chart the shift from the East to West Coast in the technology that separates the two regions. Terman pushed Bill Hewlett to start his company. He persisted even as Hewlett took a job with GE and Hewlett-Packard found a business address in Stanford's research park, a development site that, fittingly, Terman helped develop for new technology ventures.

An especially noteworthy trend in Silicon Valley is its turbulence, as indicated by the turnover of firms. In over thirty years, the valley has prospered despite the turnover. Bahrami and Evans (1995) explored this matter, describing the valley as an inherent ecosystem where ephemeral firms contribute to the overall economic climate—not necessarily because they survive. In fact, they don't. What they do is mutate into other companies: one firm replaces another. The authors call this flexible recycling. The whole area appears stable, though it is unstable at the individual-firm level.

On the point of research university ties to technology launches, there is no doubt they exist along Route 128 and in Silicon Valley. However, all high technology businesses have academic ties wherever they are located. The engineers and scientists who staff high technology firms are supplied by colleges and universities. So the consequential question is how much above and beyond what higher education does is evident by the Boston area and Silicon Valley schools? Here we find differences. There is a more profoundly interventionist side to the California schools compared to the Massachusetts schools. Stanford University provided a site for technology firms. High technology firms such as Hewlett-Packard sponsor advanced education at Stanford for their engineers. Physical proximity to the campus is closer on the West Coast than on the East Coast.

The relatively close research university–technical business startup ties are the distinctive marks of this third phase of high technology history. During the war phase, sponsorship of the development work came from the federal budget. In this contemporary phase, much innovation results from academic and large industry spinoffs. The federal impetus has declined, at least in terms of funding commercial technologies.

The differences between the two regions must also be recognized. On the East Coast, there was an existing technology infrastructure built up during the 1940s with World War II weapons research. It was a nearby infrastructure, as well. When the war ended, engineers could go back to their neighborhood MIT or obtain jobs at nearby defense firms. On the West Coast, the infrastructure was more distant. War spending built up southern California mainly, not northern California.

In Southern California, a rapid leap into space high technology was made in the 1960s. It was fueled by "Space Race" NASA funding. Again, there was an academic linkage. The academic linkage to Cal Tech in Pasadena came into full fruition in the 1960s with the eminence of the Jet Propulsion Laboratory and its lead in the *Ranger* and *Pioneer* moon probes. Earlier, Cal Tech played a role in the development of jet engines at Edwards Air Force Base.

The purpose of the expenditures were also in a different realm. Around Boston, government-sponsored research went to electronics and radar, while in southern California it went to airplanes. The largest component of high technology is in computers and hardware, so it is not difficult to understand how electrical engineers in and around Boston took up computers while aeronautical engineers in and around Los Angeles made both conceptual and physical migrations to do the work of Silicon Valley. Put in different words, the engineering talent base was in place in the East and had to be grown in the West.

The same holds true for the capital base for the two coasts. Around Boston there was a stronger and richer network of investment bankers and venture capitalists than there was around Silicon Valley. In the 1980s, there was some migration of venture capitalists from the East Coast to the West Coast who, along with indigenous San Francisco investment bankers, built an investment base, but they were stylistically different from eastern venture capitalists. West Coast engineers (who have seen both) can easily separate the money suppliers based on dress and demeanor. For example, they talk about the hard edge and numbers orientation of the East Coast investors.

There is also the matter of culture. Engineers themselves contrast the perceived rigidity of Ken Olsen of the Boston-based DEC with the informality of Apple in its early years. Engineers we have spoken with tells us about this work style difference. It is quite common in other occupations, but it exists in high technology as well.

The residency of high technology on the two coasts, and the many other areas that will be described in Chapter 16 is certainly a defining characteristic of the third stage of high technology history. There are other distinctions, too. The third phase is multifronted, continuing with electronics but also drawing in biotechnology, automation, and other activities. So the third phase is similar to the first in its broad sweep, but dissimilar because the innovation has come from the private sector rather than from government programs.

There is something, too, about the tone of this most recent phase of high technology. It is directed toward individual needs more than the earlier forms. There can be no better indicator of this than the "personal" computer which introduced the stage. A *knowledge age* or *communications age* or *information age* means enabling individuals to acquire and use information more than ever before. We can see how most technology is making this turn to individual needs, whether through gene therapy, custom manufacturing, computer based micro-marketing, personal energy generation, or other trends. The large-scale technology forms are still there—space exploration, transportation, advanced materials but the slant is now on people.

Summary

The history of high technology is part of the history of technology. Our starting point was the idea that high technology emerged from an enduring human preoccupation with tools and devices, and that modern technological developments have deepened and accelerated that imprint on humankind. High technology was organized technology. High technology businesses were those using science and engineering to produce products and services. There was frequent usage of the term *high technology* in the 1970s to describe this activity, and it has been widely used ever since.

But there were differences in the orientations of technology in the last half-century. In the 1940s, the emphasis was on warfare, while between 1950 and 1975 there was more broad-scale development in a number of fields. The dawn of the PC in the mid-1970s further broadened the base for high technology and new technologies entered. A timeline of major constituent technologies of what we call high technology shows the duration of older forms and the emergence of newer forms.

Not only have the older technologies remained, but newer forms have come into being. So how can we measure the impact of modern high technology in the postwar era? One way is to refer back to our definition of the subject and its SIC base. In doing so, we can see how the period resulted in the creation of industries that did not exist before the war. Computer and data processing services, guided missiles and space vehicles, as well office computing and accounting equipment and plastics materials were mainly new postwar industries. This is but a seventh of the Group 3 high technology industry groups, yet it shows quite plainly how high technology has introduced new forms of doing business.

Our brief history of high technology businesses originated in people's

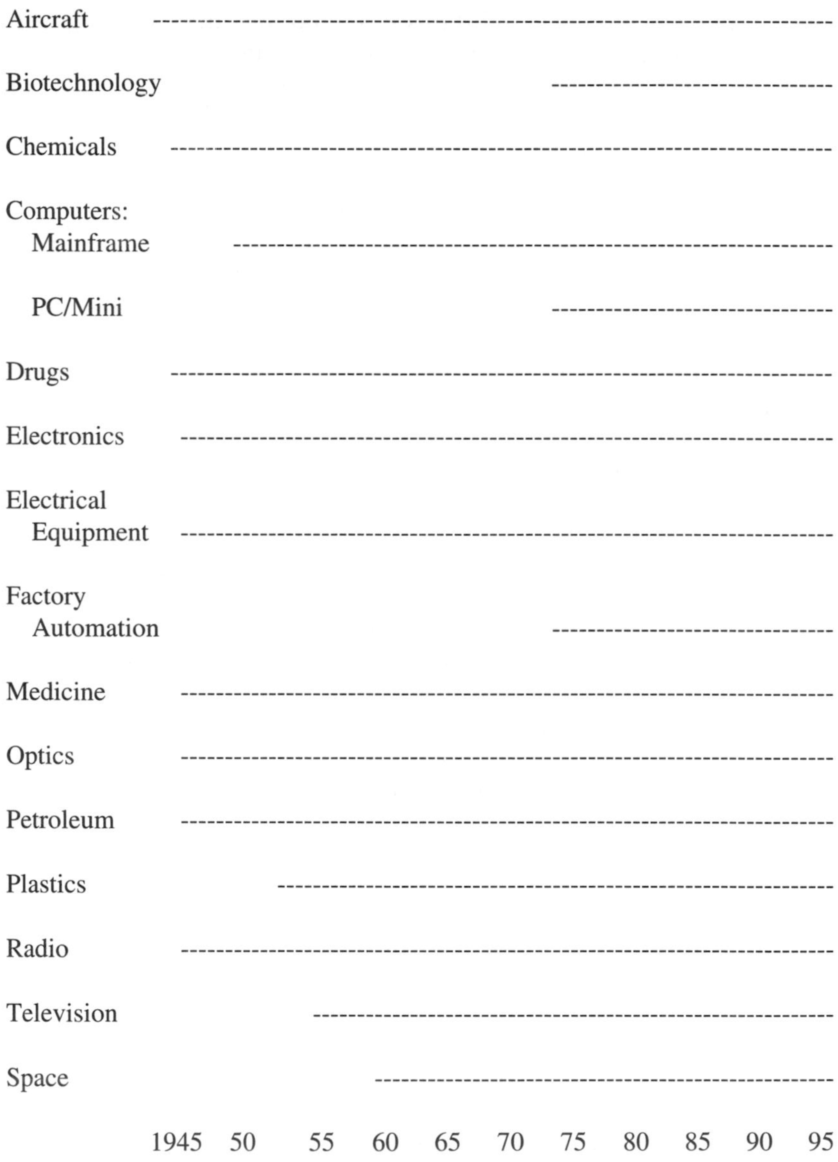

	1945	50	55	60	65	70	75	80	85	90	95
Aircraft	---
Biotechnology						-----------------------------
Chemicals	---
Computers: Mainframe	--
PC/Mini						-----------------------------
Drugs	--
Electronics	---
Electrical Equipment	--
Factory Automation						-----------------------------
Medicine	--
Optics	--
Petroleum	--
Plastics	---
Radio	--
Television	---
Space	---

Figure 2.1   Emergence and duration of major technologies since 1945

first quest for tools. Yet we also noted that it has taken particular forms at particular times, from collective civil engineering projects in ancient times to the individual genius of worldwide industrialism, back to collective war-directed technology aimed at improving weapons, to a period of private-sector worldwide innovation on many different fronts.

What is clear is that there has been a shift but never a stoppage of high technology. In fact, it is an enterprise that has accelerated social change. That change can be traced to many factors: "population growth, urbanization, the shifting proportions of young and old—all play their part. Yet technological advance is clearly a critical node in the network of causes; indeed, it may be the node that activates the entire net" (Toffler, 1970, p. 428). Themes of social interaction with technology have been raised by others as well.[2]

This history of high technology has demonstrated the subject's divergence and hence the importance of determining what makes it stick together. Could it be that the high technology version of biotechnology has less in common with computer manifestations of high technology than it does with fundamental biology? Can the same be true of the other versions of high technology?

These are central questions for this book. Thus far, we have detected a common lineage for high technology businesses. But that common lineage is separated by the centrifuge of rapid and diverse knowledge growth.

When we look at when the term *high technology* entered popular discussion, we see the early state of research, development, and technology and note how much has changed since then. New technologies have been added. There has been cross-fertilization between technologies such as optics and computers in visual scanning and computers and medicine in medical electronics and instrumentation. It is not unexpected that those in high technology not only invent devices but also invent themselves. The new technology fields require new training, especially in the hybrid fields where inventors need to learn enough about a new field to be effective in it.

In closing this chapter, we emphasize that high technology business is not a new invention. It is an activity that evolved slowly from technology and human organizations. At the same time, it has become its own juggernaut for both good and bad. In the next chapter, we'll see just how dynamic that field is by stepping directly into the daily work of high technology.

Notes

1. Note here as an example that Bill Hewlett had to leave Hewlett-Packard to serve in the armed forces during the war years, a fact that Packard considers a brief interruption. It was an interruption, but it was one that put on hold an extremely successful commercial partnership.

2. As an example, John Naisbitt in *Megatrends* (New York: Warner Books, 1982) envisioned new technology as being "high touch," designed with human users in mind and made friendly for them. Peter Drucker considers the interaction between technology and humans and concludes that technology needs to be understood as work—how and what people produce, not just the tools they use. These comments are in "Work and Tools." Drucker's article in M. Kranzberg

and W. Davenport, *Technology and Culture: An Anthology* (New York: Schocken Books, 1972).

References

Bahrami, H., and S. Evans, (1995). "Flexible Recycling and High Technology Entrepreneurship." *California Management Review*, 37: 3, pp. 62–89.

Cross, G., and R. Szostak, (1995). *Technology and American Society*. Englewood Cliffs, NJ: Prentice-Hall.

Foster, R. (1986). *Innovation: The Attacker's Advantage*. New York: Summit Books.

Gansler, J. (1981). *The Defense Industry*. Cambridge, MA: MIT Press.

Lampe, D. (1988). *The Massachusetts Miracle*. Cambridge, MA: MIT Press, 1988.

Scherer, F. M. (1970). *Industrial Market Structure and Economic Performance*. Chicago: Rand McNally & Company.

Snow, C. P. (1962). *Science and Government*. New York: Mentor Books.

Toffler, A. (1970). *Future Shock*. New York: Bantam Books.

3

THE DAYS OF
ITS LIVES

In the last chapter, we outlined the emergence of the discipline and business known as high technology. At this point, we launch into the contemporary world and enter the doors of high technology as observers of real businesses in, as close as this medium can get, real time. The purpose of this chapter is to spend some time examining high technology businesses and learn from their practitioners what is done, who does it, and how it is done.

For reasons of having so many different technology companies now and practical publication considerations, we cannot explore all of what is called high technology. Our excursion is limited to four very different companies, each having a special character yet each representative of firms in its field. We consider a true biotechnology company, a customized machine firm, a huge computer and instrument maker, and a small integrated circuit design company.

In every case, our procedure was roughly the same. There was an on-site visit with the manager, interviews with engineers and scientists, and observations. We observed work in progress with all its attendent problems and successes, and our intent here is to capture the verité of high technology. The interviews were also compared with the survey research results to further characterize these businesses.

An Endeavor of Stealth

The first destination is a biotechnology business. Biotechnology is a major high technology field, but it is a mysterious one. Of the four high technology companies we look at, what is done in biotechnology is least known. It is also different from computers and electronics. Biotechnology sales are not made directly to the consuming public, as are computer sales. The electronics industry has been around for two hundred years compared the brief life of biotechnology—this is a business in its infancy compared to electronics.

We see more cross-fertilization between electronics and computers than with biotechnology, although there are connections between electronics and biotechnology in such things as using television or computers to perform gene manipulation. There is also the use of microprocessors to automatically line up preprogrammed strains of nucleotides, saving days of manual work. This biotechnological application of computers is called a gene machine.

In all, though, biotechnology is its own field. It has its own kind of people and its own science base, that of genetics. It is also more laboratory oriented than the other businesses that visited.

There are even some problems calling this business line biotechnology. The term itself is young, so it has not settled into our lexicon to the extent that electronics or computers has. But the other problem is that it is only a recent SIC-defined industry. The category has captured many true biotechnology firms, but it may have also netted vaguely related businesses, as well as laboratory equipment suppliers. It can also include process sellers, those biotechnology firms that develop a laboratory technique and then sell that technique to firms without ever having used the process to do actual biotechnology research. Contract biotechnology firms may be included, but their work varies greatly with each contract.

There is very little that is tangible in the biotechnology business. Inventory is mostly an alien concept, save the laboratory equipment and occasional bag of genetically hardied seed. Investment is in trained workers, not large-scale production. Companies with biotechnology arms tend to recognize that their biotechnology units are not factories of the present but cultivators of the future. They require patience and investment willingness.

Biotechnology is an endeavor of stealth. Its work is not a matter of explosive insight but of careful testing and retesting. Confirmation of discovery comes in reams of statistical tests, not shouts of "Eureka." And it takes time to see the effects of genetic engineering, showing up in progeny through different generations. The impact of innovation is not immediate in genetic engineering, though the benefits of drugs produced through biotechnology may be—but that is after much nonhuman laboratory testing.

Biotechnology deals with something as common as water in a living organism. The molecule, DNA, is not only invisible but almost unpronounceable in nonabbreviated form, yet is the main focus of the entire business.

Biotechnology is also the most recent planet to be drawn into the high technology orbit. Younger than computer technology and far younger than electronics and telecommunications, biotechnology is our newest technology permutation. But it is also its own universe. It has its own science and professional base. It deals with the submolecular world, so there is a physically small dimension to it. And there is not much crossover to the physical sciences, as there is between physics and electronics. The businesses, too, are different, as will be shown.

Biotechnology requires the most explanation of the high technologies we profile, so part of our attention is to this purpose. A general description of biotechnology is followed by a review of day-to-day life in a typical biotechnology firm.

The Business of Biotechnology

The heart of biotechnology is to understand and make use of the genetic elements of living things. Its pursuit is both a science and an application of science. Discovery about the exact construction of DNA is the aim of that science. The famed double helix DNA is the origin of the genetic universe, and this structure and its molecular composition are the draw for scientists and bioengineers worldwide. It is pure science, but one that quickly passes into the application side. Biotechnology engineering makes use of recombinant DNA to manipulate separate genes. Thus the ability of engineering is to change the genetic combinations for better biology.

Biotechnology is a field of molecular biology, a field itself just twenty-five years old but with a remarkable history that certainly qualifies it as a high technology endeavor. Already, hundreds of genes in living organisms have been tagged as responsible for particular traits. Genes have been cut and reattached to produce new living traits.

Molecular biology research promises to map the genetic field of human beings, the objective of the human genome project, eventually producing specific genetic maps for individuals. As in the Age of Exploration, when maps and navigation revealed new continents for Europeans, human gene mapping should yield reasons for basic and complex human traits.

So how do technology companies made of the stuff of molecular biology differ from other technology firms? The answer seems to be in the different ways biotechnology companies make products. Some biotechnology companies want to be the first to make a discovery and then patent the process. Others may not seek to be first, but want to do it better—by lowering input costs for cloning for example. Others may have directed approaches—they may want more disease-resistant wheat for seeds that they market. Though all different, these outputs are similar that they do not have the same degree of commercialization that a PC manufacturer seeks. Biotechnology businesses aim more for *ways* of doing things better—say, ways of growing better seeds—than for a greater number of commercial products. The quality of the discovery is what counts.

The origin of biotechnology explains much of its present direction. Like the other forms of modern high technology, these businesses had their start in academic research. Funding for biotechnology research stemmed, once again, from basic wartime research grants to universities and colleges, although these grants were not aimed at creating weaponry. In fact, there was not much progress for a while. The breakthrough was in the post-1960s, with the ability to make new combinations of the basic DNA molecule—to clip here and there on the code of life to create very different types of organisms.

Biotechnology firms that have emerged recently use genetic engineering to create new forms of drugs and other results from these discoveries. Additionally, the new firms have developed live disease-detection tools. Almost invariably, the firms are started by people with PhDs in genetics or related sciences.

Genetic engineering is the obvious, well-used tool of the biotechnology business. Particular forms of a protein may be produced for pharmaceutical use. For example, a gene that produces a certain type of beneficial protein could be transferred from a live human to a bacterium. That bacterium could then be cloned and induced to produce the protein outside the human being. A genetic engineering company is thus born as a live factory. An agricultural example is the development of a genetically altered microbe that can decompose manure more rapidly.

Most would regard this business form as still in its early life, with large possibilities looming. Advances are likely to be in many different medical and biological sciences—the pharmaceutical and agriculture industries have been only the early beneficiaries. In fact, there are two main streams of development in biotechnology. One is the creation of products using genetically modified organisms and the other is the creation of genetically altered organisms themselves. If cast in the setting of traditional organizational design, then both the output and the transformation process are continuously renovated by biotechnology. Products that have already been created are growth hormones, interferons, and neuropeptides. New antibiotics are also the result of the new biology machines. As for creating genetically engineered organisms, there is more promise now than ever before. Reproduction of particularly disease-resistant forms of plants has been achieved. The genes that produce such disease resistance have been introduced into plants and scientists have seen these genes passed on to later generations of plants. While this has always been the basis for applied genetics, modern biotechnologies involves much more precise gene manipulation for more specific results.

There are at least 200 biotechnology companies in the United States. Some estimates are as high as 1,000 companies worldwide. In reality there are two kinds of biotechnology firms. One is a collection of small startup ventures with few or no products; the other is a much larger firm, often spun off from a related company such as an agribusiness firm establishing a seed research company. Midsize companies are few. For every Amgen, there are dozens of newly minted biotechnology starts. In talking with many founders of these companies, we have noticed that their scientific orientation surpasses their business orientation, particularly when thinking about competitors. Founders largely ignore or downplay their competitors, yet this is a business field that competitors are invading.

To spend a day in a biotechnology company is to discern a mere hint of this wildly growing field. Yet not spending time with these people is to leave the business shrouded in mystery. What we have done, then, is spend time with a biotechnology company to glimpse what goes on. This on-site visit was needed much more for biotechnology than for electronics and computer companies, since most of us have a realistic vision of what to expect in these companies. The same can't be said for biotechnology where our images lapse to bad science fiction or demonic human eugenics a la "Boys of Brazil". There seemed too much of this "playing God" business to make us comfortable with the biotechnology discipline.

Even being on-site is no basis for deciding if a biotechnology firm is working well. These companies have different objectives. Judging their performance is as problematic as precisely describing what these companies do. Because the output of a biotech firm is rarely a tangible, marketed product, surrogate performance measurement has to be the recourse. The surrogate most often used is patent production, and that requires more scrutiny.

As some observers have noted, patent acquisition is an imprecise indicator of biotechnology success.[1] All of the biotechnology managers we talked with bristled at the thought of having patent performance as a goal. They had other standards, though—usually experimental milestones that the company closely manages.

What kind of strategies do biotechnology firms employ? Apparently, cooperative arrangements with other firms are popular. As Shan (1990) reports the chances for such cooperation are greater if the firm's competitive position is different from its rivals. That is, the market follower is more likely to seek alliance than the market leader in product commercialization, but firm size does not correlate with such cooperative arrangements. The main use of cooperative arrangements is in the commercialization of new products in foreign countries.

While cooperative arrangements are possible, most firms we know of are aiming at growth objectives, wanting to do the right science that will lead to product commercialization in areas where demand will be huge, as in crop improvement or drug therapy.

Our biotechnology business example has such a growth orientation, which will be described later by its manager. The business is named Agracetus, and it is a midsize, well-established firm. Agracetus is a unit of Monsanto Life Science Company. It began as an independent company, but was acquired by W. R. Grace and then by Monsanto.

Biotechnology Live

Take Highway 14 west from Madison, Wisconsin, through Middleton toward Spring Green and the Wyoming Valley homestead of Frank Lloyd Wright, and you'll pass the beacon of contemporary biological architecture, the Agracetus biotechnology company in Middleton. At night, the brightly lit greenhouses are especially noticeable, though at any time the pale two-story complex stands apart from the more traditional businesses in that sprawling industrial park.

Agracetus does agricultural biotechnology for the most part, and fits in well here because corn and soybeans are grown very close by on the farms of western Dane County. These are crops that Agracetus seeks to fundamentally and permanently alter. And the company soon hopes to have a far greater impact on crops around the world.

Within the Agracetus building there are about 60 people. One-third are molecular biologists or other researchers; another third are specialists who put genes into plants; and the last third care for the plants in the greenhouses, maintain the facilities, or are administrative staff. Of all the sites we visited, Agracetus most appears "high tech." Technicians wear white lab coats and there are large rooms with long lab benches—the marks of the science business.

Dr. Robert B. Horsch is director of technology and general manager. He confirms the nature of the work. "It is almost all laboratory research," he explains, dismissing a business orientation for the place. Monsanto's plan for Agracetus is "to nurture it as a site of innovation, productive work and to make sure that what's done well up here is shared with the rest of Monsanto and vice versa." When asked about the level of control by Monsanto exerts, Horsch answers even more deliberately: "It is neither hands on nor hands off. All of Monsanto is trying to organize around the projects and products that will be valuable to customers. Some aspects of what goes on at Agracetus are almost completely self-contained and the rest of Monsanto has not been involved in those areas, so that continues on a fairly autonomous basis. There are other aspects of the work here that are integral parts of work that Monsanto is already doing and that have to be integrated into the larger picture." Monsanto provides financial support mostly. "Agracetus is an investment in research and development, not a cash provider. In the long run, we want to create new businesses for the future as opposed to continuing a history of doing a fair bit of contract research," Horsch further explains. When Agracetus was independent, it did not have the resources to register or sell products directly. Horsch believes that was why Agracetus sought a link with a major firm.

The predominant research stream is in agricultural biotechnology. "A large portion of what's done here has to do with the technology of putting new genes into soybeans. Scientists at Agracetus invented a special type of gene gun to shoot gold particles carrying genes into living cells of soybeans and other crops. This was a tremendously valuable invention," Horsch explains. Cotton improvement is another major research project. It's promising because scientists have succeeded in making a cotton hybrid fiber—one with polymer containing fibers that have better insulating properties.

Since there is a considerable lack of awareness about the day-to-day work of biotechnology, Horsch offers a detailed explanation of what happens at Agracetus:

"There are two basic areas of work. The more visible one is working with crop plants. The end point is plants which produce seeds in the greenhouse. The seeds carry the new genetic information. The plants, grown from those seeds, exhibit the traits, conferred by the new genetic information, that the farmer or food producer or manufacturer would want. If you back up from there, those plants are derived from tissue cultures that are sterile and grown on defined medium that looks like Jello in little plastic petri plates. These little plantlets have been grown from cells or tissue cultures that have been bombarded with gold particles that contain genes that confer new traits.

"There are a lot of people here who deal with those petri plates under sterile conditions—where the plant cells are bombarded with the gold coated with genes and then selected and regenerated into tissues and little plantlets and then are transferred to soil then out to the greenhouses where they grow and make seed. So we really go from a plant cell in a sterile environment to the whole plant in soil as the main flow of work."

That's the more visible part. The less visible is also detailed by Horsch: "The genes that are on the gold particles that go into the cells come from the molecular biology part of the lab. Molecular biologists deal with cultures of bacteria that are the workhorses of recombinant DNA. They alternate between test tubes where DNA is treated like a chemical and microbes where DNA is alive and replicates and expresses proteins and traits. The molecular biologist alternates between genes as a chemical and genes in living cells to do all the recombinant DNA manipulation. So, you'll see them with little pipetters squirting microdroplets of materials around and with cultures of bacteria that contain the genes in a live state. They work back and forth between the two."

The essence of the work is in the invisible DNA. Straddling between the world of the living and the world of the inanimate, the microbiologists are constantly learning about its character: "DNA is a very long molecule, much like beads on a necklace. You can read the order of the beads and you shear the long strand into fragments and separate the fragments by size. It's a double strand of pearls instead of one strand. These strands are complementary so that they recognize each other. You can melt them apart by heating the liquid they are in. As the liquid cools, they'll stick back together. It would take hours to go into it, but there are all kinds of simple chemical and biological techniques that can be used to manipulate and recombine genes."

Horsch then points out the differences between his work and other technology work: "Biological research has a lot of tried and true practices, but our starting material is usually very poorly characterized. Somebody in manufacturing like Silicon Valley manufacturers know the properties of silicon. We don't know the properties of most of the genes we work with when we start. But the techniques to cut, paste, and assay the genes are standard."

What is the most difficult part of biotechnology research? As Horsch puts it, "The infinite number of possibilities, narrowing it down—which experiments do you do? Which genes do you go after? In what configurations do you put those genes in out of the infinite possibilities? You can experiment on only a tiny fraction of what's possible." That explains the hundreds of thousands of petri plates that end up in the garbage each year and the dozens of inconclusive experiments. Strict adherence to the experimental method is part and parcel of this business.

The Miners with a Heart of Gold

Casting himself into a holistic view, Horsch adds, "There's a field of activity that's been coined as bioprospecting. Everybody knows what it's like to dig holes in the ground and see if you can find gold and diamonds. If you look at the diversity of nature, there is a huge wealth of diversity that's almost completely uncharacterized. In fact, most species that exist on earth are unnamed and undiscovered. We know that when we go into a new area and look at the species that are there very throughly, we find species we hadn't known existed before. We only looked at what was on the surface, at what was abundant, common. By extrapolation, if we can find new species in a tiny area, there must be huge numbers of species on the whole. There is this huge diversity of organisms, genes, traits that exist in nature. Bioprospecting is looking for traits of value that are out there in nature that can be used in corn, cotton, soybeans, or for human pharmaceuticals."

Horsch talks about plants being a factory today, a factory that can be made more efficient, one not needing as many expensive fertilizers and insecticides and other inputs. "Plants produce starches. They produce oils. They produce feed and they produce food. By either increasing their yields or increasing the portion of what's harvested in the higher value proteins and oils instead of starches, the plant becomes a better factory." He later amplifies, "The concept for the future is that plants are a renewable feedstock, whereas petrochemicals and things that are mined are not renewable in the long-term sense. They'll be depleted and we'll have to go someplace else." Horsch then used a *Harvard Business Review* (Magretta, 1997) interview with Monsanto CEO Robert Shapiro to elaborate the point.

Returning again to the theme of biotechnology as a business, Horsch notes that Monsanto wants to produce and market its products, the little green factories. "As part of Monsanto, our clear objective is to sell products, not sell contract research. Patents have come from Agracetus but numbers aren't important . . . as opposed to the importance and value of patents." Competitors of Agracetus come to his mind quickly and Horsch says, "Being first is important, but its going to be a complicated, differentially weighted mix of who's there first, how good their product is compared to need, what their patent estate looks like, and how many other options the marketplace provides to fill customers' needs. There's no reason for a farmer to buy a biotech product if he gets something that works better and is cheaper from chemistry or crop rotation. We're trying to be the substitute of choice in the marketplace."

In the downstairs laboratories there is a lot of what appears to be microgardening going on. A technician clips off tiny parts of seeds. Another is sorting seeds. Mostly there is laboratory equipment but few computers. Everybody is busy doing something.

As a final observation Horsch says, "It is important that the people who work here understand that what they are doing is critical to develop real products. It's the difference between laying bricks and building cathedrals. When you get right down to it, the day-to-day work for most people is little cultures in little test tubes and gels and lots of samples in complicated machines and lots of petri plates and cutting up little pieces of plants and transferring things to soil and harvesting seeds and record keeping. What makes that something more than laying brick after brick after brick is that we're building products that are useful and worthwhile. Products that will make people's lives better. That's the atmosphere that keeps the day-to-day work interesting and exciting and challenging."

Machines with Fingerprints

We could not better span the expanse of high technology than to move from the live to the inanimate, from the miniscule to the massive, and that is what we have deliberately done to show how different high technology businesses are. In this section, we go into the business of customized machinery, a wide band of industrial technology covering individual robots to computer-integrated manufacturing. First, though, a review of the evolution of this business is needed to understand its most recent developments.

The great American industrial epoch of the nineteenth and twentieth centuries sounded the beat of the repetitive machine. Metal was stamped the same way, so many per minute, to form appliances and car parts. Demand kept growing until our national population stabilized and real income growth stalled. From that new reality emerged other values: quality and convenience. Tools, vehicles, rails, roads, wings, pipes, structures, buildings, presses, looms, and all the other markings of the industrial age that were all mass produced began to change. Shorter production runs and less tolerance for defects became more common in the 1980s. The result is that our postindustrial age has an altogether different beat—more atonal, far less uniform, and meant for the ears of different audiences.

That's plainly the case in the original orchestra itself, where the old, steady machines have given way to specialized, customized machines as part of flexible manufacturing systems. Almost all of this was made possible by computers. Now, to at least some extent, all factory environments are computer controlled.

This did not happen instantly. The first phase of automating factories, in the 1970s, was an isolated affair. Manufacturers introduced programmable logic controllers (PLCs), which were connected to single machines and controlled their actions with built-in microprocessors. These would be programmed by the maker or buyers and installed on the factory floor. Often, there would be many PLCs running on the production line. The next step was interconnection to achieve production coordination. PLCs had interconnection limitations, so computers were used instead. This phase marked the 1980s, and we now have examples of complete computer-integrated manufacturing where minicomputers are running the production operation. In truth, the wide variety of computer-based methods are still found, but there is no doubt that manual labor has largely left factory floor production coordination.

Machine tool design itself has been revolutionized by computer-based design systems, the Computer-aided design (CAD) systems present in most machine shops being an example. The revolution has also been in testing. Boeing did all the design and most testing of the 777 aircraft within the world of microprocessors. That shows the depth of the computer revolution in machine making.

The business we elected to examine uses computers extensively, even if the final products are not always computer-run machines. As we show, this business has found a successful way of building machines for customers that involves both computer technology and some special business practices.

Isthmus Engineering and Manufacturing Cooperative is a small, uniquely structured and successful customized machinery and automated manufacturing firm located in Madison, Wisconsin. It is found in what could best be described as a complex of light manufacturing businesses in the southeast corner of the city. There's nothing special about this neighborhood.

What makes this business especially interesting is not only what is done but also how work is organized. There is little in the outside appearance of the business—a large building, indistinguishable from its neighbors—or its inhabitants that reveals this organization as it truly is: a business world utopia, a place where 44 people design, fabricate, and ship customized machinery. Of these, 32 people have earned their way to being members of the decision-making board. They usually meet several times a month and democratically determine how business will be done, from major strategy decisions to day-to-day operational issues. This true cooperative emerged from its founders, who had been stung by what they believed were the unfair business practices of their former employers, which ran a more capitalistically oriented organization. These founders' strong bond to egalitarian principles is heavily imprinted on this business cooperative, a seeming oxymoron that has nonetheless produced a growing and successful enterprise. As a cooperative, it must live under laws dictating that each member has one vote, that all profits go to members, and that all members are workers.

New projects are the lifeblood of Isthmus. As explained by Gary Dalgaard, who is the de facto business manager, "We are in the machine tool business. Most of our business is in machining systems or assembly automation. We seem to be doing one thing one year and another thing another." He describes a few projects—an automobile shift lever subassembly, a machine that will assemble 3M tape dispensers—and the point is made.

Typically, Isthmus is making entire machines, not subparts, and the value of the projects cuts three different ways. In a year, there about six major projects in the $1 million and above range, another six between $250,000 and $1 million. There are another 15 to 20 projects being worked on with values less than $250,000. The projects are in various stages of development, some run over multiyear periods. This calls for efficient project management, so all the significant projects have project managers.

Projects are ushered in and out the door by the 14 mechanical engineers, 4 electrical engineers, 10 assemblers, 5 machinists, 3 electricians, 4 administrators, and a few shipping clerks and maintenance people. Noticeably absent is a sales force. Concerning this, Dalgaard says, "Jobs come in the door here." Well, not really. He adds, "We have no sales people, but we have a sales effort with a lot of customer contact. I maintain a relationship with customers such as 3M. I maintain contact with their purchasing department. 3M is far more sophisticated than others. They catalog companies based on their expertise. We're in that catalog, and we have good relations with their purchasing department. An engineer in a plant in North Dakota may look in the catalog and put our name on his request. Or he'll just send it to purchasing, say 'I don't know who is good at this. You choose who and send it out.' Purchasing puts our name on request for quote. Then we get the request for quote with the usual: description, deadlines, etc."

Then, with some detectable pride, Dalgaard continues, "Because of the number of repeat customers, we are often in the position of not competing with anybody. Then it's a matter of clarifying what we're proposing, what they're they are getting, and justifying it or not. We prefer long-term, lasting relationships with a few customers. That makes it simpler for us. We think of ourselves as extensions of their engineering departments. More and more companies are reducing staff, cutting back on their in-house capabilities. We think of ourselves as replacing that capacity. Ideally, we like to get involved in projects in early. We like the idea of concurrent engineering." He is referring to joint engineering, where both supplier and customer are doing design and development together. He adds that Isthmus gets 80 percent of its repeat business from 20 percent of its customer base.

In summarizing the organization's basic marketing strategy, Dalgaard says, "See, our ability to be creative in a number of different areas, to prove capable with different machines, is our forte. We don't have a niche per se. We purposely didn't go that way. We look at ourselves as a very broadly based capability company. We think of ourselves, too, as ethical people, a cut above most of our competitors. When we develop something, say for Ray O Vac, we don't just jump on a plane and try to market it to Eveready or Panasonic. That negates the possibility of our developing a market for a product. We want to develop the ultimate level of trust and that's the way we have chosen to do it."

What difficulties are faced by aiming for repeat and long-duration customers? "As they work with us, they relax the standards because they know we don't. We have to monitor this. We like to do different jobs. With John Deere we have been designing as we go; not everything is known but we start designing anyway. Then we give them a price with unknowns thrown in. A decision is made to continue, and if it is yes, we'll continue and they'll get a fair final price." comments Dalgaard.

Designer Wendell Hottmann is sitting in front of his large-screen workstation on which a CAD program is running. What does he think the major challenge is? There is no equivocation or delay in his response: "Making it work is the biggest problem," he says as the three-dimensional static picture in front of him beings its path to reality. "Deadlines and specifications are matters of communication," he adds, dismissing the more often cited but less fundamental problems. "If we are planning well, it works. When we plan, we give ourselves alternatives and figure for possible add-ons. Every job is different, so we can't put it all down on paper. And we don't have a formal diagnostic process." Isthmus helps itself in this regard by building in front-end engineering cost loading for projects.

Dalgaard elaborates: "Tolerances have become even more tighter. If it used to be thousands of an inch before, it's half thousandths of an inch now. Process control–supplier standards are more important. You'll see what I mean later. When we design, we look at all the possibilities and design for events. If we find four ways out of a problem, we then find two of the best ways out. If it doesn't work here on the floor, there is no choice but to go back to engineering and start all over again."

Dalgaard also introduces us something he considers an important asset—the library, where the descriptions and specifications for thousands of parts are kept updated. "This cost us a lot (it comes with a librarian), but we need and use it a lot." Many projects start at the library, go to

the CAD workstations, then hit the back end—a one-acre, two-story ma-
chine-shop–looking fabrication and testing area. In this area there are
both machines and assembly lines. One looking very assembly line is an
oil pan assembly line for a GM supplier. The line will be peopled at GM,
and much of what Isthmus does on its assembly incorporates ergonomics,
reducing lifting and holding by workers. Ergonomics is frequently mentioned as we pass the machines in progress.

At one location, there is a simple, apparently completed machine
about the size of a pair of refrigerators. But looks are deceiving. The
machine isn't ready. "This is what I talked about earlier. Looks like it is
ready to go but it isn't. It hasn't been run yet. We're just at the point of
finishing the controls. The power will go on in a week, then the real
challenge begins," he warns. And what is that? He holds up a simple-
looking pump for what might be liquid soap pump. There are 10 parts,
but the customer has them configured several different ways. It still
sounds simple and that's not the source of possible doom. It's the fact
that the pumps will be machine gunned out the machine at a rate of 120
per minute, one every half second.

As this is said, the machine's sole designer, Randy Kohlhardt, clad in
a marathon T-shirt and running shoes, is using a penlight to peer into
the control system cavity. The race is about to begin, and this is a machine
that must always outrun its inventor.

The Venerable Venture

The high technology company considered next is quite familiar to anyone
who follows technology companies and probably familiar to most others.
It is a former venture that made it big. It is an example of a large technology firm—something we have not profiled but one that remains distinctively high technology in spite of its size.

The Hewlett-Packard Company is one of the anchors of high technology in the United States. Not only has it grown from the small garage
(more like a garden shed) in a Palo Alto backyard to a huge firm of
112,800 employees in 29 cities worldwide, but it has done so in the
heyday of high technology, defining high technology and moving it along
as the company itself grew. A plaque outside the garage calls the site the
birthplace of the Silicon Valley, something we have little disagreement
with.

David Packard and Bill Hewlett tinkered in the garage and, in the
now well-known story, managed to sell their first products—eight audio
oscillators—to Walt Disney for the production of *Fantasia*.

Since this first sale in 1938, nearly 60 years of business expansion has
occurred. That expansion has happened because of product development,
acquisitions, and ventures. Product inventions were the high-speed frequency counter, medical instrumentation, analytical instruments, their
first computer (the HP2116A), calculators, desktop computers, inkjet

and laser printers, and computer systems. Acquisitions and ventures intertwine with product development, starting with acquisition of F. L. Mosely Company (graphic recorder manufacturer) in 1958 and, in succession, the Sanborn Company, F&M Scientific Corporation, Apollo Computers, and Convex Computers. As described in *The HP Way* by David Packard (1995), the company has always set and has been guided by its corporate objectives and values. Those values are individual trust and respect, a focus on high achievement and contribution, business with integrity, teamwork, and the encouragement of flexibility and innovation. All this eventually translated into a $38 billion enterprise that has grown out of its initial electronics components start to become an international consumer electronics giant. We'll look now at the headquarters location and discover more about the people of Hewlett-Packard.

A visit to the Palo Alto Hanover Street corporate offices allows us to catch the flavor of the high technology giant. In most ways, it does not look much like a high technology company. The lobby is open but neither spacious nor imposing. It is a place where vendors are meeting managers. It is more the business part of the technology business that is immediately observable. Hewlett-Packard is also in several other locations in Palo Alto and in the many other cities noted. Manufacturing sites and R&D activities are outside Palo Alto. In fact, about 95 percent of operations are outside Palo Alto. CEO Lew Platt strongly believes in decentralization.

The Hanover Street location has none of the campus flavor of Microsoft's Redmond, Washington, headquarters or even the festive, banner-draped Apple headquarters. Its demographic skew is older as well. The engineers are older and many have been at Hewlett-Packard for years. It is a business-slanted technology firm and that is clearly evident during our visit.

Since the company offers some 20,000 products, it is probably easier to say what Hewlett-Packard is not doing in high technology than it is to say what it is doing. The only area missing is pharmaceuticals. However, most of the revenues originate from computer products and services and peripherals, making computers and electronics a core business.

Although Hewlett-Packard has moved several times since being at the Stanford Research Park (there are still about 10 buildings there), it is still very close to the campus. With 31 percent of the workforce being scientists or engineers and 1,300 new university recruits every year, the academic connection remains important. This connection sets the evolution of Hewlett-Packard. The company keeps close ties with many universities around the world, and Hewlett-Packard employees take classes subsidized by the company, not just at Stanford but also at other sites. That explains the fields of bicycles at Hewlett-Packard locations.

At HP Laboratories, also in Palo Alto, much more of the research and development intensity is found. The company spends 7 percent of net revenue on research and development activities (.5 percent at HP Laboratories). The laboratory is less businesslike and the workforce appears younger with more men than women. In a way, it seems more disconnected with the headquarters and considerably less structured. It is more like what would be expected, with laboratory benches connected by the ubiquitous glowing workstations.

Laurie Mittelstadt is a materials scientist at HP Labs. She joined the company as a technician in 1980 with a M.S. in Physics and Astronomy. Wanting to get more real-world experience, she earned an M.A. in Materials Science and Engineering at Stanford while working for Hewlett-Packard. Mittelstadt provided information about what day-to-day work was like at the company, its challenges and rewards, why HP is different, and what brought her to the company (which could be subtitled the power of Lucile Packard's home-baked cookies).

On the last point, she remarks, "I grew up in the Bay area and it's one of the choice companies in this area. I had interviewed at other companies where there were many security clearances involved, like Lockheed and GE. There were a lot of government contracts and Defense work, which I did not want to do, so HP was very attractive. It was known as an open company where people didn't have passwords to get into the door. That was one of the things that made HP different. They have this reputation for caring about the employees—not hiring a bunch of people then laying them off like an aerospace company might. We used to always have cookies and coffee every morning. When I was hired, I was actually attracted by the doughnuts, a tradition that goes back to when Lucile Packard baked cookies for the employees."

The printing technology department is where Mittelstadt spends much of her time, although she has another intriguing assignment we will describe shortly. Thermal inkjet technology was invented at Hewlett-Packard and the company continues to explore the process in the printing technology department. "My particular work is on laser ablation, which is an ultraviolet laser that drills nozzles in plastic, which are in products now. I work on refining that and trying to improve the way we make print heads," she explains. Concerning the challenges and difficulties of doing this she adds, "We're always trying to make the print quality more uniform from nozzle to nozzle. The difficulties are that it's a materials process and there are always physical limits to what you can do. Our job in the lab is to identify the limits of the technology. How far can you push it before you have to look at some other means to get what you want? With a thermal inkjet pen, how can you make something of the highest quality that is pleasing to your eye, like a *National Geographic* picture?"

It's not all machine making for Mittelstadt. A surprisingly large part of the job is research oriented. As she describes, "Much of it is scientific investigation. What's the best thing to be working on? When do you start looking for some other printing technology? What do the customers want? And for the HP divisions that make the pens, we're constantly working with them on what they want. It's a mix of science and application orientation. It's finding things that are useful to work on. We don't do things just for the sake of science. We can't, because we're a small group with limited resources. As we get more into printing technology, it's become much more capital intensive. You can't just make one of the printheads in the lab anymore. So we aim at doing the lab work that is useful." Mittelstadt notes that it's especially important for the corporate lab to be future focused because Hewlett-Packard divisions have their own research and development teams. With this comes opportunity, however. Mittelstadt elaborates: "In the lab, we're expected to set our own agenda. Each engineer is expected to keep abreast with colleagues and the divisions with what is going on and helping to define what to work on."

The way engineering work is done at Hewlett-Packard has both conventional and distinctive aspects. Project management is used, but engineers are not required to reach strict milestones in the early investigative phases of a project. As Mittelstadt puts it, "We usually have some project we are working on for which the scope is a year or two in most cases. There are no real sharp deadlines and the project involves making something in the lab, usually some kind of film or drilling some kind of material or assembling something and testing it. We also have to keep abreast of all the other work going on. Any day, though, there's not a particular thing you have to get done. Engineering is different from any other profession in that we don't have to bill our time to any particular account." There are review and controls, however. "For projects, we have reviews at least twice a year where we have to present our results to the divisions, our customers and managers. On a shorter-term basis, we have project meetings once a week and there are people from the divisions drifting through once or twice a week," she states. The engineers work with technicians on a daily basis doing software programming, measurement, and instrument control tasks. There is also time spent preparing presentations, using voice mail, and searching the Net for information that supports projects. Overall, Mittelstadt sees her time as breaking out among design work, communications, and professional development. There is not much repetitive work. "If there is, there is something that is the matter." she adds wryly.

With the large emphasis on self-motivation and performance, what happens when projects fail? "Some people have made big mistakes, but for most of the projects we can eke some kind of learning out of it," she states, offering the example of a project that made optical fibers only to have it canceled. The people were kept and transferred to two other important departments.

Mittelstadt calls herself a special case, meaning she also has a half-time rotational assignment with the director of personnel for HP Labs in addition to the printing technology work. She is quite enthusiastic as she describes the additional work: "I've been involved in a lot of networking activities for women engineers in the company. Through that, I'm working with the human resource area and looking at ways of helping employees look at where they are in the company and how they can help steer HP labs into the future."

This is an evolution of her involvement in working with the company's women engineers. Although there have always been women engineers at Hewlett-Packard, their number reached a critical mass in the late 1980s, when they first came together and networked. The Hewlett-Packard women engineers gravitated to each other at an external Society of Women Engineers Conference in Kansas City. As Mittelstadt explains, "They thought, 'Isn't this kind of stupid that we have to come here to meet each other and talk about business,' so they decided they would approach Corporate with the idea of a technical meeting for women engineers—to get together and share energy and meet each other and some high-level corporate speakers and technical papers like IEEE conferences. So we did that back in 1988 and it was extremely successful." Mittelstadt adds that the conference dispelled the ideas it would be a coffee klatsch or gripe session. The last conference in 1995 had grown to over 3,000 women engineers. Mittelstadt says that the Technical Women's Conference provides networking opportunities and highlights role models. For herself, she appreciated "just walking into a ballroom with 800 women engineers. It was kind of a turning point experience for me because, after sitting through all my college life and most professional societies and conventions which are 90 percent men, my perspective just turned and made up for the long haul. I gathered a lot of energy from being around the women, especially women engineers." Mittelstadt adds, "Women have said that by seeing other women who have made it in the company and talked about their experiences as they took different risks, it inspired them. One woman said she turned down a job as an R&D manager, but after hearing from HP's own Carolyn Ticknor, vice president and general manager of HP's LaserJet Solutions Group, and how Carolyn said the hardest decision she made was to take the first step as an R&D manager, this woman was encouraged to do that, too."

In summarizing her views, Mittelstadt could not think of a better place to work. "What we found in some of these large events we had for the labs is that what people really want is a feeling they are contributing something to the world and aren't just punching in their hours. A lot of people spend a lot of hours here, but it's definitely by choice. More than that, though, is that people want to bring their lives into their work." Lucile's cookies have plainly hit their mark.

If anything can summarize Hewlett-Packard it is that the company is the enduring mark of high technology business success in the United States, having a long, uninterrupted chain of technology products as a testament to Hewlett and Packard's working in a small shed in 1938. When industry giants faltered, as did IBM, Wang, and others, there were

Hewlett and Packard plugging away, showing that growth could be planned when quality and customer support were part of the vector. To a large extent, that growth was achieved by the kind of employee practices we observed in Palo Alto.

Integrated Circuit Design

The electronics industry is the oldest form of a technology industry, with roots back to such inventions as Schilling's electric telegraph in 1825. It is deep mine for innovation and one that continuously sends up new nuggets. Electronics is essential for many businesses—telecommunications, utilities, entertainment, printing, manufacturing—and an ubiquitous part of all others. It must be considered a linchpin for all high technology.

The evolution of electronics has been dramatic and well told. We do not deal with it here except to pick up the evolutionary stage as it stands—a point when small circuit design firms have the capability of producing integrated circuits.

Our look now is at one small part of an exceptionally broad industry. The firm of interest is a small entrepreneurial company—Reining International—located in Madison, Wisconsin.

Reining International creates application-specific integrated circuit designs for customers such as Dairy Equipment Company, which is looking for reductions in product size, briefer development times, and lower costs—all achieved through the properties of silicon. Reining International also does electrical and computer engineering and product development services. The 10 employees (8 scientists and engineers and 2 administrative staff) produce annual sales on the order of $1 million.

A third to a half of the company's customers are medical firms, owing mostly to the background of the founder, Bill Reining. The other customers are in process control, dairy, agribusiness, and a mixture of other industries. Circuit design is a big part of what this company does, but Reining International prides itself on the extended range of what it does for customers. That includes conceptualization at the start and preproduction phases and product at the end. Typical design firms offer more limited spans.

This company combines many of our themes—entrepreneurial, university spinoffs, mixing software and electronics, and technological intensity. It is a microcosm of what we might find in young, closely held electronic design houses. We'll explore the workings of this firm and have the owner and president, Bill Reining, explain his business.

Like most of other visits, we find Reining International hidden in an unexpected location—a business park in Madison, Wisconsin. But inside the door, it's plainly high technology ground. This is the land of very small technology, mostly chip design and computers doing that work. A device with probes only fractions of microns in diameter is poised to work on circuitry once an operator arrives to inspect an integrated circuit through the attached microscope.

Inside a business-style office with no laboratory, on the other side of a large but nondescript dark wooden desk, is Bill Reining, who clears the edge of the desk of small odds and ends. We talk briefly about the University of Wisconsin Space Astronomy Laboratory where we both worked in the 1960s. It was there that Reining started the Blinding Flash Award, which was given to the person who destroyed something and created both light and smoke in the process. The award had an etched plate with the fictitious first recipient, who had a shock-induced expression of perplexity on his face as clouds of smoke dissipated from something vaguely recognizable as electronic equipment crumbled in his hands. Winners were Dymotaped on the back, a back that can no longer accommodate the thirty years of awards, since the tradition is still practiced by the Space Astronomy Laboratory.

Reining, too, remains mischievous and lanky though somewhat steeled by his own efforts at technical entrepreneurship. Reining International did not pass smoothly into the present, as we'll now discover.

Perhaps the most notable aspect of Reining International is the resilience of Bill Reining himself. He is a career entrepreneur, having been through previous technology enterprises (one prophetically named Phoenix Engineering) before forming Reining International in 1989. It is privately held with Reining the major owner, an offshoot of Reining SC, a holding company. Before that, Phoenix Engineering was in operation—a classic freelance university-inspired business. "That was back in the Space Astronomy days, when we were looking for something to do at night. Phoenix was formed by me, John McNall, and Terry Meidner. We offered services to UW departments to help connect their experiments to computers they had laying around. In 1966 it became a corporation. We had short-haul projects. There was no marketing at all. The projects would go into someone's basement and come out. I did a project for Central Colony (a state institution for children with disabilities) for free-roaming children; it used telemetry for tracking. It was becoming obvious to me that the medical world needed telemetry," Reining explains.

Then disaster struck. The young, gregarious, red-haired Scottish partner, professor of electrical engineering John McNall died. Reining bought out McNall's share and went ahead with Phoenix. Then disaster struck again. "In 1988 an employee basically falsified some test results on a product. As a result, we delivered something that didn't work. We ended up in a giant legal battle, something between us and the client, us and the employee. Before it was all done, the lawyers got all the money and Phoenix Engineering was broke," reveals Reining. A note was held by Reining and in the ensuing bankruptcy he was able to get Phoenix's assets, which he used to build Reining International with the financial help of a client. The present business arose from the singed feathers of its predecessors—it is the company that deserves to be called Phoenix. Reining comments on this: "The road to success is persistence—not money, not location, but persistence. Persistence is the one thing that if you want to be an entrepreneur you had better have because it's a lot of fun, but there are a lot of problems you have to get through."

What, then, were the steps taken that secured a future for Reining International? Other partners were brought in. The company remains privately held. The company also sensed an opportunity outside the United States and acted on it. "Up until Reining International came into being, the focus was totally domestic. We made a conscientious effort starting in 1993 in international coverage. That effort was begun because the economies (U.S. and European) were countercyclical. As consultants, we do well when the economy is lousy—because companies are down-sizing, getting rid of internal engineering staffs. They still need products. We ride that wave, but it is a hard thing—three quarters up and three quarters down. It's very difficult for staffing—to keep them busy during quiet periods. That's kind of why we started this international thing. But there are still two-to three-year lags between us and Europe," elaborates Reining. He estimates that a quarter of Reining's present business comes from Europe, where he concedes doing business is different. "We've never had a marketing program in the United States. It's all word-of-mouth. People would hear about us from friends. In Europe, however, it's different. We come in cold. No one knows us. So we're attending shows. We've got representatives over most of Europe and we're making a very concerted effort to develop that business," he states. He mentions a few customers in Wales and Denmark.

As Reining explains it, another thing that has kept his business going is its special nature. "The big thing is we design integrated circuits—a very special kind, a mixed-signal integrated circuit. This is a circuit with both analog and digital functions embedded in the same piece of silicon. So you think of and can get many different functions in it—embedded microprocessing, embedded software, big microprocessors. We call them Crays in a chip, quarter-Crays, half-Crays, etc."

Having an idea of what the silicon can do, Reining International knows how to carve it for the customer's needs. "Normally, the people come in with a concept. It may be a formal concept or a nebulous concept—most projects are betwixt and between. We will take their concept and refine it to a specification. We'll design, simulate the design, proto-type, get them into production by working with whoever is chosen to do it. We almost always do prototypes but no production, but we work with production people on the first run until everything is working smoothly and then we're kind of out of the picture," explains Reining.

For Reining himself, the engineering part of the business is but a passing enterprise. He now spends a third to a half of his time doing sales and marketing (mostly sales) and a quarter of the time in adminis-tration. It is one of his laments.

We are introduced to Dr. Maha Jaffer in the circuit design area of the business. She is working on the physical arrangements of high-voltage multiplexers on an integrated circuit. There are important considerations here—bus length reduces speed, maximize the number of microprocessors but make room for physical connections, and so on, yet she has an advantage that circuit designers of old never had. With her computer, she can observe the design down through layers of circuitry to the gate level and redesign as she needs.

As we loop around the office, we see an electronic technician working in what looks like a more commonplace electronics shop. There are circuit boards on the wooden workbench and tools and soldering irons, but no high-tech instrumentation. In another area a technical writer and engineering vice-president are chatting about business. In a way, these people are silicon sculptors whose artistry cannot be detected by what they make but by what it does. Their work is in the area of the invisible, but the results are in the realm of the real.

Summary

The four companies profiled in this chapter have demonstrated just how different high technology companies can be. They have different business objectives and different ways of operating. Even their structures differ substantially. If there is a common thread, it is that the founders and managers all have an optimism about technology and a desire to make it successful commercially. As much as science is the base for these companies, business is their raison d'être.

In the next chapter we will turn exclusively to high technology ventures and devote the next five chapters to that subject. And we will start at the start, when the venture is a gleam in the eye. Our subject will be how the venture idea becomes a venture with resources—specifically, the financing of high technology ventures, an unavoidable step toward actualization that separates idle dream from business operation.

Note

1. Joan O'C. Hamilton, writing in the April 22, 1996, issue of *Business Week*, argues that although Japanese companies captured half the patents for human DNA sequences between 1981 and 1995, such attributions play quantity over quality. Further, she notes that there was a similar and earlier fear that Japan would take over the electronics business through patent creation and acquisition. Japanese firms filed many electronics patents, but they were for narrow scope products. Biotechnology moves even slower than electronics, and patents may be no advantage because they are lockstep with manufacturing processes that can easily become outdated during the long march to commercialization. When a drug enters testing, it cannot be changed even in the way it is manufactured because changing the manufacturing may compromise safety or purity.

References

Magretta, J. (1997). "Growth Through Global Sustainability: An Interview with Monsanto's CEO, Robert B. Shapiro." *Harvard Business Review*, January-February, pp. 78–88.

Packard, D. (1995). *The HP Way*. New York: HarperCollins.

Shan, W. (1990). "An Empirical Analysis of Organizational Strategies by Entrepreneurial High Technology Firms." *Strategic Management Journal*, February, pp. 129–139.

II

NEW
TECHNOLOGY
VENTURES

4

FINANCING
VENTURES

In the previous chapters, we provided an overview of the business of high technology. That included defining high technology businesses, discovering the history of high technology, and putting ourselves in the workday world of high technology. There is something important missing, however. It is the spark for the kindling—that initial supply of money that puts a venture into action by paying for people and materials. That money is often provided by venture capitalists, and this chapter is dedicated to examining the practices of venture capitalism in high technology launches. As we discuss later, some other sources for venture funding are family loans, other partners, conventional business loans, and personal savings. But venture capital is one of the newer and more intriguing sources, and so it is our focus here.

Specifically, this chapter discusses the role of venture capital and venture capitalists in high technology ventures. It also presents the views of a number of venture capitalists on the subject of high technology ventures. The process of obtaining these opinions and the content of the interviews are described in this chapter, as well as the results and the contributions these results make to our look at high technology ventures. We then draw conclusions about the venture capital connection.

Something Ventured: Something Gained

Much of what is said about venture capitalism is said by venture capitalists themselves, and most of that is said by venture capitalists who make technology firms their trade. Including the opinions of venture capitalists specializing in high technology is essential. Their perspective helps enrich our understanding of the entire issue of high technology ventures because these figures have a genuine stake in the formation and maturation of high technology companies. That stake cannot be equaled by reporters or researchers who are not directly involved with venture success.

Put into its most basic form, venture capitalists supply money and expect more of the same in return. They are not in the business of social largesse, though they have considerable goodwill for the practicing entrepreneurs they fund. Venture capitalists are not grant givers, either. Grants for technology ventures, few as they are, originate from government sources or private benefactors, not venture capitalists. That is an entirely different kind of operation, one usually associated with financing basic research rather than business practice. In Chapters 17 and 18 there is more detail on state and federal support for high technology.

Where can the venture capitalists be found if one is looking for them as a funding source? They are somewhat elusive, expecting to be found by the searching entrepreneur rather than openly publicizing their presence. Advertisements for their services are in the specialized trade publications, not in mass-circulation publications. In their operations, they deal mainly in capital that is obtained and passed on to promising firms. They usually do not do highly visible venture promotions. Venture capital firms resemble investment banks in general function, but not in the physical sense. Their setting is more officelike than institutional. In this environment, there are the typical office administrative functions observed anywhere: incoming phone calls, mail distribution, staff meetings. But at the core of venture capitalism are frequent meetings with prospective recipients, the pouring over of spreadsheets done by investors in private offices, and the hours of scrutinizing business plans. In larger firms, staff members do the analysis of the business plan and make recommendations to a decision-making partner or officer. In smaller firms, there is naturally less division of labor and the partner does the work. Venture capitalists are in a quiet, organized, but crucial business niche in which their ordinariness belies their significance.

The Venture Capitalist's Role in High Technology Venture Development

Venture capitalists are paradoxically the most intimate and most distant party to the venture experience, at least in terms of how they are commonly understood. High technology founders themselves put it best when they call venture capitalists derisively "vulture capitalists" or reverently "guardian angels," depending on their own experiences. Their references to skyward intervention capture the idea that venture capitalists can grace the nascent venture with first-capital infusion or they can pick at its equity bones with the dispassionate mechanics of scavengers. Venture capitalists will keep their distance as the venture struggles, yet their eye on it remains keen in case they need to pick up the pieces.

These are stark extremes and not the only modes of the venture capitalist, but as extremes they leave a dramatic imprint on us and the founders who partner with them. From the technical entrepreneur's perspective, the choice of a venture capital agent is one of the most crucial decisions that can be made. The history of high technology demonstrates just what that can mean.

A few of the connections made between venture capitalists and high technology are legendary, none more so than the launching of Digital Equipment Corporation (DEC). This founding marks the formalization of the venture capital influence on technology ventures. When Ken Olsen and Harlan Anderson started DEC in 1957, they contacted General Georges Doriot for funding. Doriot was head of American Research and Development Corporation (ARD), a venture capital firm that had been recently started by a cadre of MIT alumni. Pioneering venture capitalism and pioneering computer designs went hand in hand in this instance.

Doriot was a Harvard Business School professor and the consummate Continental. Patient with his ventures, not expecting instant investment returns, Doriot considered his favored ventures as progeny. He asked Olsen to meet his wife before he decided to invest in DEC. Doriot saw the strength of a marital partnership as important as a business partnership. He taught his Harvard students how to marry the right kind of spouse, one who could endure the hardships of entrepreneurship.

The association between DEC and Doriot was to last thirty years, one of the most enduring and successful venture capital and venture combinations of all time. Though Doriot extracted a very large equity share from DEC, he was patient with his corporate client and didn't expect the instant and plentiful returns that some of his contemporaries did. The bond between Doriot and DEC remains a classic and now legendary nexus in the business world.

There are other companies that have been aided by venture capital. Intel received the venture capital sponsorship of Arthur Rock. Apple Computers was founded with a $350,000 sum from Rock and others. Microsoft, Lotus, Software Publishing, and Digital Research all obtained some level of venture capital funding. What's remarkable is that the initial investments were so small in these star companies. As Doriot did, these companies took their investors for the equity ride to the summit. Aside from the famous names in computer technology, many more less well known firms are sponsored by venture capital. After all, these companies were not renowned when the venture capitalists were paying for them. That may not even be the expectation of the investor, who may instead more reasonably expect a target ROI from a known technology rather than an astronomical ROI from an unproven technology.

As noted before, venture capitalists have the primary function of providing technology companies with capital. The performance of that invested capital is very closely related to the performance of the venture capitalists. Venture capitalists have more than theory at stake in their decision on which ventures to fund. They have their firm's capital and their own reputations at stake, along with the prestige and status of their venture capital firms. This is especially true of well-established venture capital firms such as those surveyed for this chapter. The individuals who make the investment decisions are usually compensated on the basis of how successful their decisions are. As a general observation, venture capital firms tend to reward success more than failure. The compensation component based on contribution to firm profit is the major income source for almost all venture capitalists.

Even a brief review of the venture capital business will show how sizable and important it is. Capital supplied by venture capitalists for all types of ventures, including high technology ventures, amounted to $3.8 billion in 1994, a large jump from 1993 but not nearly the record $4.2 billion of 1988. This year was the high water mark for venture capital in the past fourteen years. Although the amount has ebbed and flowed recently, venture capital has generally grown, though not nearly as much as during the great boom of 1975–1985. In this period, venture capital investment was exponential, starting with $250 million in 1975 and nearing $3 billion by 1983.

In considering the different stages of growth of a firm, venture capitalists put very little money where it might be expected—in the formation of the company. Expansions and bridge loans are the stages that take up most investments. Startups and seed funding accounted for only 20 percent of capital in 1993.

What is true is that high technology companies attract a considerable portion of venture capital dollars. Software and services dominate investments, with 203 firms and 21 percent of dollars. Nearly three-quarters of the firms funded in 1993 were high technology firms, using a liberal definition of the term *high technology*.

The Lure of High Technology

What attracts the venture capitalist to high technology? There may not be a simple answer. It is not a matter of being a safe bet in technology as opposed to other types of businesses. Technology firms do not stand out as safer investments compared to other businesses. Although precise definitions of high technology are not used in the Dun & Bradstreet reported business failure rates, broadly defined high technology groupings have comparable failure rates (by business failures per 10,000 firms). Finance, retail trade, utilities, mining, and agriculture all had fewer failures in 1994.

The attraction of high technology for venture capitalists has to do with potential return on investment. High technology is a growth and profit opportunity, as we have shown earlier. But the appeal also has to do with the longevity of investments, as much as with the intensity of returned investments. More than one venture capitalist has told us that he is "in it for the long term" and that he is "always interested in companies that will be around in the future." Venture capitalists are known to make repeated investments in technology startups, even though they may have been burned by earlier failures. This is a characteristic they share with failed entrepreneurs, forming a camaraderie among the doggedly loyal.

Also mentioned by the venture capitalists are the challenge and attractiveness of the sector. There are venture capitalists who don't understand, or even want to understand, the science behind high technology. They accept the challenge of working with the unknown and try to make their investment decisions work despite being hobbled by a lack of scientific knowledge. There are venture capitalists who are scientists and engineers, but these are few compared to the finance masters and MBAs who make their way into the field. To other professional investors, the essence of the investment decision is much more prosaic. They think putting money into high technology beats putting it into a refuse disposal business. There is at least a modicum of glamour in this business compared to some others.

Our going to the venture capitalists themselves helps validate or dispute theories about how they behave. The inclusion of their views here also helps build on the knowledge base of high technology ventures as constructed by high technology theorists and researchers. For instance, the incorporation of practitioner insight could help modify popular versions of venture capitalism, as well as verifying or countering theory and research. In that sense, it offers a real-world test for high technology venture theory and research.

Funding Considerations

The rationale for considering the views of venture capitalists originates from the premise that high technology ventures make use of venture capital firm services. Venture capitalists do not fund all high technology ventures, but the ventures that are funded depend greatly on this outside capital source. In looking at this aspect of high technology, we see that venture capitalism can offer information on an often crucial dynamic: the internal management of the venture and the external supply of capital. Put another way, it is not just the infusion of capital that sustains new high technology ventures but also the ability of venture management to determine when and how much capital is required to operate the high technology venture. The supply of outside capital and the capacity of the venture manager to acquire and use it correctly are what this dynamic is all about.

It is commonly assumed that venture capitalists are the sole suppliers of money for the technology venture. That's understandable, considering how the spectacular stories—DEC and Apple—have been publicized to the point they obscure the ordinary, more modest failures and success in the industry. But it's certainly not the case that all technology money comes by way of the venture capitalist. There are a variety of ways money comes into a new venture and the venture capital route is only one of many. Venture capital is, in fact, not the major component of high technology venture funding. It is part of the array of personal savings, private loans, or government funds mustered to create technology companies.

Several researchers have looked at the role of venture capital in business starts and have found it to be a varied and crucial function.[1] In addition to actually providing startup money, venture capitalists offer consulting services for new ventures. The KPMG Peat Marwick firm, for example, offered educational and consulting programs for venture managers and prospective venture founders. Spaienza and Timmons, in a work cited in note 1, observed that venture capitalists provide sounding board, business advisor, financier, and coach-mentor functions as primary ways of supporting new ventures.

Venture capitalists as a whole serve to screen and evaluate as they consider venture proposals and funding requests. It is these functions that have particular application in this study. The evaluation function allows the researcher to investigate how practitioners establish and weigh the various factors they believe will lead to a successful venture and the consequent satisfactory return on their investment.

Venture capitalists are in the position of being independent and self-interested assessors of potential venture success. They usually don't have a bias for or against the *what* the venture proposes to do, or are at least more restrained than the venture founder's unbridled enthusiasm for it. Venture capitalists are also well positioned in terms of timing and can make decisions most often before the debut of the venture, in the critical pre-startup phase. They are also involved as the venture begins operation and as it grows because there is often a need for additional funding.

This timing and the venture capitalist's importance suggest that the role of the venture capitalist can be crucial for the high technology venture. The methods used by venture capitalists come as close to providing a real-world model for decision making about successful high technology venturing as is available. Indeed, it is the first real-world test for the new technology venture.

What Researchers Say about the Venture Capitalists' Role

There has been some recent research on venture capitalists, and much of it has emphasized their importance in venture founding. Venture capitalists play a role very early on in the technology organization.[2] The most significant and directly related study of venture capital in high technology was done by Edward Roberts.[3] In examining the initial capital needs of 156 technological firms spun off from MIT departments and laboratories;

Roberts found that capital needs were provided by personal savings for small ventures and by groups for larger ventures. Roberts cited research that noted how a small group of venture capital firms provide most of the capital for technology firms seeking such funding. Twenty-one venture capital firms constituted fewer than 5 percent of the 464 firms in the database, but they accounted for nearly 25 percent of the highly innovative technological venture investing.

Venture capital can be viewed in the context of financial support for technology ventures. A significant article by Freear and Wetzel (1990) looked at the financial histories of 284 new technology-based firms founded in New England between 1975 and 1986. The aim was to discover where these firms obtained their funding. It was found that private individuals other than the funding management team and its relatives were the most prevalent source of equity financing. The second most common form of equity financing was venture capital funds. A small number of companies received funds from nonfinancial companies and public stock offerings. The authors concluded that individual and venture capital funds play complementary, not competing roles.

The importance of venture capital funding is evidenced in the practical side of high technology as well. A conclusion from KPMG Peat Marwick's 1988 document entitled "Building the High Technology Business" on business planning for technology ventures is that high technology firms require more initial capital more quickly than do conventional businesses. A guide to developing a high technology business plan prepared by Arthur Anderson & Company entitled "How to Develop a High-Technology Business Plan," published in 1983, suggests that venture capital firms are an appropriate capital source when money is needed quickly. Indeed, high technology investing is a special case of business investing, in which the capital lifeblood is needed quickly and sufficiently. This is recognized by prominent consulting firms, as well as individual venture capitalists.

Survey of Venture Capitalists

In order to determine how high technology venture capitalists look at prospective ventures and what they look for, we contacted 14 venture capital firms.[4] The firms were all identified as top venture capital sources by their assets, as reported in the former *High Technology Business* magazine in the March 1988 article by F. Paul entitled "Where Venture Capital Is Investing Now."

It is appropriate to consider the top venture firms because these firms presumably have sufficient experience and resources in dealing with many high technology ventures. They see a lot of proposals and they invest selectively—their investment portfolios bear this out. It also seems reasonable to expect that these firms have assessed a large number of high technology business plans.

Four firms agreed to provide detailed responses from the original fourteen firms. The responding firms replied with detailed information on how they characterize successful high technology ventures, as well as giving their general approach to funding new high technology ventures.

The individuals interviewed represented venture capital firms with $200 million or more in assets. Their investment amounts ranged from $10 to $100 million. Typically, the percentage of the investment in high technology firms was 75 percent of their assets. Each of the venture capital firms had a special interest in some type of high technology venture. For example, one firm specialized in biotechnology ventures while another invested in electronics. Firms from both the East and West Coasts were represented. Although these venture capital firms had investment preferences, they were also open to other investment opportunities. A majority of the high technology ventures funded by these firms were founded by sole entrepreneurs who had technical expertise and some level of managerial experience. The ventures considered for funding were mainly in the startup phase and were well diversified in terms of representing electronics, biotechnology, and computer ventures. The four responding firms were generally representative of the top 14 firms.

Interview Results

Each participant was asked three identical questions: (1) What distinguishes a successful high technology venture from an unsuccessful venture? (2) What criteria do you use in deciding to fund high technology ventures? (3) What has the firm learned in funding new ventures? If the venture capitalists developed distinctions between successful and unsuccessful ventures, they were asked to give specific examples.

The respondants showed strong agreement on the distinction between successful and unsuccessful high technology ventures. One East Coast venture capitalist, respondent A, remarked, "They say you should pick the venture leader over the venture, and I'll agree with that. But it is not easy to pick the best managers. Past record isn't enough. There is always a big risk in new ventures, even with proven managers."

Another East Coast venture capitalist, respondent B, also emphasized the role of the manager stating, "Clearly, top management is what defines successful ventures. The managers must be resourceful and able to direct the entire company to a goal."

Still another East Coast venture capitalist, respondent C, replied to the open-ended question by commenting, "Executive management is the key to the high tech venture, the owners, the drivers. Putting it into one word, it is management. If I had to put it in more than one word I would say management, management, management."

Respondent D, representing a West Coast venture capital group, remarked, "We always look for good people. Then we consider the good idea. That is the sequence we follow and it is the right sequence. There are plenty of good ideas out there, but fewer of the right people to get them to work."

Two of the venture capitalists elaborated on their answers when asked to consider specific examples of high technology ventures that succeeded or failed. Two other respondents did not wish to reply. Respondent B stated,

> The difference between success and failure is always one or two things or a combination—the management team and how early in the process you move the critical risk factors. If the risk is a technological hurdle, before you spend any money on sales, marketing, administration—anything—did you remove that risk? Get it under control? Jump over it? See the light at the end of the tunnel? Whatever, is that risk now managed? In a biotech company, is there the breakthrough you thought? In a computer company, did the chip work? It's the management of the risk, not necessarily the risk, that is key.

Respondent D replied,

> As far as exact instances are concerned, or comparisons between one success and one failure, if I knew that we would have all made a lot more money. The simple problem is that it varies from situation to situation in some profound ways. It's very hard to generalize on this question. You hear a lot of stories in the venture world about it involving the quality of the management effort or their experience in a particular industry or the competitive environment. They are all factors. Most of the industries we invest in are moving rapidly. There is a lot of change going on, and by the time you get your product in the market or you're hoping to generate revenues, the entire landscape may have changed and it's no fault of anyone—the management or the people who financed the companies. You're in rapidly changing industries. Sometimes things fail for technical reasons. You're attempting to take the next step and the next step doesn't work.

Do these reasons for funding technology ventures bear up if we look at the reasons why ventures are not funded? Apparently so, because the main reason given for rejecting applications for venture capital finance was "management weakness." It accounted for 71% of rejections in a survey of venture capitalist practices.[5]

In responding to the question about the criteria used in deciding on funding a high technology venture, the responses were varied. A West Coast venture capitalist, respondent D, stated his firm's criteria as,

> We do a personnel evaluation, look at the characteristics of the organization, see if the market niche they are trying for is real. We ask if what they produce people will buy, and we say no to ventures that are trying to create a market. We find that evaluating the technical risk is hard to

do. The risk is that somebody else will invent a clone. A cheaper clone can put the original out of the market.

Respondent A offered, "There is not a single way but a number of ways. We'll look at their managers and their market. But there is no formula for evaluation or some kind of test that we apply to everyone."

The comments from respondent D were, "The innovation level of the product is right up there, but it's not the only thing. It's the whole package that gets our evaluation, maybe a subjective one but it is one we are comfortable with."

The criteria cited by the venture capitalists were essentially subjective, in that none listed quantitative techniques to assess the personnel, market, or product. None of the venture capitalists referred to industry rating standards either, and it appears they have developed their own informal evaluation methods. This is not to say that quantitative methods are not used at some stage of evaluation, but apparently they are not preeminent as a decision-making reference point.

Within the sample, then, there was no uniform approach to evaluating these ventures. The practices appear to have originated from the venture capital individuals and organizations. However, the venture capitalists did provide other sources for their decisions to fund a venture. For one, they examined business plans. According to respondent A, "We'll look at the business plan for sure and the resumes of the principals. If they have it, we'll also look at product information, brochures, spec sheets for example."

All the venture capitalists used extensive questioning in assessing the proposed venture. Their questions dealt with reasons why the venture was founded, how sales goals were developed, how the company expected to meet those goals, and what resources it had, along with many other questions.

Two of those interviewed remarked that they viewed funding a venture as continuous involvement. They expected the venture founder to come back to them for financial support during several stages of development. As respondent D put it, "I know I'm in for continuous involvement no matter what happens. If it takes off, they are often back for more money. Even when the venture doesn't work, I keep in touch. One founder has been through a couple of other failures, but I still work with him, share ideas and such." This observation was generally supported in the other interviews, as respondent A suggested, "High technology ventures are different than conventional business investments. Technological risk is involved and the venture can be doomed or it can succeed on the basis of accurately reading this risk. We spend a lot of time with those companies, get to know them and the risk they face. And we tend to stay with them too."

The other side of risk is gain. As noted by respondent B, "The return on investment can be quite good in the high tech field—much higher than in conventional businesses. That happens in the computer field, for example—that's a good counter to the big risk."

Only one of the venture capitalists elaborated at length on what was learned from his experience in capitalizing high technology ventures. According to this venture capitalist, respondent D, "The successes are great, but I really build relationships with some of the more dramatic failures. These people really have a lot on the ball, but the venture just didn't work out. They stay in the technology business. One guy tried twice in the software field. I keep talking with them because I think they will eventually make it, even if we don't fund them."

It is also noteworthy that the four venture capitalists who were interviewed considered themselves to be more than providers of venture capital. Two mentioned their consulting function in advising those with ideas whether to seek investors. Respondent D explained: "Consulting with the entrepreneur is a big part of all this. It's working one-on-one, almost like a business partner that often takes up the time. Consulting is as big a part as getting them money." And Respondent C added, "I like putting together [money from] a number of investors and [my] own firm's capital in exchange for a major equity interest. It is larger scale, like deal making instead of just funding."

The remarks by respondent C are an expansion of the customary view that venture capitalists decide how to invest the assets of a single firm.

Other Views and Analysis

In order to provide perspective on the role of the venture capitalist in the high technology venture, we held two additional interviews. One was with the president of a software venture who had worked with venture capitalists, the other was with an experienced venture capitalist and former head of the Wisconsin State Department of Business Development. These interviews were done immediately after the initial interviews.

The president of the software company offered an entrepreneur's view:

> Venture founders should not give away too much equity initially. The reason is that the venture usually has to come back for more funding, and the venture founder could end up giving away controlling interest in the venture. If any outside investor holds more than half ownership in that venture, the founder is relegated to employee status, and he or she cannot not fully pursue the concept that originally motivated them.

The results of these interviews, including that with the software company founder, were then summarized and discussed with an experienced venture capitalist and former head of the Wisconsin State Department of Business Development. The purpose of this was to have an overview of high technology venture capitalist views from another informed source. The reaction of this venture capitalist to the practices of the high technology venture capitalists was that their informal evaluation methods were similar to the practices of venture capitalists who sponsored conventional businesses. However, he also noted that there were special characteristics of high technology that were both difficult yet necessary to assess: technical competence, technological risk, identification of a market, and degree of product innovation. He concluded that venture capitalist activities in the high technology area merited a distinct, specialized approach within the broad field of the venture capitalism.

The two additional views did not contradict the opinions of the four venture capitalists interviewed. In fact, the two additional observations corroborated the informal evaluation styles, decision-making approaches, and desire to obtain equity on the part of high technology venture capitalists.

Venture Capitalist Interviews in Review

The results of these interviews form the basis for a set of practical criteria for studying high technology venture performance. Among these criteria, venture management is a crucial aspect. Venture capitalists hunt for management experience and reward it when they find it. They also use informal and varied venture evaluation techniques.

The interviews also add a new dimension to our understanding of high technology venture performance: quality of management as a consequential factor. Relatively better performing ventures will have more experienced, trained management and better management systems than will lower performing ventures. Although the venture capitalists suggested rather than definitively linked performance with management factors, their strong endorsement warrants consideration.

The venture capitalists interviewed uniformly agreed that venture leadership is crucial to performance. Apparently they would select an effective leader over an effective idea, if given a choice. Nevertheless, at least one of the venture capitalists argued that the idea must be carefully and perhaps equally evaluated, since there is such high risk associated with the business. The ramifications of this dominance of leadership over concept or product was not elaborated on by any of those interviewed.

The venture capitalist is ideally positioned to assess risks in relation to rewards in a technical venture, but may not be equipped to assess all aspects of those risks. Such risks may be technological obsolescence, duplication of the technology by another firm, failure to grow through timely infusion of capital, technical labor requirements, and unforseen costs. This limitation has been conceded by the venture capitalists themselves. Indeed, venture capitalists offer insight but not omniscience about the art of technology venturing.

We must factor the views of these venture capitalists into how we look at these special kinds of ventures. The venture capitalists view themselves as important to the venture. They are particularly sensitive to the leadership versus venture concept conundrum and recognize the difficulty in correctly weighing these considerations. But none of the venture capitalists went beyond leadership, risk, and the venture idea as reasons for deciding to fund or not. Since the interviewees were asked to elaborate in an open-ended question, they had the opportunity to suggest other reasons. They may not have done so because of time constraints or perhaps they saw no reasons to do so, given their resounding and immediate endorsement of the managerial factor in high technology venture success. We need to note that of all those interviewed for this book—venture founders, governmental officials, community leaders and others—it was the venture capitalists who seemed most pressed for time.

Venture capitalists also appeared to see themselves as involved in the venture at different stages and not solely at startup. All expressed the view that they expected continuous involvement in the venture, and even when the venture failed, they continued to maintain interest in the activities of the founder.

The interests of these venture capitalists were not solely on the fate of invested funds, either. Virtually all of them expressed an understanding of the efforts required by the founder to make the venture successful and most indicated a willingness to share both the good and the bad side of the venture's fate.

They also articulated a dependency on the venture founders and on their technical proficiency and managerial expertise. They believe that ultimate success of a venture is in the hands of the venture leader. Two of the venture capitalists stated that it was difficult to find the attributes of technical proficiency and managerial skill in a single individual, yet they depended on those attributes to make the venture a success. As expressed by respondent D, "It is very rare to find somebody who is on top technically and can still run a business—not really possible to get someone like that. So I try to make sure the weak side, technical or managerial, is not too weak."

The same two venture capitalists felt more qualified to evaluate the managerial aspect of a venture than its technical side. In spite of such concerns, the venture capitalists were not reluctant to make investment decisions about prospective ventures.

It is important to note that, from the point of view of the venture capitalist, the venture leader or founder is synonymous with the venture manager. This was implicit in the interviews as they described the same individual as having founding and managing responsibilities. For purposes of this research, venture founders and managers are one and the same.

There were only four venture capitalists interviewed, and this is a limitation of our approach, yet they represent the largest of the venture capital firms and can be seen as bellwethers for venture capitalist activity. Additional in-depth interviews would be needed for a comprehensive look at this specialized field.

Additional Venture Capitalist Perspectives

Do venture capitalists see things the same way as venturers? The answer is not always yes, as a result of past research. Dean & Giglierano (1990) examined the problems and differing perceptions of venture financing. Based on interviews and a questionnaire, it was found that venture capital firms reduce risk by showing preferences for certain stages of financing or for funding single versus multiple-round funding. The preferences appear to be idiosyncratic. Therefore, entrepreneurs must be careful to deal with venture capitalists whose tendencies are consistent with their goals. A major problem faced by entrepreneurs seeking such funding is the effort required for the search. It cuts into day-to-day management.

At startup, management and leadership are one and the same.

Another feature of recent venture capitalist practices is that investment now spans venture founding and continued operations. After the 1987 stock market crash, many high technology venture capitalists moved away from funding high-risk initial stages and to later-stage business operations. So as we move on to consider established high technology companies in the coming chapters, we can be assured that venture capitalists will remain with us, ready to fund the needy and deserving operation, as well as advise it and perhaps profit from it.

Summary

The views of venture capitalists are important determinants in what ventures will be funded. Our interviews revealed considerable informality in evaluating ventures. Management was a consistent criteria, however. There are numerous sources for capital for the technology venture—private savings, business loans, and grants—but venture capitalists play an increasing role in financing high technology ventures. For an entrepreneur to be successful with a venture capitalist, they must demonstrate strong management. In the next chapter, we will explore what strong management means, as embodied by the venture founder. In short, we will discover how much the founder matters.

Notes

1. R. Oakey reported in *High Technology Small Firms* (New York: St. Martin's Press, 1984) that 30% of new technology firms founded in the Bay area of San Francisco were established by venture capitalists. Peter Drucker noted that a viable high technology business sector requires access to venture capital, among other requirements. See P. Drucker, *Innovation and Entrepreneurship: Practices and Principles* (New York: Harper and Row, 1985). H. Spaienza and B. Timmons, in their 1989 *Academy of Management Proceedings* paper on "The roles of Venture Capitalists in New Ventures," discovered that both entrepreneurs and

venture capitalists believe the venture capitalist role is important in general ventures.

2. An empirical study by H. Spaienza and J. Timmons (see note 1) examined the role of venture capitalists and chief executives of ventures. The authors found that the venture capitalist plays more than the role of financier; the consulting role is also rated as important. In addition, early-stage ventures required a higher level of venture capitalist involvement. While there have been other significant studies of the role of the venture capitalist in the successful venture. See N. Fast, *The Evolution of Corporate New Venture Divisions,* a 1976 Harvard Ph.D dissertation, and A. Dingee, L. Smollen, and B. Haslett, "Characteristics of a Successful Entrepreneur," in *How to Raise Venture Capital* (New York: Charles Scribner's Sons, 1982). The subfield of high technology venture capitalist evaluation systems and methods has not been explicitly investigated, although the studies mentioned provide information potentially useful for high technology ventures.

3. Contained in "Initial Capital Formation for New Technology Enterprise, *IEEE Transactions on Engineering Management* 37 (May 1990): 81–93, as well as his book, *Entrepreneurs in High Technology* (New York: Oxford University Press, 1991).

4. Venture capital firms contacted were: Accel Partners, Princeton, NJ; Allen Patricoff, New York, NY; Brentwood Associates, Los Angeles, CA; Burr, Egan, Deleage, Boston, MA; Institutional Venture Partners, Menlo Park, CA; Kleiner, Perkins, Caufield, & Byers, San Francisco, CA; New Enterprise, Baltimore, MD; Norwest Venture Capital Management, Minneapolis, MN; Oak Investment Partners, Westport, CT; Sequoia Capital, Menlo Park, CA; Sutter Hill, Palo Alto, CA; T. A. Associates, Boston, MA; Welsh, Carson, Anderson, & Stowe, New York, NY; and Warburg, Pincus Ventures, New York, NY.

5. The Organization for Economic Cooperation and Development report *Venture Capital in Information Technology* (Paris, 1985), referred to as *The Databook of Venture Capital,* reports that other reasons for rejecting applications were "project contrary to investment philosophy" at 20%, "poor business plan or analysis" at 13%, "non-feasibility of project" at 12%, "insufficient or uncertain market" at 9%, "insufficient return on investment" at 7%, and "lack of uniqueness" at 6%.

References

Dean, B., and J. J. Giglierano (1990). "Multistage Financing of Technical Start Up Companies in Silicon Valley," *Journal of Business Venturing* 5: pp. 375–389.

Freear, J., and W. Wetzel (1990). "Who Bankrolls High Tech Entrepreneurs?" *Journal of Business Venturing* 5: pp. 77–89.

5

HIGH TECHNOLOGY
FOUNDERS AND BUSINESS
PERFORMANCE

Any workday, some 100 million Americans travel to work. Almost all of them do this for others, not solely for themselves. They do it for their family and their bosses, or perhaps they have learned to do nothing else. A tiny number of these people—perhaps 100,000—do the same thing, only they do it for themselves and their families. What they do differently is arrive at, work through, and leave an uncertainty so fundamental that it is hauntingly unspeakable, even with spouses in the supposed refuge of home. Unlike most of the 100 million, they live and work knowing it may be the last day at the job because it may be the last day of the job.

These 100,000 people are the technology entrepreneurs—men and women whose daily struggles we celebrate as the twin virtues of independence and technical acumen, a particularly "American" attribute that translates to a particularly "American" job description. They are our astronaut cowboys, our technology adventurers.

There is no escape from the adventure, however. Every day there is another problem to be solved by the technology entrepreneur. The circuit board is still shorting. The memory is impenetrable, even when using machine language. The equations never balance. The new polymer still doesn't hold up during the testing. These are the laments over the chess game in which Venture Death is the opponent, an endlessly repetitive scene straight from Ingmar Bergman's *The Seventh Seal.*

The problems are not just technical; business problems are equally troubling. Why can't I get a handle on my cash flow? Am I going to have to give up even more equity to buy the new building? Business and technical problems loom as the Scylla and Charybdis of the technical entrepreneur's voyage, randomly ensnaring the entrepreneur focused on one problem when another problem also demands attention.

Yet technology adventurers are not a brooding lot. They are a buoyant bunch whose optimism is evident, albeit tried by the challenges of entrepreneurship. They clearly need to be optimistic, since so many things can wreck their businesses.

How do people come to create technology companies? There are a few common ways. Technical entrepreneurship is a game that some players learn from their fathers; others have it thrust upon them when their "safe" jobs in big companies suddenly end; and others have discovered it as a by-product of anger with a company that simply would not listen to a good idea.

There are many sources of technical entrepreneurship, and not all of them are noble. Some entrepreneurs want to put parent companies out of business. Others want to place their genius on the market for others to admire. Although there are far fewer entrepreneurs than there are workers in general, their personalities, capabilities, and backgrounds are about as varied as for any other population. The difference may well be motivation: technology entrepreneurs actually do something about their ideas.

Although human motivation itself is an elusive and complex subject, we can look into some specific background factors that could temper motivation for the person wanting to start a software company, or a Global Positioning System mapping company, or any other technology variant. This needs to be done to do justice to explaining why entrepreneurs act.

This chapter concerns the matter of who starts high technology firms. It also presents findings on high technology venture performance in relation to these founders' backgrounds and prior work experience; education and prior work experience are the specific factors constituting the background experience of founding entrepreneurs. We also look at some possible explanations of how founders may affect venture performance, and draw conclusions about the relationships between entrepreneurial characteristics and business results.

Information for this and all the chapters on high technology ventures was collected through a self-administered written survey. Using the "Venture Economics" database of high technology ventures, detailed information on venture operations, the environment, strategy, and performance was obtained. One hundred and one ventures participated in the survey. More specific information about the survey is contained in the appendix.

Put yourself in the place of someone thinking about starting a high technology business, as we did when we interviewed successful founders. You may be surprised. We found that the creative spark was not always as expected. Very often, it was not the invention or innovation itself but the market need for some kind of improvement—the conditions for the spark, not the spark itself. This was typified by an entrepreneur who said "We saw all these paper and pencil spreadsheets and all these accountants. What a backwards way of doing things in this day and age—paper and pencil. That's what started it all for us." This entrepreneur's general ledger software was launched with a view of the market first; then he wrote the software that made the business. This type of entrepreneur doesn't rush something to market. He sees the market, then develops the product. Indeed, if there is one common characteristic of failed technology enterprises, it is the founder's assumption of a market to support his innovation.

What is the role of the entrepreneur in creating a technology business? Is it a solitary or a many-peopled process? Does the technology entrepreneur do all the business functions or just some? In the chapter we discuss what has been said about entrepreneurship in general and apply that to technical entrepreneurship. We also want to know if some of the factors of entrepreneurship make a difference in the performance of the venture. The chapter is devoted to these subjects.

The Entrepreneurial Personality

In looking at the entrepreneur and his technology enterprise, we enter a world that has been better studied than venture strategy and origin. Almost all studies of entrepreneurship have concentrated on the founder. It is a long tradition of inquiry as well. Why do some people start businesses and others don't? That question has launched a thousand articles based on talking with the captains—a necessary but not sufficient resource for discovery.

The tradition of investigating the nature of entrepreneurship by looking at the entrepreneur has been around for at least thirty years. Economic theorizing about entrepreneurship has had an even longer heritage, articulated most notably by Adam Smith more than two hundred years ago. In calling the entrepreneur a risk bearer and capital provider, Smith saw a role for the intervention of thought in the forces of economic production. Modern technical entrepreneurs probably think little about Smith, Say, and the other theorists as they write their business plans, but the theorists anticipated them. The fact that we can trace mention of the contemporary entrepreneur back two hundred years is as much a verification of Smith's theory as it is validation of some imprinted code in the entrepreneurial mind for starting new forms of economic organization.

Many of our ideas about entrepreneurship have come to be both a definition and a value, as suggested earlier. This is especially true of technical venturing. Entrepreneurship is something valued by writers who observe it. Peter Drucker calls entrepreneurship the heart of development of businesses, as important as management is.[1] Others have taken the view that entrepreneurship is of such value that it should be made a permanent feature of the modern business organization. Doing entrepreneurship within a company has even come to be its own term—*intrapreneurship*. Such variation on a theme may strike some as antithetical to organizational cohesion, since few large companies find real advantage in encouraging employees to act like entrepreneurs. Nonetheless, the language is easy to use even if the culture is difficult to assimilate. Drucker's remark is a strong statement from someone who is largely credited with defining and showing the importance of modern management.

The technologists who want to become entrepreneurs will find that their personalities have been diagnosed, at least if entrepreneurs bother to look at what has been theorized about their reasons for creating their organizations. Psychologists have dissected the venturing technologist. They have taken three different approaches in doing this: the content approach, where entrepreneurship is explained as the fulfillment of inner needs; or the process approach, which explains why people make particular choices; or the reinforcement theory, which proposes that behaviors occur in accordance with resultant punishments or rewards. Any of these streams will give you the reason entrepreneurs act. Anchored at the content end by Maslow and at the reinforcement end by Skinner, the spectrum includes major theorists with differing ideas about human motivation.

The psychological explanation is the predominant way that entrepreneurship has been understood, but there is a problem in finding single, generally accepted theory of motivation. There are, after all, at least the three divergent schools of thought just mentioned. There is, however, a psychologist frequently referred to in studies of entrepreneurship whose view of motivation has been tested and used by several entrepreneurship theorists. That psychologist is David McClelland, (1961) and he formulated the idea of the need to achieve (which he noted as *n-Ach*) as a condition for entrepreneurship. This need to achieve is contrasted with affiliation and power needs, which are stronger among nonentrepreneurs. In fact, the need to achieve is dominant among most entrepreneurs, yet it is found in only 5% of the population. McClelland saw the entrepreneur as one who seeks innovation rather than routine. He is one of the few theorists who define the entrepreneur in these terms.

Entrepreneurs seek to succeed in business or to excel among their peers. Noteworthy is the fact that individual power needs are not great, nor is the need to affiliate. This last point has some interest for us, owing to a pronounced pattern of using teams for venture founding. Teams of like-minded technologists are often found in high technology companies. We can speculate that, among their venture partners, McClelland's entrepreneurs are less concerned with affiliation or power than with innovation. Some might even say the high *n-Ach* entrepreneur could threaten the venture team if he insisted on his way without valuing the affiliation with other founders.

In McClelland's depiction, the high technology entrepreneur could be well described as a "self-starters," "driven," "bull-headed," "ambitious," "hard-working," and "independent" Though not formally tested here, these descriptions seem to apply to the group of founders we interviewed. These are personal qualities that prospective technology entrepreneurs might want to inventory before starting a company.

The tradition of content psychologists such as Maslow and Hertzberg can be used to explain why entrepreneurs are motivated to take risks to start new businesses and decline the purely monetary rewards of regular jobs. But the application cannot be precise because motivations change and cannot be clearly deciphered from the enigma that is human personality. Nevertheless, the content psychologists appeal to us as especially appropriate in looking at technical entrepreneurship because they emphasize the "fire within" the entrepreneur. The technical entrepreneurs we met all had quite varied environmental circumstances. Some had money from previous successes; others had only second mortgages. Some had happy family lives; others were divorced or getting divorced. All, however, seemed to have an inner compass that kept them heading in a direction they set themselves.

Why Entrepreneurs Start Businesses

Because there has not been much work done on how people are motivated to start technology businesses, we look at why entrepreneurs start firms in general and then make some comparisons with high technology startups. This should prove helpful as we turn again to high technology entrepreneurship.

On a hypothetical basis, it would seem reasonable to expect that founding entrepreneurs carry with them a body of experiences and educational development that has an impact on venture performance. For general ventures, there is usually not much specialized technical training, but virtually all high technology ventures are founded by technically trained people who are actively engaged in the initial management of the ventures during their crucial formative stage. The founder or founders (as the case is in many high technology business launchings) tap both experience and education. Past experience exposes the founder to good and bad business practices. Technical education builds the knowledge base of science or engineering that is necessary to produce a valued technical innovation. It may not be an original invention, though. The technical education may solely access the innovations of others. Engineers talk with one another and read trade publications. Information on new and better techniques spreads like wildfire. This print and oral communication about new advances may be enough to trigger the latent entrepreneur.

By and large, technical venture founders start businesses that tap their technical education and convert it into a product or service they are familiar with. The Roberts (1991) study of high technology entrepreneurs found that technical entrepreneurs tend to work in development areas rather than research, have about ten years of work experience, and are often supervisors. Another characteristics he found was what he called "an entrepreneurial heritage": technology entrepreneurs were frequently the sons of self-employed fathers, although the fathers were nontechnology entrepreneurs.[2]

The question arises about the sufficiency of these two factors—education and experience—as founder relatedness factors. Isn't entrepreneurship, especially technical entrepreneurship, a much richer activity—much more faceted and origin dependent—than the two factors of education and experience? There are certainly a considerable number of conditions that could be regarded as venture relatedness and accordingly could influence performance. For example, socioeconomic status of the founder, sex, culture, beliefs, personality, attitudes, values, family circumstances, individual ability, motivation, and geographical location are some factors that could be examined. And there are plenty of well-known and respected theorists who speculate about the influence of these forces.

We have limited ourselves to the factors of education and work experience for three reasons. First, these factors were believed to be the most directly related explanations of the nature of the technical undertaking. That conclusion came from preliminary interviews with technical entrepreneurs before this book was even planned. Second, both education and prior employment are factors that have captured the attention of others. They have been looked at before by researchers and managers, including venture capitalists and university researchers. Consequently, there is other work we can reference to find out the connections between the venture and its founder. Third, education and work experience are measurable factors. Other factors, such as motivation and extrinsic psychological conditions, may be measurable but have not been tested in standardized ways for a large group of technology entrepreneurs. The studies we would need simply are not there.

Thoughts on Entrepreneurship

The classic development of the entrepreneurial function depicts the entrepreneur as the driver of the venture. Entrepreneurial organizations do not simply evolve from proto-organizations, ameba-like and aimless. They are deliberate structures that would be nothing without human intervention. So both general ventures and technology ventures are not inevitable creations of economic processes; they are the conscious result of motivated people. Economists have seen the entrepreneur as intrinsic to the creation and functioning of a venture. In a way, the entrepreneur is how humans touch market forces. The technology entrepreneur then brings together the rings of technology, entrepreneurship, and economics. The technology entrepreneur must hold together more for longer, and there is perhaps something admirable in that.

Among all the modern economists, there is one who comes closest to depicting how technological entrepreneurship works today. Joseph Schumpeter (1950), and in various other works and commentary, described entrepreneurs as individuals who carry out new combinations of the means of production. His writing points out that this is not synonymous with the definition of a capitalist. The functions of the capitalist are distinct from the functions of the entrepreneur. In high technology operations, this distinction is manifested in the separation of the capital-generating functions from the technical production functions. Although this situation does not occur in every technology launch (some founders provide their own capital, as noted), it happens in sufficient numbers to be a notable characteristic. With this useful distinction provided by Schumpeter, we can appropriately focus on the purely entrepreneurial functions of the high technology founder. Schumpeter provided further insight by defining the venturing function and the venture capital function. These are modes of operation that have become their own professions. One is an entrepreneur; another is a venture capitalist. The prospective high technology founder need not think he must self-fund his business. Capital infusion can be external.

The development of entrepreneurial theory has been rich and lengthy, but the question of concern here is if there is a special nature to technical entrepreneurship and, if so, what its relationship is to venture performance. Do technology ventures "act" like general ventures? More contemporary views on entrepreneurship will help us consider high technology entrepreneurs and their related business performance.

Most research on new ventures has pursued this speculation, but despite its substance and duration, the results have been mixed. As examples of opposing research conclusions, Hoad and Rosko found that the venture business founder who was both educated and experienced in a similar business achieved the highest performance for his venture, compared to weaker background-related founders.[3] Sandberg (1986), however, drew a different conclusion: that founder experience was not related to successful venture performance. With respect to technical ventures, Cooper produced some of the earliest findings, discovering that more successful high technology ventures were associated with founders who had significant related technical experience.[4] Cooper's research is positioned with those who argue that entrepreneurial relatedness is a factor in technical business success. Although not directly related to the performance-founder factor, an investigation of high technology company formations in the Austin, Texas, area by Susbauer (1969) revealed that technical founder characteristics such as age of formation, educational attainment, and previous work experience were similar to those of the population of engineers and scientists as a whole. In other words, technical entrepreneurs were like technical workers; nonetheless, founders who had a role model for entrepreneurship were distinguished from nonfounding engineers and scientists, even though this factor did not sufficiently explain venture creations—testament again to the nature of these studies as descriptive rather than explanatory. Susbauer did not study venture performance and entrepreneurial characteristics, concentrating instead on the venture founding process.

Based on interviews of 30 high technology CEOs, as well as a larger number of other CEOs, Maidique and Hayes (1985) endorsed the role of the entrepreneur as critical for success, with strong leadership cited as a reason for high technology success. Indeed, leadership is implicit as a characteristic of the entrepreneur. The authors asserted that successful high technology organizations foster an entrepreneurial culture and their top management is directly involved in day-to-day business operations.

This view is contrasted by Drucker (1985), however, who suggested that the founding entrepreneurs of high technology firms do not know when to step aside and let professional managers take over. The founders of these troubled firms continue to manage day-to-day activities beyond the point of effectiveness. They lead too long, micro-manage, and do not trust the managers they have hired.

All this leaves the practicing entrepreneur with no obvious path to successful venturing. This is even more true of the technology fields, where there is even less research. In spite of this, endorsement of the entrepreneurial value has been constant and recent. Gates (1995, p. 231) states, "Entrepreneurship will play a major role in shaping the development of the information highway, the same way it shaped the personal computer business." He calls the few successes and many failures of businesses in this area as

> the hallmark of an evolving entrepreneurial market; rapid innovation occurs on many fronts. Most of it will be unsuccessful, regardless of whether it's attempted by a large or a small company. Large companies tend to take fewer risks, but when they crash and burn, the combination of their sheer ego and the scale of their resources means they wind up digging a bigger crater in the ground. By comparison, a start-up fails without much notice. The good news is that people learn from both the successes and failures, and the net result is rapid progress.

So is there anything special about the high technology entrepreneur? On this question writers have summarized their observations rather than formularize high technology entrepreneur characteristics. Burns complied a listing of important characteristics of successful R&D enterprises that is applicable to the technology entrepreneur: First, adequate technical training, ability, and experiences; second, the freedom to establish the entrepreneur's own mix of activities and work regimen; third, personal dedication; fourth, satisfaction, which means a sense of urgency by the individual; fifth, diversity by each individual in several functions; and last, the provision of feedback.[5]

This listing gives us many of the psychological dimensions of entrepreneurships discussed earlier, but it adds the technical training feature.

This addition makes technical entrepreneurship somewhat special, at least in our view.

What are high technology entrepreneurs like? Can you walk into an engineering convention and pick out people destined for entrepreneurship? Not likely, in our view. Based on in-depth interviews and our personal associations with 30 high technology entrepreneurs, we found no visible commonality except that they were men. Only one was a woman, who was actually a co-founder of a systems integration company. The technical backgrounds of these 30 men were evident, but they were not insufferable about it. We had heard a lot about engineers being the most conspicuously self-congratulatory group of all professions. "Passed my PE exam years ago but I remember . . ." was an often repeated comment. Yet our sample were not braggarts. A few even claimed to have no technical expertise and thought of themselves more as decent businessmen than as technical wizards. All of them worked somewhere else before starting their companies, but the age range was considerable—from the mid-twenties to the fifties. If they were put together in a room, there would be nothing especially distinctive about how they looked.

Talking with those entrepreneurs was a different matter. Two characteristics came out among all of them: conviction and articulation. A sense of earnestness about what they did emerged and their ability to articulate their business purpose was evident. Their articulation spanned other subjects as well, including excursions into politics and economics. In a way, this worldliness was not part of our expectations; we had viewed them as being narrowly focused and driven. An interest in the future and a fascination with the computer revolution were also woven into what they said. This was totally expected, however, since these entrepreneurs were engaged in the revolution and needed to be as wary of a new competitors as they were alert to new technologies.

An especially interesting aspect emerged when we talked with a husband-and-wife founding team of a young systems integration company. The wife had the technical background with an MS in computer science. The husband had no technical background and taught history of science as a university teaching assistant. When the two met with prospective customers, all the technical questions were directed to the husband but the wife answered. Some customers caught themselves and others didn't. "We worked with them anyway," the husband commented.

The technology venture founders also had their share of hardships. A software company founder who began with a partner told us that his partner became mentally ill and he had to take over as both engineer and business manager. This ordeal (for both) lasted for several years, until the incapacitated partner was bought out by the other owner. Another founder talked about the economic difficulties he and his wife faced trying to raise 12 children and launch a database service for libraries. For these entrepreneurs and most others, the story was the same: there were unanticipated problems that came from unexpected sources. In many cases, these problems were discussed with far more fervor than the separation angst of leaving the original company. Cutting the economic umbilical cord and setting off to the new frontier with only self and a concept seemed a distant and minor inconvenience compared to losing a partner, meeting the demands of family, and beating the creditors.

The tougher side of R&D entrepreneurship was noticed by Schrage (1965), who took a group of technology entrepreneurs from 22 companies, assessed and rank-ordered their success, and used psychological tests to measure behavior. Business "success" and the psychological test results were then compared. Schrage reported that the most successful entrepreneurs had anxiety and self-awareness. Unlike McClelland (1961), whose body of works explored and developed human motivation achievement needs, Schrage did not find a high need to achieve in this group.

From the data compiled in our sample of 101 high technology ventures, we found that founders create ventures that stem from their educational backgrounds or business experience. This tendency differentiates them from nontechnical venture founders, who more often begin businesses less related to their academic disciplines or past business experience. It is also consistent with Roberts's (1991) own discoveries on high technology formations around Route 128 in Massachusetts, in which survey respondents had initiated ventures consistent with their education and business experience.

How much of each factor—the academic background and business experience—separates the successful from the unsuccessful high technology venture has not been determined. Karl Vesper (1979, 1983), a long-time venture researcher, offered his opinion on a number of occasions that technical training up to the BA level is important in high technology startups. Roberts found that his group of technology entrepreneurs had master's degree, usually in engineering. Vesper also stated that work experience is important, and that education plus work experience are the most promising background. Vesper (1983) envisioned the venture as the outcome of pushes (negative aspects of a current job) and pulls (opposite factors such as independence and personal reward). These pushes propel the entrepreneur toward action.

In all, though, we don't have a good idea of the relationship between the technology entrepreneur and performance. Our own understanding comes from case studies of a small number of entrepreneurs or from applying research on nontechnology ventures to our subject. As a result, we have an opportunity to devise and test our own ideas on the matter. And that is what we do in the next section.

The Connection between Entrepreneur and Performance

In our research, we developed and evaluated five main ideas based on the possible connection between the entrepreneur's experience and education and the outcome of the venture. Each idea separately captures the technical education and the business experience. A combination of these factors was also done in a single proposition, the effect of which was to facilitate a complete look at the relationship.

Idea One

In the first idea, we wanted to know if technology venture performance was positively related to the business experience of the founder. Does more experience produce better performance? The duration of business experience—in number of years—was compared to performance. In cases of numerous founders, the average number of years experience was used. Then performance measures were applied for the first five years of the venture operations. These measures were average sales, increase in sales (difference between initial sales and year-five sales), employee staff increase, pretax return on investment (ROI), and market share (indicated as the percent of sales divided by total industry sales as estimated by the survey respondent).

In four of the five measures, the results did not show a connection between more founder business experience. There were negative correlations in all performance measures except ROI, the sole area of positive but not statistically significant association.

Based on this evidence, it cannot be concluded that performance is positively related to prior experience. Is this dismal news for the prospective entrepreneur slaving away at a laboratory for the sole purpose of getting enough experience to launch his own venture? Not in our view. Experience counts, but it doesn't guarantee success. Rather like the union card, it is a precondition for going into an industry that demands technical proficiency. That's what others have found and that is where we agree. Entrepreneurs told us that past experience was critical for them, so it cannot be dismissed lightly—after all, real-life experience can be the best laboratory for theory. However, experience cannot be used by itself as an explanation for performance. The business performance of technology ventures has more influencers than what we have detected thus far.

Idea Two

In the next case, our attention was turned to the technical education of the venture founder. Our discussions with venture founders led us to expect that technical education was an absolute must and that more technical education would be an asset to the company. Our interest was in finding if there were significant differences in the performance of high technology ventures based on the extent of technical education of the founder.

The responses were reviewed to ensure that the founders did have technical backgrounds and that the technical education background was related to the venture. For example, computer science–educated venture founders were expected to begin computer and computer related firms, not biotechnology companies. These conditions being satisfied, the analysis was conducted.

Six response categories were offered to the survey respondent. They were (1) no degree, (2) BA or BS, (3) some graduate work, (4) MA or MS, (5) additional graduate work, and (6) PhD. The first three categories were placed in one classification and the remaining three categories were placed in another. Performance was compared across the two classifications.

There was some support for the idea that, although there were performance differences based on the amount of technical education, it was not possible to say that it was statistically significant. Noticeable differences were found in average sales, ROI, and market share. But the results suggest that there is no performance distinction between relatively high technical education and relatively low technical education. This finding disputes our expectation that greater technical education would generate higher performance. It prompts us to think again our suggestion that technical entrepreneurs can expect a performance payoff in direct accordance with their level of technical education. Education may be an "admission card"; the great majority of the survey technical entrepreneurs have at least a bachelor's degree. The very few who did not have a college degree happened to have gone through their entrepreneurial phase 20 or more years ago, when a college degree was not as common. Even so, the education was technical in nature. One founder had training as an electronics technician and set up his business in his home outside Dayton, Ohio, in the 1960s. He also had occasion as an entrepreneur to arrange his work so that he could seize unique experiences, such as being a passenger in a sailplane piloted by Chuck Yeager, a flight he called one of ultimate confidence in the pilot and reinforcement of the perks of entrepreneurship. He felt he would have never had this opportunity if he was punching a time clock.

Another entrepreneur without an undergraduate degree entered the field of desktop publishing, itself a spinoff from computer graphics–based human model photography. This entrepreneur fits Robert's (1991) profile of a second-generation entrepreneur. The father started a weekly advertising newspaper. The son's technical training came in through the back door, almost literally. He taught himself computer graphics by bringing in new equipment, piece by piece, into the backroom of the newspaper business.

The point is that certain types of high technology businesses more easily induce the entrepreneur without formal education. Although we have examined this idea only informally, the general area of software and computer services appears to draw most degreeless entrepreneurs. That might be because of the pervasiveness of computers, the needs of businesses that have computers, and the relatively low barrier to entry for the software consultant.

Idea Three

On the chance that the combined effects of business experience and technical education might have a proven performance impact, we looked for a positive association between a combination of the two and venture performance. At least one writer (Vesper, 1983, for example) has contended that venture success is related to these two factors, so it worth looking at the combined factors even though separately the factors did not support our expectations.

The number of years of technical experience was combined with the academic achievement level. Academic achievement level was classified as a number between 1 and 6 that expressed the highest achievement of the founder, as was done in testing the previous idea.

The results showed no positive association between any performance measure and combined experience and education.[6] This was not unexpected given previous results, and it reinforces the notion that education and experience are not determinants of high technology venture performance.

Idea Four

Thus far, our interest has been in the individuals who start technology companies. But our exposure to technology firms, as well as the studies of others, leads us to consider the impact of teams rather than sole entrepreneurs. An additional test was done concerning the venture founding team. Cooper and Bruno (1977) found that ventures founded by teams had better survivorship than ventures founded by single individuals. Roberts (1991) noted that 64 percent of sampled high technology starts had multiple founders and had better survival chances. Reich (1987) even went so far as to celebrate venture teaming. He argued that the "myth of the entrepreneurial hero" is supplanted by collective entrepreneurship, in which the rapid transfer of technology and the complexity of modern entrepreneurship make the individual hero passé.

To explore the team influence on performance, we speculated that multimember founding venture teams are associated with ventures having better financial performance. This seemed to be a logical extension of the premise that teams could bring more experience and judgment than could individuals. However, the results didn't prove this speculation. Team-based ventures did not have better average sales performance than individual-based ventures. The other four measures of performance were not used because they had too few responses.[7] The positive effects of team foundings do not necessarily manifest themselves in better venture performance, at least within the confines of our study.

Idea Five

The last part of our survey was directed toward establishing if prior venture founding experience had any association with better financial performance. The specific question explored was whether founders with prior venture founding experience would be associated with better venture performance. This idea came from the venture capitalists, who told us about technology entrepreneurs who "tried and tried again" to form successful ventures. Failed first attempts did not deter the intrepid, and the venture capitalists spoke warmly of their associations with these jaded entrepreneurs, even though the venture capitalists lost money on them.

The measurement was based on whether the founder had any prior founding experience. Once again, we didn't find support for better average sales performance for the current venture.[8]

Thus the technology entrepreneur might find eventual success in the founding of a firm in spite of these results. If there is a 70 percent probability of success on the first venture and 70 percent on the second, there is nearly a 50 percent chance of success eventually. Some encouragement can be found in multiplying the separate probabilities of success coming from multiple foundings.

Summary and Conclusions

In this initial look at any strong relationship between performance and founder characteristics, we see no support for such a relationship. This was unexpected but important in its implications. High technology entrepreneurs do have something in common with general entrepreneurs—that is, entrepreneurial characteristics are not associated with higher levels of business performance, according to the measures we used and as far as our group was concerned. This does align with the conclusions of more recent general venture researchers and also counters the earlier idea that education makes a significant difference.

The most consequential implication for those who may want to start a technology venture is that the door is open for them even if they do not have a technical education. People point to John Sculley in going from Pepsi to Apple as an example, yet the leap was not entrepreneurial, since both firms were well established and Sculley had a definite technical bent, as evidenced by his youthful invention of a color television scanning system. The message here is better directed to the formally trained technology entrepreneur, who may expect that his greater learning will lead to a better company. That is not in the cards we have played out here.

For technical entrepreneurship, the question is more properly, Who are the entrepreneurs? Multiple founders, or team entrepreneurship, is the mode of high technology formations. And teamsmanship is found not just in technology venturing but also in the whole realm of technology. In large engineering firms and the R&D labs of well-established companies, there are teams working on lead projects. It may be that technology foundings simply reflect this situation rather than being a whole new dynamic in high technology foundings where teams are demanded, not used as options. This is such a pronounced feature that it certainly deserves a closer look at some point.

Team entrepreneurship is one of high technology's most distinguishing characteristics. It may well be its clearest demarcation from older forms of entrepreneurship, where we see the imprint of the individual. The use of teams introduces a whole net set of factors—team dynamics, team size, substitution of members, complementary skills, and the like. These are all fair game for closer study.

In the next chapter, we will turn to a subject related to the entrepreneur: the circumstances in which the entrepreneur found himself before founding his new venture. Captured by the term *origin relatedness*, this is part of the folklore of entrepreneurship. Like most folklore, it has the vitality of momentum, though close examination shows its fissures.

Notes

1. Drucker describes this in *Technology, Management and Society* (New York: Harper and Row, 1970).

2. Although Roberts (1991) concentrates on technology ventures near Boston, the book is a complete discussion of the high-tech entrepreneurship business, wherever it may be located.

3. W. Hoad and P. Rosko studied venture success, as described in their article "Management Factors Contributing to the Success and Failure of New Small Manufacturers" (*Bureau of Business Research Publications*, Bureau of Business Research, University of Michigan, Ann Arbor 1964).

4. This result is in A. Cooper and A. Bruno, "Success Among High Technology Firms," in the April 1977 issue of *Business Horizons*. Cooper has produced much information on this subject, including other articles and papers on the role of incubator organizations, founding conditions, and the environment for technical entrepreneurship.

5. This listing was devised by R. Burns in his book *Innovation: The Management Connection* (Lexington, MA: Lexington Books, 1975).

6. This test shows uniform results with respect to both the correlation coefficients and the probability values. The correlation coefficients are all negative (but not strongly negative) and the probability value is not significantly different from zero at 1, 5, and 10 percent confidence intervals. As a consequence, we conclude that there is not sufficient evidence to accept the alternative hypothesis that the combination of education and experience are positively related to performance.

7. The sign test was employed in this instance on 50 cases of reported data. For each of the three industry groups, ventures with higher than average five-year sales were compared to lower than average sales.

8. The sign test was again employed in 40 cases with the data. There was insufficient statistical basis to reject the proposition that there is no difference between the more experienced and less experienced founders.

References

Cooper, A., and A. Bruno (1977). "Success Among High Technology Firms." *Business Horizons*, April, pp. 16–23.

Drucker, P. (1985). *Innovation and Entrepreneurship: Practices and Principles.* New York: Harper and Row.

Gates, W. (1995). *The Road Ahead.* New York: Viking Penguin.

Maidique, M., and R. Hayes (1985, Summer). "The Art of High Technology Management." *McKinsey Quarterly*, pp. 43–62.

McClelland, D. (1961). *The Achieving Society* Princeton, NJ: Van Nostrand Reinhold.

Roberts, E. (1991). *Entrepreneurs in High Technology: Lessons from MIT and Beyond.* New York: Oxford University Press.

Sandberg, W. (1986). *New Venture Performance.* Lexington, MA: Lexington Books.

Reich, R. (1987). "Entrepreneurship Reconsidered: The team as Hero." *Harvard Business Review*, May–June, pp. 77–83.

Schrage, H. (1965). "The R&D Entrepreneur: Profile of Success." *Harvard Business Review*, November, pp. 56–69.

Schumpeter, J. (1950). *Capitalism, Socialism and Democracy.* New York: Harper.

Susbauer, J. (1969). *The Technical Company Formation Process.* Doctoral dissertation, University of Texas, Austin.

Vesper, K. (1983). *Entrepreneurship and National Policy.* Pittsburg: Carnegie-Mellon University.

Vesper K. (1979). *New Venture Strategies.* Englewood Cliffs, NJ: Prentice Hall.

6

ORIGINS AND PLANNING IN NEW FIRMS

There is a mythology about high technology. It holds that one or more spurned geniuses hatch a technological breakthrough in a garage. They do this on their own because they cannot get their organization's leaders interested in their ideas. The innovators commercialize the invention on a shoestring budget and later, triumphantly and with Orestesian justice, dispatch the organization that scorned their insight.

Classical mythology is supposed to lead us back to natural truth—in this instance to the origin of high technology firms—and there are flashes of truth that happen in that backward excursion. The evolution of Apple Computer is testament to the myth, but there is much deviance in other cases. High technology businesses start in many different ways, in many different places, by many different people. The businesses are as individual as their founders and a few instances will illustrate the point.

In some cases, origin is through acquisition rather than invention. The habit that large agricultural firms have of snatching up young biotechnology companies is well known. It happens in software as well. Bill Gates acquired the software that was the basis for the hugely successful MS-DOS operating system that was later adopted by IBM for its PCs. Gates masterfully positioned MS-DOS so that it became the logical choice for IBM in a now well-documented episode of business acumen.

Contrast that with the origin of another equally spectacular venture—the Lotus Corporation, whose founder Mitch Kapor actually wrote the spreadsheet program himself. Here, there was close paternity by its creator of the main corporate product. Both creations—Kapor's invention and Gates's fortuitous acquisition—fall in the realm of high technology venture operations. Yet the essential innovations of each company took very different paths to market.

These high technology giants had different management styles as well. Kapor recognized that he was a programmer, not a business leader, and he turned over the reins of his company to professional managers. On the other hand, Gates increased his control over Microsoft during its introduction and growth—far longer than Kapor kept Lotus to himself. That Gates still offers no sign of relinquishing Microsoft further demonstrates the founder differences.

There is also the matter of the manger in the garage. The examples are few but famous, and we have already described Hewlett-Packard's shed start in an earlier chapter. But these examples distort the fact that most high technology ventures emerge within corporate walls as specially established venture divisions or as products of full-time employees who plan the venture and then depart as a new firm. Even in federal research laboratories, the germs of impending technology adventures are constantly forming. All these sources far outnumber the garage births. As the myth is pursued, it often turns out that there was considerably planning, even before the garage stage. That happened with the Apple founders, as it did with Hewlett-Packard.

There is another way that ventures come about. It is more on the unseemly side, but it does happen and that is through theft and other acts. Because there are few who admit to this, the evidence often comes from history. Glenn Curtiss sneaked into a closed display of the Wright Flyer to discover how the brothers mastered rudder and aileron controls. The software business today is rampant with industrial theft, part of which makes its way to a new business.

Back to the moral mainstream, we can say that the virginal technology startup is virtual fiction. It is rare to find a technology venture that was launched by an entrepreneur fresh out of school, without any previous employment. In the ventures we looked at, only 2 percent of the founders said the venture was their first job. Almost all founders had prior experience in a related field—so much so that most researchers assume that technical entrepreneurs have all been one-time corporate bedding plants.

Where the Ideas Come From

The big idea and the idea to commercialize it almost always crystallize while the entrepreneur is working for another company. That makes it worthwhile to consider what else, besides the entrepreneur, was transferred to the new technology venture. It seems plausible that technology entrepreneurs take laboratory techniques, and in the dire view of the former manager, perhaps equipment as well. Software skills, potential customers, and supplier contacts may also be taken.

This is not simply idle speculation. Technology entrepreneurs have told us that this has happened. One was reluctant to divulge anything about his venture because his main competitor was the firm he departed, and he made very similar instruments—devices he developed for his former employer. The former firm was located only ten miles away, and he fretted about offering any information because he thought his former firm would use the information to do him in. He might have had good reason for this fear because his own marketing staff called his machines clones of the origin company. This was one example in 101 ventures we examined, but such direct replication is evident.

By and large, though, high technology venture founders were willing to talk about what they learned in their earlier jobs and what they applied to their own ventures. The activities of the founder in the former organization is an important element of origin relatedness. Most of the tone of the founder was fondness, not contempt. As one entrepreneur put it, "My idea came from there. I would talk it up and get feedback from the other engineers. That was the nice thing about that company. They let you brainstorm on your own at times. I let it percolate for quite a bit before going off on my own. Even though I had a winner idea, I couldn't go off on my own right away. I had to test and plan and I did a lot of it there. They helped me launch my dream."

Another entrepreneur who came from a university research laboratory said, "We would take turns working on an outside project. The people who worked in the lab didn't really know how smart they were until we worked up some of the stuff and sold it through a separate company. The customers really were the ones who saw how good it was."

Incubation may be the most apt way of characterizing this episode for the entrepreneur. There is no sudden departure from the former organization—calculated cultivation is more like it. These two entrepreneurs not only got their original ideas from the incubating company but were also able to develop them and do a little test marketing on the side.

Unlike general ventures, high technology entrepreneurship is skill-intensive in very specific ways. Other innovations have to be well known to beat them in the market. That is, to make the better mousetrap, you have to know the other mousetraps. As noted in Chapter 1, technology launches are often not done by management generalists but by engineers and scientists with supplemental significant business experience. The reason for this could be that exposure to the technical innovation that would drive a new venture occurs in the domain of science and engineering rather than in general management. High technology business are fundamentally technology-knowledge businesses, so knowing the science is essential. The high technology venture refugee must take the right things, not all the things, from the incubator. That information can come in a narrow range, though—the right way to make integrated circuits or the right place to alter the DNA molecule.

Origin Relatedness

The connection between prior work experience and the new venture is called *origin relatedness* by researchers. In our terminology, this phrase refers to the degree of similarity to an incubator company in industry structure, product, organization, and processes used. When we consider origin relatedness, we tackle the question of how different the new venture is from the organization its founders may have come from. We also look at what was transferred from the prior organization to the new organization.

Origin relatedness is a matter of degree rather than totality—horseshoes and hand grenades instead of pregnancy. One would be hard pressed to say that nothing came from the prior organization. Sometimes the negatives of the prior organization are themselves sufficient to spur the new venture. A technology venture founder can be so disgusted with the practices of the prior organization that they will use the firm as an example of what not to do. The relatedness here is a dialectic one, but it is relatedness nonetheless.

More often, though, origin relatedness is a matter of emulating the operations or products of the departed organization. Peter Drucker in works such as *Innovation and Entrepreneurship: Practices and Principles* is among those who has observed that this as a viable way for new technology firms to complete. Imitation is less expensive than paying for research and development work.

Origin relatedness is manifested in several ways. One is when the venture founder uses experimental procedures and laboratory techniques acquired in a previous job. We've heard of pharmaceutical entrepreneurs who picked up the latest FDA testing protocol from the firm when they departed and immediately used it in their new business. Origin relatedness might also be a particular quality assurance test for silicon semiconductors in the electronics business. We consider these as examples of skills origin relatedness.

Another place origin relatedness might be seen is on the production line. For example, the new venture may duplicate the production techniques used by the incubator organization. If a computer was manufactured by the new venture in essentially the same way, with the same kind of machinery, as was used by the incubator organization, we would say that there is a high degree of production origin relatedness.

Still another form of origin relatedness is if the venture product itself is similar to the product of the incubator organization. This can be considered product origin relatedness. Put in terms of business nomenclature, it is cloning. High technology business is a rich repository of this practice. It is so pervasive (e.g., uninterruptible DC power supplies) that electrical engineers have added completely useless gates on circuit boards for the sinister pleasure of seeing them mindlessly replicated on cloned versions produced by foreign competitors.

Yet another example of origin relatedness is in marketing, as when the venture competes in the same market, or nearly the same market, as the incubator organization. This market relatedness is particularly perilous, especially when the new venture enters the same market with the same product against a better established incubator organization. The product may be different, but the customers are identical or similar.

These different forms of origin relatedness are prevalent in high technology—so much so that certain high technology businesses have achieved distinction within their fields as spawning shallows for generations of other high technology launches. The Fairchild Corporation is one such firm with a genealogy of startups that former Fairchild employees have initiated. High technology firms created by these former Fairchild employees are themselves breeding grounds for new ventures. It's a begetting of ventures that reads like Old Testament scripture to the technology historian.

Despite what seems to be a prevalent practice in the high technology field, origin relatedness has not been extensively studied. Past studies are mainly descriptive and do not connect venture performance to origin relatedness. This is curious because it seems that relatedness can have a real influence on new ventures. We look at the connection between origin relatedness and performance later in this chapter.

In a study of 161 California high technology firms, 64 percent of the new firms were related to their incubator firms (Cooper, 1985). Cooper also found 84 percent of the newly founded firms began operations in the same geographical area as the incubator firm. For Cooper, an incubator was the organization that employed an entrepreneur before he founded his own venture.

Incubator organizations are both technical and nontechnical organizations in the cited study. It may be a software engineering firm or a paper manufacturing plant—there appear to be no hard-and-fast rules about high technology emerging exclusively from technical companies.

In subsequent research on high technology origin relatedness and performance, Feeser (1987) explored the possible connections among size of the incubator firm, technology and markets of the incubator, proximity (relocation of the founder at venture launching), and type of venture (either for profit or not for profit). Size of the organization was measured by sales amounts and number of employees. An assumption of the study was that the founder worked at the prior organization; as noted earlier, however, that may not always be the case. The test of proximity relatedness was to establish if the venture location required the founder to move his residence. Here again a cautionary note must be introduced; a residence change is not likely to be an indicator of proximity relatedness. The domicile of the venture in relation to the incubator organization would be a better measure. Founders may reside anywhere, and that would be immaterial to venture location and incubator location. The founder may even live in a different state from the venture or incubator location.

Aside from this caution, the Feeser study found that high-growth computer ventures were incubated in larger organizations and their founders had more venture startup experience than founders of slower growth computer ventures. Dollar value of sales was the growth measure used.

High Technology Incubators

What can be said about the high technology ventures we looked at with respect to origin relatedness? Were high technology ventures highly related to their incubating firms or were they less related?

There was a surprise here. We found that high technology ventures have less origin relatedness than might have been expected. To reach this conclusion, we looked at the four different types of relatedness described earlier. We compared technologies of the present venture to the prior organization. We examined the extent product development technologies were transferred from the former organization to the new venture. We also looked at the extent the market was similar for the incubator and the new venture. Lastly, we tried to discover if marketing skills and techniques acquired in the former organization were used in the new ventures.

Only 38 percent of our 101 ventures had the same or very closely related technologies. The rest had some or little or no relatedness. Similarly, only 26 percent had considerable (same or very closely related) product development technologies. Ventures with considerable market similarity were 30 percent, while 29 percent had considerable transferring of marketing skills and techniques. There was more technology transferring than the other categories. Nonetheless, less than a third of the ventures had substantial relatedness in the other categories, so overall there was not much relatedness between the venture and the incubator. These ventures had much more independence than might have been originally thought.

There were differences when the separate high technology business classifications were considered. A typical computer venture had some technology relatedness but no product development technology transfer, market similarity, or marketing skills transfer. Electronics ventures had more marketing skills relatedness than either computers or biotechnology. Biotech ventures had more product technology relatedness and more technology relatedness than the other groups. In all, there were differences among the different types of ventures as far as relatedness is concerned. It is also true there was only a modest level of relatedness in the surveyed high technology ventures.

Another facet of this discovery is that if a venture had a particularly strong degree of relatedness in one category—say, technology relatedness—that did not carry over to the other categories. Far more often than not, a venture with the same technology had little or no marketing skills transference. This might be thought of—in particular by the incubator company—as robbing the cupboard but not the kitchen.

High technology ventures may be related to the incubator organization in other ways as well. Our consideration was by type of incubator and the size of the incubator. The ventures we explored were categorized by incubator type: public firm, private firm, university, or government. For the size of the incubator, annual sales amounts and employee numbers were obtained.

Among the computer ventures, the incubators were predominantly public or private firms, with each accounting for 42 percent of the total for a combined 84 percent. Only 14 percent were university incubations and 2 percent were government incubators.

Were the incubators large or small organizations? Excluding university and government incubators, fewer than half (45 percent) of the incubations were in small businesses when small businesses are defined as having 500 or fewer employees. (The 500-employee demarcation point is commonly used by the Small Business Administration to identify small businesses.)

The incubation story is somewhat different for electronic ventures. Here, 66 percent incubated in public firms, 24 percent originated in private firms, and 5 percent were in university or government environments. The size of the incubator is also different for the electronics ventures. Only a third of the ventures sprang from small business firms.

In this look at relatedness, the last group considered are the biotechnology ventures. We found that biotechnology ventures were evenly split three ways by type of incubator: a third were private firms, a third were public, and a third were university based. University-based incubation is much stronger for biotechnology ventures than it is for computer and electronics ventures. This biotechnology link to universities was expected and has been borne out in our sample.

Slightly more than half (55 percent) of the biotechnology ventures were launched by entrepreneurs who came from small businesses as we have defined them. This is different from both the electronics and the computer incubations. The typical biotechnology launch emerged from a public or private small business or university background.

Based on our look at origin relatedness, we have seen significant differences in the type of incubator and the size of incubator among our three chosen high technology businesses. We'll continue to explore differences and similarities among the three groups in this book.

Despite the differences, it is striking that all three venture groups show considerably more origin relatedness to larger size businesses than might have been expected. Of the over 20 million businesses that file taxes, only about a tenth have more than 500 employees. Yet in our group, a much greater percent had incubations in large businesses.

As President Coolidge put it, the business of America is business. The derivative corollary is that the business of business is small business.

It seems that many large businesses beget many small technical businesses, at least in our group.

Explanations abound for this. A larger firm may have more advanced technology, it may have a larger human resource base to tap, or it may have better initial capital sources, customer contacts, or other advantages.

Venture Performance and Origin

Does origin relatedness really matter to the high technology venture? To answer this question, we generated a total relatedness score for all the high technology ventures in our sample group. The group was then split on the basis of relatively higher or lower relatedness. The same five performance measures were used as described in Chapter 5 (average sales, increase in sales, employee growth, ROI and market share).

There were no statistically significant performance differences found between the higher and lower relatedness groups (although average sales and increase in sales came closest to showing a performance difference). What we end up with is the conclusion that the two degrees of origin relatedness—high and low—are not linked to the performance of the ventures in our group.

To Plan or Not to Plan

Besides origin relatedness, a possible consequential factor worth investigating is the use of written planning documents and procedures in the high technology venture. We turn now to this subject and draw some conclusions about "putting it in writing."

The business of starting a business is, at times, a paper industry itself and high technology is no exception. Like any business, the high technology firm is required to do tax filings, corporate record filings, personnel record keeping, and many other governmental compliance procedures. Many of these documents are needed by multiple units of government, at the federal, state, and local levels.

Connected to these are the nonpublic internal planning documents of a business. Business writers and consultants urge entrepreneurs to develop business plans. Mancuso (1983) is among the noted business advisors who stress plans. KPMG, a firm which did considerable technology consulting, provided technology entrepreneurs with guidelines for organizing their businesses, the gist of which is that business plans are necessary for high technology ventures. The theme of this advice is that it is better to plan and document practices in advance and adjust them later than it is to go into business without written plans and procedures.

In this respect, high technology business plans have much in common with business plans of other types of business ventures. That is, the structures of the business plans are similar, borne out in the half-dozen plans we saw. The customary business plan contains a summary of the business, an organizational chart, a production plan, functional area plans, a personnel plan, marketing and sales activities, and a budget. There are permutations for high technology business plans, however. For example, functional area plans might pay special attention to the research and development function because it is the organization's lifeblood, while less attention might be paid to the production function, since product output may not be immediate. Since much high technology is an intellectual industry, focus is often on plans for the development and protection of patents and copyrights. Another area often addressed in detail is equipment, with clear interest in acquiring and replacing computers and laboratory instrumentation frequently for both tax depreciation advantages and technology currency needs.

There are two other types of documents often found in high technology business launches. One is a personnel policies and procedures manual and the other is a marketing plan. We have seen personnel manuals even in eight-person ventures and marketing plans for high technology companies with less than $200,000 in sales. These firms put the information on paper not just for the present but also to anticipate growth.

Written documents for personnel policies and procedures must, by law, have information about retirement plan coverage. Fringe benefits are typically described, as well as cash bonuses, stock plans, and savings plans. High technology companies also add educational procedures, recognizing that it is in their best interests to keep employees abreast of developments. There was quite a bit of variation in the level of detail contained in the documents we saw, from sketchy 3-page personnel "manuals" to comprehensive 50-page documents.

The marketing plan is quite understandable, since marketing is perhaps the most critical function for the new high technology venture. Investors almost always insist on having a marketing plan.

Again, there is considerable variation and many models have been used. For the most part, they contain at least the following: (1) a marketing strategy; (2) a description of the product or service; (3) an indication of target customers; (4) promotional, pricing, and distribution considerations; and (5) sales goals and marketing budget.

These are the elements of a minimal high technology marketing plan. They are variations on the standard marketing plan that might be used by a retail firm. There are differences in the way high technology firms do marketing, however. We present much more about this subject in later chapters, but the relevance here is that marketing practices may be transferred from the incubator to the new firm.

In all, there is strong endorsement for written plans and procedures. But what about the technology ventures themselves? We found that only 15 percent of our sample high technology ventures had written policies and procedures for employment, purchasing, and other functions. A much larger number, 81 percent, had business plans; and 51 percent had marketing plans. So, our typical high technology venture has a business and marketing plan but no written policies and procedures for personnel.

It is important to note that our interest was whether the ventures had these documents when they were formally organized. These percentages would very likely increase as the ventures moved into actual business operations. It is, in fact, very difficult to imagine a viable high technology venture operating without plans to at least some degree. The ventures seemed to value such plans, but they mainly valued doing business plans. Marketing plans are a hit-or-miss proposition as far as plans are concerned and the other plans are rare.

Written Planning, Procedures, and Performance

Does having written plans and procedures make a difference in the performance of high technology ventures? Our venture sample was split into two groups. Those ventures that had relatively more written policies and procedures (based on a total numerical score made up of points for a business plan, market plan, and written policies and procedures for employment, purchasing, and other functions) were compared to those that had fewer plans.

There was no significant difference in performance on the five performance measures between those with more planning and less planning. The closest we got is in the ROI measure, but information is insufficient to assert a difference even in this measure. As a consequence, there is no basis for saying that high technology ventures with written plans do better than those without written plans.

Although our research results do not tie venture performance and planning together, for other reasons that have been cited, the practice is warranted. Business planning is an inevitable companion for organizational development. Established high technology firms have business planning functions, even if they did not start with them. There is some point at which written plans and procedures become crucial, and many ventures discover that it is better to do it now than later.

Business planning practices are also important because they are used by capital providers to evaluate the venture. Venture capitalists review business plans at different growth stages. Rapidly growing technology ventures often outgrow their nests and need the infusion of capital from many sources to continue growth. So the business plan gets scrutinized continuously. In fact, we know of no instances where there was substantial funding of a technology venture when the venture lacked a business plan.

The existence of business plans, policies, and procedures is also seen as an asset by potential purchasers of the venture. The buyout is a stronger possibility for the entrepreneur if he can deliver documents that show a plan of operations and the methods to get work done. Successful high technology ventures do get purchased by public firms or private owners; business plans keep that option viable.

Lastly, the planning exercise and documentation can help demonstrate if the venture is truly feasible. Sometimes, the founder needs the most convincing of all. And sometimes the convincing can be done only when pen is put to paper, especially when the venture is a marginal business proposition. Creating a virtual venture on paper is preferable to investing sweat equity and bearing the *Sturm und Drang* of venture extinction. Business planning can be used to identify possible problems. If the problems are insurmountable, the paper venture can be crumpled and tossed before the same happens to the founder.

This is not intended to frighten the prospective founder, even if it were possible to deflect a truly committed technology pioneer. There is reason for hope. It is axiomatic that no venture succeeds without trying, so the attempts must continue. It's also true that the pure numerical probability of success or failure of one venture is independent of another, though we are inclined, on the basis of observation, to believe, that a founder who has succeeded in one venture has an edge in successive tries.

Our remarks on business planning followed our survey research results and are based on observing high technology ventures in action. Observation helps because we do not have much research in this area. Putting these observations in a motto, what we have is: to plan is to prevail in the high technology venture business.

Summary

In this chapter we described origin relatedness and the planning, policies, and procedures of high technology companies. The ventures we explored did not have a high degree of origin relatedness, and there were no evident performance consequences from this lack of relatedness or of policies and procedures. Nonetheless, business analysts have stressed the relevance of these factors.

In the next chapter, we will look at a fundamental aspect of technology venturing—business strategy—and explore possible performance implications. For the technology venture, strategy choice and execution may well be key to eventual success.

References

Cooper, A. (1985). "The Role of Incubator Organizations in the Founding of Growth-Oriented Firms." *Journal of Business Venturing* 1, pp. 75–86.
Feeser, H. (1987). *Incubators, Entrepreneurs, Strategy and Performance: A Comparison of High and Low Growth High Tech Firms*. Doctoral dissertation, Purdue University.
Mancuso, J. (1983). *How to Prepare and Present a Business Plan*. Englewood Cliffs, NJ: Prentice-Hall.

7

STRATEGY AND THE NEW TECHNOLOGY VENTURE

In the previous chapter, we considered the origin of the high technology venture. Our search continues here for a connection between business performance and venture operation factors. If performance isn't linked to entrepreneur characteristics or idea origin, then perhaps strategy, environment, organizational structure, or business practices makes a difference. In this chapter we focus on business strategy.

We look at the new technology firm's strategy decisions, its access to the resources needed to grow the venture, and some organizational features that might have an impact on the high technology venture's initial business success. These different dimensions are explored and their impact on performance is established.

In Chapter 12, we will consider strategy for the established company, but for now the concern is with emergent firms and formative strategies. The result of examining strategy at these two different points should be a better understanding of technology strategy and the evolution of such strategy. It also gives us a continuum for a strategy perspective.

A Strategy Frame for Technology Ventures

Theorists and strategy researchers have long stressed the importance of strategy for new businesses, that it is consequential. As described by Singer (1995), new ventures suffer more difficulties and business failures than do established firms. That is clearly true, but it becomes merely a premise for the rest of Singer's explanation. There is a pattern to the mortality, with most newly founded businesses lasting only a few years. The explanations for this have traditionally centered on poor management and inappropriate strategies. Singer approaches venture mortality as similar to the patterns of development of an inherently hazardous process. These hazards are strategy related and consist of generic entry barriers, density of development hurdles, amplification, amplification of maturation error (not fixing things early enough), and sequence and control in development. This contemporary view of strategic error, placing emphasis on the identification and reduction of business risk, serves as our launching point for a discussion of strategy and the new technology venture.

Note initially that there has been scant research on high technology venture performance; most research concerns ventures in general. In a recent study, Chandler and Hanks (1994) sought to identify variables that should be related to venture performance. It was hypothesized that both market attractiveness and resource-based capabilities are directly related to new venture performance. Additionally, specific resource-based capabilities were hypothesized to be related to the competitive strategies chosen by a firm. It was also proposed that the fit of strategies to resource-based capabilities was related to venture performance. The results were that the conjectures tended to be supported. In short, strategies are consequential.

The strategy for an emergent technology company flows directly from the entrepreneur and the circumstances of the founding. Strategy is making the decision about the general direction for the business. It is a constant exercise for the entrepreneur rather than a one-time event. The business strategy must be concocted before the business opens it doors. It must also be modified in the face of new realities as the venture matures.[1]

Strategy cannot be postponed for a viable venture, although it may be hidden in the mind of the founder. A technology entrepreneur with a product and no strategy is immobile inertia personified, but when the product is put in motion with some kind of strategy, the business becomes real. It leaves an imprint. The direction of the venture may not be apparent to those not privy to the entrepreneur's strategy thinking, even if it is in motion. These observations demonstrate the permanence and obscurity of the core strategy question—its own espirit in the entrepreneur's noble quest.

Strategy and the entrepreneur are closely aligned. First comes the entrepreneur and then comes the strategy. It is the business strategy that is the first complete expression of the entrepreneur's business sense and his vision. It can be a complement to his inventiveness or its sad dénouement. Entrepreneurs told us how much they dreaded the strategy-making step, perhaps knowing how consequential this action is. But they also knew that they had to take that step.

Strategy Origin

The new firm's business strategy may not originate from a single founder. In most of our sample, the technology ventures were founded by multiple members who presumably crafted their strategy together or at least agreed upon a basic course for their joint enterprise. The strategy itself may be a single stream, but it flows from several tributaries.

We now look at the core, day-to-day functions of the venture, where the decisions are made that can unravel a dream or forge a firm. There are no moments of truth but there are epics of uncertainty. The effects of strategy decisions are not immediate. In fact, this ability to deal with uncertainty has been one of the definers of entrepreneurship.[2] This is an important part of venture development, and although the constant decisions about daily operations are routinized and minor compared to the big thinking of venture conception, they are just as consequential. (Those who are considering launching a technology company will be especially interested in this chapter because what is discovered could have a great bearing on their entrepreneurial efforts.)

Few ventures blunder their way to success. As expressed in the line in *Casablanca* about the Allies' "blundering their way" into Germany during World War I, the successful technology venture is not a blunder or happenstance but the fruit of a well-planned orchard. The most successful and sizable technology firms all had clear initial business directions. IBM, DEC, and Hewlett-Packard serve as examples. These companies changed their original business strategies and markets, but they nevertheless had articulated strategies in place before moving ahead. Those strategies consisted of a view of the market and a vision of success. IBM seized the office machine market and DEC paved its way with the minicomputer. Hewlett-Packard saw opportunities in the professional laboratory by building electronic lab equipment for scientists and engineers.

Defining Strategy in Entrepreneurship

Strategy for the emerging technology company is how the venture intends to arrive at its initial objectives, whatever they may be. But there are so many possible business objectives and so many paths to those objectives that clarification is needed here.

The objectives are not the strategy, though that is a widely held view. Turning an initial profit by five years may be the first objective; the strategy is how that will be done. The profit objective may be attained with the introduction and production of a wind shear detection device, but strategy is this venture's path to profitability. Strategy is not an end in itself, so to test whether something is a strategy, ask if it is a mechanism for a further end. To test for an objective, ask if it is more of an end—a result of a strategy. In a way, this is the classic means-end distinction for the most part, with objectives or goals being the end and strategy the means. There is more discussion of this in Chapter 12, on business strategies for established high technology companies.

Business-level strategy for the high technology venture is probably the least visible part of its operations. Goals are established and proudly articulated by founders. "We want to be the market leader" or "Total customer service" are the goals prominently featured in marketing materials. The founders themselves are also visible—they have to be. For the most part, the only thing a new venture has is its founder's visibility. But strategy is less so. It lurks in the pages of the business plan or is muttered in response to probing questions. Our interviews with entrepreneurs led us to the conclusion that some business founders have vague ideas about what strategy is.

Researchers trying to see if strategies have an impact on performance have determined that strategy does matter.[3] They have worked with general ventures in mind, however, not high technology ventures. These researchers have studied the strategy-performance nexus, either imposing their own obstacles or encountering the obstacles of others. To obtain results, they narrowly drew the studies to control for externalities or have been troubled by the great challenge of netting in the complex array of factors in a testable model of high technology venture performance. Neither of these approaches has provided a fundamental understanding of strategy and performance for the high technology field.

Prior glimpses at high technology have been in a narrow spectrum— in one instance it consisted of only computer ventures. These high-growth computer ventures had strategies of sticking to an initial product focus.[4] The finding is useful because it does address a high technology business sector, however—one of the few studies that do.

Some older studies suggest that high technology ventures with closer proximity to resources experience better performance than ventures more distant from resources.[5] The older (1970s and 1980s) studies of high technology new businesses tended to look at location factors and founders in accounting for business performance. The later studies enlarge the perspective to include more environmental factors, such as industry structure.

Strategy Choices

The freedom of infinite strategy choice is an irresistible force for a company that has not yet come into being and is confronted by the immovable objects of heavy industry competition and the sobering statistics of business failure. Such straddling of opportunity and threat has been addressed by Porter (1996), who calls strategy development in these circumstances a "daunting proposition" because of the uncertainty of customers, products, and services. In developing industries (as we have called technology industries), imitation and hedging are quite commonplace, as firms don't want to be left behind. Under these conditions, enduring companies are those that start early creating a "unique competitive advantage." Other high technology firms pursue imitation far too long, adding features and cutting process to the point where they doom themselves. For Porter, this reinforces the point that high technology strategy must be done by the old rules and that claims about technology ventures as special cases in a new era of competition are not true.

Not all strategy choices are possible for the new technology operation, even imitation. Resource limitations and sunk investments may preclude profitability and imitation strategies, at least in the short term. The technology entrepreneurs who we interviewed had second mortgages on homes and very little left to fall back on. Yet they still had to wait it out, slowly building market share and then achieving profitability. Very commonly, they did not feel comfortable that the new firm was up and running, even after five years, of operation and they were still looking for their first significant profits from the company at that time.

Strategy choice for the new entrepreneur can be translated to questions such as,

> What market should I go into? Should it be something I know a lot about even though it may not be a high-growth sector, or should it instead be a high potential market?
>
> When do I go in? Do I wait and let others take their lumps first, or do I go in early and learn along the way?
>
> How about the scale of entry? Do I go in with everything at once, or try a small market first?

These are not small matters. But not only do they have to be answered, they have to be answered together, since timing is at the core of success. As Burgelman and Sayles have noted, timing is part and parcel of entry strategy.[6] Choosing an appropriate strategy and executing it when industry conditions are right are both necessities for emerging technology companies.

As the range of strategy possibilities diminishes, one form of strategy emerges more than others. That is a niche strategy. A niche marketing strategy is evident in the organization that decides to serve certain market niches or a single niche rather than the whole market. Niches are often ignored by the major technology players because they may not be large enough to be cost-effective or their customers may be far too demanding.

The niche strategy was the most popular choice among the ventures in our sample. About 70 percent of those ventures use what they indicated was a niche strategy. Broader-based strategies involving entry into more than one market and the use of a main marketing strategy in a undifferentiated approach were far less popular. So, although niche marketing has been the most recently developed of these three approaches, it has found a receptive audience among technology founders.

The reason why the niche strategy was so popular may be its conceptual clarity compared to other choices, or the business acumen of these founders. There are certainly other possibilities, but our subsequent personal contacts with a smaller group of these 10 venture founders indicated that these founders had good understandings of niche strategies and deliberately chose this form for its expected benefits.

Note that niche marketing is depicted as a marketing strategy. That means the marketing mix of promotion, distribution, price, and product are directed to meeting marketing goals. Niche marketing strategies have achieved (if that is a proper description) a singular overuse. Devoid of context, a niche strategy has come to mean almost any marketing strategy in which the goal is not total market domination but conquering a section of it.

The most appropriate context for a niche marketing strategy is selection from the set of marketing strategies that have come into popular use. They can be categorized in four descriptions, from the niche player to the market follower, to the market challenger, to the market leader, each with progressively larger market shares and different competitive concerns. Challengers attack leaders. Leaders defend themselves. Followers live off the remains and nichers are selective feeders.

High technology ventures have little choice of being anything other than nichers or followers in most cases, unless they are grand-scale spinoffs carefully nurtured by a beneficent sponsor. An essential distinction needs to be made between the market follower who takes their cue from leaders or challengers and the nicher who take their cue from themselves. The follower will imitate market initiatives of larger firms while the nicher will operate more independently. Essentially unconcerned with the fray about them, the nicher serves neglected sectors of the market or serves served markets better.

In high technology, a true niche strategy could be a PC troubleshooting business for used equipment only in a certain geographic market. This is a sector of the market normally ignored by manufacturers, yet it could be a profitable niche where there is concentration of PCs and a climate for reselling and using old PCs.

Part of new technology venture marketing strategy is the development of new products. New ventures introduce new products for the most part, although a venture can market old products acquired from another firm. The ventures in our sample had their own products or services. They also had some form of a new product development process implanted in their overall strategies.[7]

We cannot leave the subject of strategy and entrepreneurship without observing something about the changing nature of strategy conceptualization. The most recent development has been attempts to incorporate chaos theory into strategy. Smilor and Feeser have written that chaos theory looks at systems that behave chaotically and that long-term forecasting of results is unreliable. Yet the theory offers a way of understanding systems that do not reproduce results, such as the evolution of entrepreneurial companies. They state four factors are necessary for a new high technology company: talent, technology, capital, and know-how. The diversity of these and other founding factors results in a system that never exactly repeats itself. Public policy can influence these factors, however. Some principles of chaos theory can help explain the nature of risk in entrepreneurial ventures, according to these researchers.[8]

What the Ventures Revealed

Our attention now turns to the reported results from high technology ventures. We consider the types of strategies used, as well as the consequences of those strategies.

Does the niche strategy make a difference compared to other strategies? In our venture sample, those that followed a niche strategy were put in one group and those that followed either a multisegment, broad-based approach or an undifferentiated broad-based entry strategy were put in another group. Performance was checked on the same five dimensions as described in Chapter 5. On none of the five performance measures was there a significant performance difference between the nichers and the nonnichers, although we came close to establishing a difference on the basis of growth in staff.

Even though there was no indication that a general niche strategy might work better (in our sample, nichers actually had slightly lower sales increases), it could be that certain types of niche strategies work better than others. That was the next idea we tested. The results are described after we review what the different niche strategies are.

A high technology nicher could cut its price, offer higher quality, provide broader or specialist product lines or market segments, or improve services or its distribution system. These are the types of niche strategies customarily used in marketing. But our results show that there is no clearly superior type of niche strategy. In our sample there were no significant differences in performance among the ventures that carry out any of these six types of niche strategies.

What this leaves us with is the conclusion that niche marketing strategies for high technology ventures have not demonstrated any inherent superiority, nor do the types of niche strategies sort themselves out as preferable, one over another. This isn't to say that niche strategies aren't appropriate over other forms. But they are not, in and of themselves, proven performance producers in these circumstances.

What of the other broad-based nonniche strategies? Not all high technology ventures use initial niche strategies, so we looked at differences in performance during founding between the ventures that went into more than one market segment (as opposed to nichers, who went into a single segment) and those that used an undifferentiated strategy. The undifferentiated strategy meant using a main marketing strategy for all markets.

New ventures that used an *undifferentiated strategy* had better average sales performance than ventures that used a differentiated strategy. The differences between ventures with differentiated vs. undifferentiated strategies was not deemed statistically significant, however.

Another important consideration for ventures is *timing*. When do you launch the business? There has been some support for the idea that an undifferentiated strategy makes a difference, but what about the time effects? The key to time effects is the industry life cycle. The industry life cycle is the concept roughly analogous to the biological life cycle, in which there is rapid growth, maturation, and decline. Industries have been looked at this way by economists and financial analysts. Some strategies would seem to work better in the early stages of the industry rather than in the later stages.

Since the early stage of the industry life cycle is a struggle for identity for new technology companies, we theorized that differentiated strategies would produce better performance. That was contrasted with the idea that undifferentiated strategy would not be as linked to better performance early in the life cycle. Our findings offer some evidence that while it makes a difference to enter the industry early with an undifferentiated strategy, there were only a few cases in which this happened, so we don't conclude that there is a difference.

In all, then, we found that initial strategy choices do not have a clear, demonstrable effect on performance. Technology entrepreneurs should bear this in mind as they launch their businesses. There is no single, prescriptive strategy for the technology business launch.

Product differentiation is one way the high technology venture might get a foothold in the market. This is a typical marketing strategy that bears elaboration here. Differentiation of products is important when there are many products available, when the core product cannot be distinguished by customers on the basis of superior features or benefits, and when customers make buying decisions on the basis of additional features rather than the product itself. Marketing practitioners make careers of doing product differentiation. A typical high technology product differentiation could be found in spreadsheet programs, where the core program is essentially the same. The differentiation is in the number of platforms it runs on, special graphic features, and so on.

Our look at product differentiation was done on the basis of quality, value, and product support. After combining these dimensions into a score, we speculated that ventures with more product differentiation than their competitors had better performance than ventures with lower degrees of product differentiation.

So what were the results? The ventures in our sample that had more differentiation experienced no better performance than those that had less differentiation. Only average sales came close to showing a difference.

Another common marketing practice of companies is to conduct *market segmentation*. In the high technology field, this occurs as the entire possible market for a technology firm is divided or segmented into different groups. The different groups are homogeneous in certain characteristics and heterogeneous from other groups. The criteria for segmenting the market may be predetermined by the organization, or they may evolve from naturally occurring commonalities. High technology segmentation is potential customer segmentation; that is probably the most common way it is done. For example, some potential customers might be much more computer literate than others. This could be a possible basis for segmentation, though there are many others.

Segmentation is only the first step in devising a marketing program. What follows is the decision about which segment to target. The targeting decision is where the niche or more broadly based strategies come into play. The last step of this process is to position the firm's products and services in relation to those of other companies. That completes the marketing program, at least in its theoretical manifestation.

What can be said about segmentation in high technology ventures? First, it is surprisingly popular. Over two-thirds of the ventures in our sample used segmentation. Indeed, the concept and application of segmentation is very popular among emerging high technology companies. Does it make a difference in performance? Apparently not, because when our sample was divided between segmenting and nonsegmenting ventures, there were no significant performance differences between the two groups.

Our results may be due to the fact that most of our ventures entered in an early phase of the industry life cycle (as they reported), where the necessity to segment (and reap the performance benefits of doing so) and target marketing are less pronounced than in later stages. Using market segmentation is pronounced in mature and declining industries, where there is severe competition for customers and natural market growth has diminished.

The last of the marketing strategies considered is *distribution channels*. These are the channels to customers. In high technology, they may be manufacturing reps or value-added resellers. The channels are not considered as retail establishments but as the different paths to retail establishments. In a product's life cycle, the early stages of market entry have single or few distribution channels. In the more mature phases, multiple channels are usually employed. Our expectation was that differences in the number of channels would have consequences for high technology marketing, since distribution is such a critical aspect of marketing (considered the *P* for Place in the 4P model). Our further expectation was that ventures with single or few distribution channels would have better performance than ventures with multiple channels. Fewer channels do not tax the resources of the venture as greatly. Fewer distribution channels also adhere to the "recommendations" implicit in the product life cycle.

Our sample ventures provided some limited support for the idea that more distribution was better. With more distributors, the ventures had slightly better sales growth, average sales, ROI, and employee growth. However, high technology ventures that have very few or single channels had better market share performance than ventures that had many channels. The market share result came close to being statistically significant, but the other measures did not.

Our results on distribution channels run counter to expectations, but they emphasize the point about marketing acumen in emerging technology ventures. Distribution is an important marketing issue because it is how new companies establish a market presence and how established companies compete successfully.

Is Being Close Better?

One possible consequential performance difference is customer and supplier proximity to the venture. A formation-stage decision about who the customers are and where they are located is truly strategic. The same applies to supplier proximity.

High technology ventures can obviously avoid transportation costs by using nearby resources, but there is more to it than that. They may be corporate spinoffs and share nearby customers with their sponsoring firm. Intellectual resources may reside with nearby customers and suppliers. Our venture sample liked to "check back" with colleagues in firms they left to get valued opinions. One technology venture founder needed to keep in touch weekly with his engineer friend in the original company just for feedback on an idea or two. The post-1970s American auto industry view of having close suppliers has had quality improvement implications. This was a lesson from Japanese Total Quality Management, and it has not been lost an high technology companies.

Before going into the subject more fully, we should say something about the locations of high technology firms. Are these companies concentrated or diffused? We'll provide more details on location in Chapter 16, but here we need to consider the extent of concentration somewhat because this is a defining characteristic of the business sector itself. It also directs us to answering the question of proximity and performance.

Giese and Testa (1988) compared the national average of high technology employment to census geographic divisions and devised a concentration index for high technology. They found a not unexpected greatest concentration of high technology industries in the western states of California and Washington and in the eastern states of New Hampshire, Massachusetts, Connecticut, and New Jersey. Employment concentrations are higher in the eastern states, but the western states have a greater percentage of the largest high technology businesses. Between the coasts, there is a wide scattering of high technology businesses and low concentration. Only Utah, Arizona, Colorado, Kansas, Missouri,

Minnesota, and Indiana show average concentrations of employment. All other states have below-average concentrations of high technology businesses.

Although the technology businesses tend to be concentrated, the question remains if their customers and suppliers are clustered near them. At least one author has shown that local resource access is important in innovation success.[9]

Another aspect is the matter of where the entrepreneur locates the high technology firm. The availability of professional and technical resources has dominated the location decision.[10] Location of operations is presumed to be a strategy decision for the entrepreneur. We also consider it a rational decision, although there are psychological dimensions to it. Most traditional ventures are grown in founder hometowns, and the same appears to be true for technology ventures. There are compelling reasons for homebound entrepreneurship: family, community, convenience, and transportation costs that can overshadow some purely rational aspects of this crucial strategy decision.

The location decision has more permanence than other parts of business strategy because there is extremely little relocation of going technology ventures. While organizational form evolves as strategies are modified for new environments, location seems to be as rooted as the mightiest oak in the community of the company's.

These considerations help introduce our findings on the location and performance issue. Turning to the question of whether nearby customers and suppliers make a difference in venture performance, we can say that ventures with closer customers and suppliers had better performance on two measures, average sales and sales increase. This was somewhat significant at the 10 percent level. On the other three measures, there was insufficient statistical proof but general agreement that closer proximity means better performance. To reach that conclusion, we compared the ventures for whom a majority of customers and suppliers were located within ten miles to the ventures whose customers and suppliers were located more than ten miles away.

What can be pulled from the resource access idea is the conclusion that a new technology venture could very well benefit from locating its main business operations close to its suppliers and close to its customers. The seemingly obvious here may not be so obvious to the overly ambitious entrepreneur who simply assumes that the whole world is his oyster and that customers are everywhere. It seems appropriate to discover potential customer concentrations and locate the operation near them. The same holds true for suppliers, be they labor suppliers such as engineering schools or raw material suppliers. The truly optimal combination is to have both customers and suppliers nearby.

One of the unique aspects of high technology venturing is the presumed need to have marketing and research and development functions in close contact with each other.[11] At the venture stage, research and development skills and marketing skills are both very critical. Research and development traditionally is the source for new products, and marketing is the source for customers. If these functions are not in concert, the technology venture can quickly falter.

The frequency of communications between these two departments is one indicator of cooperation. Another indicator is joint planning operations. These reflect the degree of cooperation between functions and were the targets of our questions about performance and coordination.

Our sample ventures were split between those with relatively more cooperation and those with relatively less. Using this idea of cooperation from A. Gupta (see note 11), we looked within our sample and found no real support for a connection between venture performance and cooperation between research and development and marketing. In fact, only ROI showed a slight but not statistically significant relationship. It should be added that most of our sample had relatively more cooperation and only a few—between 6 and 12 for the five performance categories—had less interdepartmental cooperation.

The prospective technology entrepreneur shouldn't interpret these results as meaning that marketing and research and development cooperation is not important. It is simply that there are no demonstrated links between performance and such cooperation.

Differences among Electronics, Biotechnology, and Computer and Computer Related Ventures

One of our tasks was to determine how homogeneous high technology really is. We wanted to discover if there were performance differences among the three types of ventures as far as strategy was concerned. Our results showed that there were notable performance differences, though nothing overwhelming emerged. In doing this research we followed the same sequence used in assessing the separate strategy issues in the earlier part of this chapter.

The use of niche strategies or broad-based strategies separated the three types of ventures with respect to average sales performance. Also, the sales increase performance measure separated the venture types in the broad-based strategy. Lastly, using the niche strategy separated the venture types in the ROI measure. These were the instances where statistical significance was found.

It might be noted that in 11 other possible outcomes (three strategy categories times five performance measures), there was no statistical support. Although performance differences were found in most of these outcomes, we could not conclude that they were statistically significant.

In all, this evidence was taken to show support for the idea that the three venture types are indeed different in the strategy performance measures and that the three ventures are more dissimilar than similar. In short, we are talking about three different animals, not three heads on the same hydra.

Customer and supplier proximity was the next matter considered. Our interest was whether there are differences in performance among the electronics, biotechnology, and computer and computer-related ventures with respect to access to resources or, as we also called it, "customer and supplier proximity." Our expectation was that there would be performance differences among the venture subgroups when customers and suppliers are located closer to the venture location than when they are farther away. As earlier, we used ten miles as the demarcation between closer and farther away.

The results provided the strongest endorsements thus far that these three types of ventures are distinct from one another. In a majority of possible outcomes, we found performance differences. Our venture subgroups had different average sales performance no matter where they were located (either closer or farther from customers and suppliers). The same was true for ROI. For the measure of sales increase, there were subgroup differences when the customers and suppliers were farther away. The same was true for staff increases.

What this suggests to the aspiring technology entrepreneur in electronics, biotechnology, or computers is that they need to take their cues from the particular venture type rather than the broad classification of high technology. A prospective software entrepreneur can't expect to learn much from the electronics venture field as far as resource access is concerned.

Turning to a marketing strategy question, we looked at whether there were performance differences based on the extent of product differentiation used. Splitting the ventures into categories of "more differentiation" and "less differentiation," we found subgroup differences for average sales, sales increases, and ROI. There were not enough cases of "more differentiation" to draw conclusions. Once again, the case has been strengthened for understanding technology ventures in their separate domains rather than in aggregate. High technology is not a monolith.

Is there a performance difference when market segmentation is used compared to when it is not used? Our sample showed that on one side, that of segmentation, there were performance differences among the subgroups. All five performance measures showed statistically significant differences regarding the use of segmentation. For ventures that did not use segmentation, there was not enough evidence to indicate performance differences. This leads to the conclusion that there are performance differences among the three venture types with respect to segmentation. Considering that market segmentation is a common marketing strategy for ventures, this result is consequential.

Next, another marketing strategy was considered: whether or not there were performance differences based on the number of distribution channels. Our ventures were split into "few and single channels of distribution" and "many channels of distribution." There were performance differences found for ventures with fewer or single channels. The performance differences were in average sales, sales increase, ROI, and market share. There were not enough cases of ventures using many channels to establish results on this side.

Our final look at venture type differences has to do with the level of integration between the marketing and research and development functions. As defined earlier, this is the frequency of joint planning and communications between these two organizational units. The ventures were separated into "more integration" and "less integration" categories.

The gap between R&D and marketing functions is more of a plague for the larger, hierarchical technology firm. Marketing departments and R&D may only be connected at the top of the organization, yet it can still reside in the newly formed venture. If a single manager does marketing and a single engineer does research and development, and they don't communicate or coordinate functions, the results can be just as devastating. It can also be a problem in the single founder–single technology business. The founder may mentally be connecting R&D with production but may miss the marketing technology connection.

In our sample, there were performance differences for average sales, sales increase, ROI, and market share (but not staff increase in instances where there was more integration between functions). For less integration, there was statistical support for differences in the ROI measure alone. This represented a split finding, which is far stronger on the "more integration" side than the "less integration" side, so the inferences can be made mainly when we find more coordination between marketing and research. Nevertheless, since technology ventures tend to have fewer employees, there is presumably more coordination between research and development and marketing, and the results are more applicable to typical venture activities.

Once again, we find that it is worth considering the distinctive aspects of these three venture categories rather than putting them together as a typical high technology marketing and research coordination operational mode. Those who are embarking on a technology venture should consider the importance of functional unit coordination, as well as the fact there are performance differences among the types of ventures. This can then be planned for by requiring coordinated planning and budgeting by marketing and research and development units.

Because of the importance and complexity of strategy choices, there have been a variety of tools devised to help entrepreneurs make their decisions. Some involve matrices in which factors are plotted against one another. Others are decision tree analyses, payoff matrices, and a host of techniques in financial analysis and strategic management texts.[12]

Conclusions and Considerations

Strategy counts for the new technology venture. Location strategy has been shown to be very important. Close proximity to customers and suppliers has business performance consequences for the technology venture. Marketing strategies do not have clearly established performance results, though there are indications that having multiple distribution channels helps in some instances. Close coordination between marketing and R&D functions has also not emerged as consequential for performance. Our three venture groups of computers, biotechnology, and electronics show differences rather than similarities on the strategy dimension, which gives support to the idea that high technology venturing is not a monolithic enterprise.

All of these results pass the "makes sense" test and are consistent with the discoveries of other researchers. Most normative accounts of strategy are based on small samples in fixed evaluation conditions, so any generalizations ought to be circumspect. We have tried to use a broader perspective involving more ventures and arrive at the position that venture business strategy requires a contingency approach. The research on actual ventures, such as work by Sandberg and McDougall (see note 2), has supported this view which translates into an equivocation of sorts. That is, no single strategy is appropriate at all times for all technology ventures.

A caution for new business founders is to not neglect the development and articulation of their path to success—the strategy they choose or assume by default—even if it is only the Spencerian survival imperative. Even grossly drawn, it is a result of entrepreneurial choice and has long-lasting consequences. Our causal observations of high technology ventures have revealed that founders emphasize personal skills over chosen strategies, an inclination shared by venture capitalists. Yet strategy has been given research endorsement here and founder characteristics have not. That is a message that should not be lost, despite the fact that strategy may be undetectable to the casual observer of a high technology venture.

Our interviews with technology company founders have shown that these individuals have an easier time coming up with the "what" but more difficulty arriving at the "how." They know and can detail what the product will be, but are sometimes stumped when asked how they will get their innovation to market. On the other hand, the founders who could answer both the what and how were the ones who enjoyed commercial success.

There are other important and possibly quite consequential factors that need to be assessed before we close the book on technology venture performance. In the next chapter, we will consider another broad issue that may have an impact on venture performance: the environmental factors that surround the technology venture. Discussion of these factors will help complete our depiction of technology ventures and allow us to look at the fully evolved venture, as we will do in Chapters 9 through 16.

Notes

1. Changing strategy as high technology businesses grow has been a research subject itself. Robert Kazanjian, in his 1983 University of Pennsylvania dissertation *The Organizational Evolution of High Technology Ventures: Impact of Stage of Growth on the Nature of Structure and Planning Process*, demonstrated that there were certain organizational patterns in some stages of high technology business growth. These patterns were characterized by different types of strategic planning questions. High technology evolution means different strategy issues, such as changing the form of the organization.

2. A study of this is Jean Schere's 1981 University of Pennsylvania dissertation, *Tolerance of Ambiguity as a Discriminating Variable Between Entrepreneurs and Managers*.

3. In *New Venture Performance: The Role of Strategy and Industry Structure* (New York: Lexington Books, 1986), author William Sandberg reviewed venture business plans and financial performance. He found a relationship between business strategy and performance, as well as industry structure and performance. Patricia McDougall followed this research stream and, using cluster analysis, discovered that a combination of new venture business strategy and entry barriers accounted for considerable market share, return on sales, and return on investment variance. She tested other venture origin, strategy, and entry barriers models and concluded that no single strategy is optimal. These findings were in her 1987 University of South Carolina dissertation, *An Analysis of Strategy, Entry Barriers, and Origin as Factors Explaining New Venture Performance*.

4. This was discovered by Henry Feeser in his 1987 Purdue doctoral dissertation, *Incubators, Entrepreneurs, Strategy and Performance: A Comparison of High and Low Growth High Tech Firms*. The research concerned the computer technology sector.

5. R. Oakey, in *High Technology Small Firms* (New York: St. Martins Press, 1984), is one such study.

6. As they state it, "Not surprisingly failure is more probable than success in initiating new business ventures. There were literally dozens of critical events that had to be worked out right and come together at the right time in order to effectively build a viable new business based on new technology." In R. Burgleman and L. Sayles, *Inside Corporate Innovation: Strategy, Structure and Managerial Skills* (New York: Free Press, 1986), pp. 7–8.

7. Others have studied high technology firms and new product strategies. M. H. Meyer and E. B. Roberts, in "New Product Strategy in Small High Technology Firms: A Pilot Study," MIT Sloan School of Management Working Paper #1428-1-84, May 1984, report differences between the most successful and the poorest performing technology firms. The most successful (in terms of growth) firms concentrated on one key technological area and introduced product enhancements related to that area. The poorest performers had undertaken unrelated new technologies and entered new product market areas.

8. In R. Smilor and H. Feeser, "Chaos and the Entrepreneurial Process: Patterns and Policy Implications for Technology Entrepreneurship," *Journal of*

Business Venturing 6 (1991): 3. David Levy adds further endorsement to the value of chaos theory as providing a framework for understanding of the evolution of industries and their player interactions. In conditions of complex and dynamic competition the outcomes are not known and chaos theory is applicable. A simulation was developed using interactions among a computer manufacturer, suppliers, and market. The simulation showed managers might underestimate the costs of international production. In D. Levy, "Chaos Theory and Strategy: Theory, Application and Managerial Implications." *Strategic Management Journal* 15 (Summer 1994), pp. 167–178.

9. R. Oakey found more innovations coming from small high-tech firms with nearby suppliers. Reported in *High Technology Small Firms*, cited in note 5 above.

10. C. Galbraith interviewed 98 high technology executives to arrive at this conclusion. In "High Technology Location and Development: The Case of Orange County," *California Management Review*, 1985, pp. 98–109.

11. A. K. Gupta studied this matter in a 1984 Syracuse University doctoral dissertation, *A Study of R&D/Marketing Interface and Innovation Success in High Technology Firms (Integration, New Product, Research Management, Development.*

12. A example of a matrix model is the three by three Optimal Entry Strategies created by Edward B. Roberts and Charles A. Berry. By putting "market factors" on the Y axis and "technologies or services" on the X axis, and scaling the axes on the basis of familiarity, the authors suggest strategies for seven different sectors, from internal development to acquisitions, joint venturing, internal venturing, licensing, venture capital, and educational acquisitions. In E. Roberts and C. Berry, "Entering New Businesses: Selecting Strategies for Success," *Sloan Management Review*, Spring 1985, pp. 3–17. This scheme is well suited for technology ventures since market factors and technological aspects of products are key factors in all high technology business.

References

Chandler, C., and S. Hanks (1994). "Market Attractiveness, Resource Based Capabilities, Venture Strategies, and Venture Performance." *Journal of Business Venturing* 9: 4, pp. 331–349.

Giese, A., and W. Testa (1988). "Targeting High Technology." *American Demographics*, May, pp. 38–41.

Porter, M. (1996). "What is Strategy?" *Harvard Business Review*, November-December, pp. 61–78.

Singer, B. (1995). "Contours of Development." *Journal of Business Venturing* 10: 4, pp. 303–329.

8

■■
■■

TECHNOLOGY VENTURES
AND THEIR
ENVIRONMENT

This chapter addresses environmental factors of the technology venture. We also explore which environmental factors may influence high technology venture performance. We step outside the immediate and controllable and look at some conditions that reside in the external tempest of the marketplace.[1] This dimension is frequently ignored by entrepreneurs, yet it may be an important determinant of success. The perspective here on the venture's environment will complete our review of technology ventures.

This chapter thus considers an important subject: the environment of the technology venture. But practicality must make us selective. We can't and won't analyze all environmental factors. Instead, we broadly suggest what they are, so that those who might want do their own analysis can do so using their own devices.

The Outside Forces

No technology venture is an island. It exists within myriad forces and conditions that can influence how well it fares. The new technology business is a planet in a larger universe. A look beyond this planet reveals other planets—competitors perhaps or friendly bodies in similar businesses. Often, the swirl of mass and motion is seen as the universe where the high technology venture resides.

This version is Copernican, but we need a view that is of the age of Hubble. The solar systems themselves are captured in galaxies that are

moving, a perspective from Hubble that lets us get even more fundamental. We move beyond the new venture in its solar system to a galaxy
of environmental forces. The entrepreneur who is just looking at the
other planets will not detect the sweeping motions that catch their venture, as well as all other ventures and established businesses, and set them
on an intractable course to an unforeseen destiny.

In the broadest setting, the new venture's chief environmental factor
is international competition in high technology. The United States once
dominated high technology innovation, but other developed countries
now compete directly with the United States. Japan, the United Kingdom, Western Europe, and Russia all have vital and innovative high technology sectors. In fact, all industrialized countries have high technology
sectors, and some countries are leading the way for others, such as with
Japan's development of robotics and Germany's work in pharaceuticals
and automotive engineering.

Another facet of the international environment is the role of government. While high technology products and services are global in nature,
companies exist in the province of the laws and policies of their home
countries.[2] Any analysis of international environmental factors must include the actions of governments to support, remain neutral, or hinder
their technology sectors. Government support produces uneven effects.
For example, close cooperation between industry and government is
manifested in Japan's MITI joint business government planning council,
while in the United States the essentially laissez-faire presumption keeps
government distant from high technology development.

These are important differences for the technology venture and established company as well, yet they are located outside the business. Close
cooperation by government and high technology business can help in a
number of ways. First, the government itself is a large lot buyer. Second,
there can be resource sharing. Third, and related to the second point, is
the risk mitigation benefit as government and business each assume some
risk of technology development. Fourth, tax and regulatory policies can
be written to favor domestic technology companies. Fifth, governments
can be strict or lax in their enforcement of imported technology patents,
a major concern for software firms.

Another environmental factor that dwells within the realm of government is regulation. The regulatory area affects high technology ventures in different ways. For ventures aimed at marketing new drug
therapies, regulation by the Food and Drug Administration involves constantly winning approval for commercialization. At the other end of high
technology, the new independent software consultant has a relatively free
rein, providing it meets licensure agreements for software used.

Technological forces are themselves frequently external to the venture's operations, a situation we note but do not belabor. Here, we are
concerned with large and sweeping technological changes that alter the
way an entire industry does business. But we are also thinking of ways

that unrelated technologies can impact particular industries. An example is how programmable controllers have entered the factory floor, an area once thought the domain of mechanical devices. We make a distinction here between external, transplanted technologies and internal, business-driven technology improvements that come from the businesses themselves, such as new airframe assembly machines or enhancements in computer operating systems.

Still another environmental factor is the sociocultural element, a potentially complex and influential element. To a great extent, cultural attitudes about technology shape its acceptance among consumers and the labor base that makes these high technology products. The parts of high technology that invade privacy or lead to a dispersion of weaponry create negative attitudes, while more positive aspects lead to better medical technologies or ease workplace drudgery and are welcomed. Technology entrepreneurs need to determine if it is ambivalence, acceptance, or rejection that awaits their efforts. In this way, social and cultural factors are clear, consequential environmental influences.

Very closely related to the sociocultural element is the factor of community. In this category fall the literal definitions of community as well as the less formal collections of professions, trades, and interest groups. Community encompasses the educational factors of training and understanding of science-based businesses, social support for technology companies, and other, more localized but nonetheless important environmental factors. These community factors are crucial in an analysis of the venture environment.

Our look at environmental factors mentions these many factors, but our analysis went into the more limited environmental sphere of economics, in several specific ways. The environmental factors assessed in this chapter are barriers to entry and industry life cycle.

Barriers to entry, though discussed in more detail shortly, are the economic concept of having, in a certain market, economic-based advantages for present competitors and entry costs for those who want to compete in these market. High barriers to entry are an advantage for incumbents and a disadvantage for newcomers. Some barriers to entry might be the costs of acquiring patent rights or capital costs. Exclusive access to a supplier might be another barrier to entry.

A corresponding but less well known concept is the barrier to exit. That is the cost of leaving an industry, which may be prohibitive. It may also be the creation of a technological advance. Perhaps the most prominent, example of a barrier to exit is the nuclear power plant in the utility industry. So much capital was invested in these plants that their decommissioning and consequent "hanger queen" status send shudders down the backs of utility executives, who fear they will recover very little of their initial investments.

By *industry life cycle*, we mean the idea that groups of similar businesses experience the biological analog of birth, growth, maturity, and decline. They do so in a nonbiological world, however. (We have discussed this in Chapter 7.) For example, certain technology business sectors are growing in overall sales while others are shrinking. Particular growth areas now are in telecommunications, while electronic components is an area of slow or no growth.

The environment for new ventures certainly does not consist of these two features alone. In the broader business universe, such domains as politics and government, or the social, cultural, macroeconomic, labor, and natural resource factors that have been mentioned must come as features in a true environmental assessment. Our restriction to the economic side is one of both convenience and promise. This study has explored business performance, and we have collected information on the economic factors potentially impacting business performance. That's the convenient part. The promise part has to do with the fact that industry life cycle and barriers to entry have been subjects of considerable research on company performance, and there is a chance to add our unique perspective of high technology performance.

For this chapter, we drew considerable guidance from industrial organization theory, which identifies, classifies, and evaluates the forces acting on the business organization. To a lesser extent, organizational behavior theory was also utilized because it provides a reference frame for understanding ventures and their environments. The ultimate purpose of our approach is to depict the dynamics of the venture, its environment and its performance, with the result of gaining an understanding of the significance of industry life cycle and barriers to entry on high technology venture performance.

When high technology founders talk about the business environment, they don't usually do so directly. Discussion is more often in terms of timing. For example, we've heard entrepreneurs talk about looking for the right time to do something. The move to introduce a new product is a timing matter. But what is really meant is whether the environmental factors are favorable for a new product. The competition is sleeping. The industry is growing. Customers are willing to pay dearly—these are the favorable circumstances embedded in the timing cover. Taken this way, timing is a combination of not only of when but also of how much and where—the essentials of any environmental analysis.

Our exposure to entrepreneurs in this study resulted in a curious split. We saw most founders as acutely sensitive to the greater environment while others were oblivious to it. The sensitive ones engage in the typical strategic management activities of environmental analysis, industry analysis, and scenario building, while the near focused plunge ahead, certain that a market is there for whatever superior machine, device, or process they have to offer. What this observation reinforces is that technology entrepreneurs are their own people, some of whom consider everything and others who consider only what they need to think about.

Related Ideas and Prior Findings

From what our sample technology founders have suggested, their appraisal of environmental factors is one of ambiguity. Many founders don't systematically assess their environments, but this doesn't mean there aren't other sources for environmental assessment. There are, quite fortunately, a considerable number of people who study the general venture environment's influence on performance. But there are considerably fewer who consider the environment for high technology ventures. Venture research by Sandberg (1986) and McDougall (1987) supported the importance of industry structure in explaining venture performance. Industry structure includes the industry life cycle and barriers to entry ideas. Our interviews with venture capitalists also suggested that timing and amount of capital are crucial for high technology venture success, as well as the stage of industry life cycle at entry. Access to capital emerged as a viable barrier to entry that was worth a closer look.

A broader net in these waters was flung by Karl Vesper, who did not limit himself to strictly economic barriers. He proposed that the venture can be either aided or retarded by forces. These forces, pushes or pulls, either positively or negatively affect the development of the venture. The negative forces are quite substantial. For him, the negative forces are a bad concept, lack of market knowledge, no technical skills, no business know-how, complacency, social stigma, a comfortable current job, time pressure, legal constraints, patents, and lack of capital.[3] The last three are addressed in this chapter while others were at least partially addressed in the preceding three chapters. All of these barriers are pertinent for high technology business founders, although Vesper did not focus exclusively on these. Thus the high technology venture environment is envisioned here as being those forces and conditions either residing outside the existing venture and having a possible influence on it or constituting a factor that will influence the emergence of the venture.

While we have looked at the environment in relation to new business, we cannot dismiss its importance as the venture becomes a thriving firm. Environmental influence is continuous. Readers need only inspect the subject of corporate venturing to see this. Block and MacMillan (1993) have sections on "Developing in-depth knowledge of customers and markets" and "Industry and market changes" that help serve the point about the persistence of environment.

These considerations move us closer to having a specific set of environmental factors to assess and the more immediate purpose of introducing industry life cycle and barriers to entry as environmental forces acting on the high technology venture.

The Impact of Environment

Our consideration of environmental factors was done in steps, by testing several related ideas on high technology business performance and the venture environment. Together, they advanced the notion that venture performance is related to the nature and extent of barriers to entry and that entry at different stages of the industry life cycle has performance consequences. The development of these proposals assumes, of course, that there are barriers to entry for high technology ventures and that the industry life cycle model is applicable.

Information from our sample founder group supports the existence of barriers to entry for new technology formations. Not a single venture founder indicated that their venture faced no barriers to entry—every one reported facing some barrier to some degree. The heads of software companies were unable to get needed licenses for software from larger companies, whom they believed were stalling just to run them out of business. The head of a well-known firm's biotechnology division reported that some patents given to the firm were considered far too broad and were being reviewed as restrictive of competition.

The same was true for industry life cycle. Our discussions with high technology venture founders before our survey was finalized confirmed that the industry life cycle concept was applicable, although there were instances when it had to be drawn out. Venture founders had a general idea of what stage of the life cycle their particular enterprises occupied. All of them could place their businesses on either the upside or the downside of the cycle, and most could further distinguish between introduction and growth on the upside of the cycle. This informal confirmation of the importance of industry life cycle and barriers to entry made it reasonable for us to then ask about these factors in the formal survey of venture founders. This was not done as a test of founder knowledge but to confirm the validity of the life cycle concept in the real world.

These two environmental factors are successful transplants from the academic world to the business environment. Industrial organization researchers and economists have successfully introduced and refined the concepts of entry barriers and industry life cycle to the point of acceptance in the practical as well as the academic domains.

These concepts have been applied to all kinds of industries and their success has been in providing a useful way to explain organizational behavior. What companies do has much to do with how they deal with economic barriers. The spectrum of what they do also reflects the range of what any business can do in the industry life cycle. Businesses in a mature phase cannot reestablish in a growth phase. If we reject the notion that barriers to entry and industry life cycle apply to high technology ventures, then we must accept the idea that these ventures are exceptional business organizations. Yet for this we have discovered no evidence whatsoever.

There is little basis for viewing high technology organizations as exceptions. Their goals, organizational forms, and operations are similar to traditional businesses. Even governmental definitions of high technology, such as those reviewed in Chapter 1, distinguish high technology from traditional firms solely on the basis of workforce composition and research and development expenditure. Governmental agencies have found no other bases to define this form of business, including industry stage. In most respects, the high technology firm is similar to traditional firms. In accordance with this, our application of barriers to entry and industry life cycle are justified.

Barriers to Entry

In classical industrial organization theory, barriers to entry "may be defined as a cost of producing (at some or every rate of output) which must be borne by a firm which seeks to enter an industry but is not borne by firms already in the industry" (Stigler, 1968, p. 67). Although not explicitly identifying high technology barriers to entry, Porter (1980) lists economies of scale, product differentiation, capital requirements, switching costs, access to distribution channels, proprietary technologies, raw material access, favorable locations, and government policies as barriers to entry. The strength of the entry barrier and competitive actions are key to Porter's conception of competitive strategy. Of these barriers, virtually all—with the possible exception of raw material access—are applicable to the high technology venture.

On capital requirements, however, Porter and Stigler take different views, with Stigler arguing that capital requirements are not a barrier to entry but a determinant of an economy of scale, since existing firms require capital as well. Bain (1956), another economist who has contributed much to refining the barriers to entry concept, approached barriers to entry as part of a discussion on conditions of entry. Bain described types of entry as easy market entry, effectively impeded entry, ineffectively impeded entry, and blockaded entry. These conditions of entry were important to us as we constructed the questions we asked our sample founders. It became a matter of the *extent* of the barrier rather than if a barrier was present or not. Bain also explicated on economies of scale, product differentiation, and absolute cost advantages as barriers to entry. That is, high technology ventures face barriers to entry because they are the new businesses seeking market entry. Blockaded entry happens when, for instance, a software design company senses the entry of a rival and tries to prevent that new firm from getting a license to use software for development purposes.

This book takes the approach that the different capital requirements for a new venture are a basis for saying there is a barrier to entry for new ventures. New technology ventures need capital to operate. Thus the capital requirement barrier has a degree assigned to it.

There is one further note about what has been said regarding barriers to entry and the pace of the technology that drives new business formations. Industrial organization theorists have considered the conditions favorable for rapid technological change. A typical view is presented by Scherer (1970), who states that barriers to entry must be modest to accommodate rapid technological change. Severe barriers to entry presumably impede technological change, and technological change is the lifeblood of high technology ventures. Taking this perspective into account, we see that barriers to entry affect the entering venture as well as incumbent organizations. This further fortifies the value of considering barriers to entry as a possible performance determinant.

Discovering if Barriers to Entry Matter

We now turn to the results of our experiences with the 101 technology ventures in our sample. The methods and business performance criteria used were the same as indicated in Chapter 5.

Our first intention was to determine any connection between performance and capital barriers to entry. We were interested in finding out if there was a difference in performance between high technology ventures that overcome less severe capital barriers to entry and those that overcome more severe barriers.

The technology firm requires capital during the startup phase, according to Roberts (1991). He also asserted that capital funds product development and equipment and provides working capital. The startup phase is within the scope of this study, so we were interested in the impact of capital requirements on performance.

Our attention was on the founders who regarded the capital requirement as a barrier to entry. If the entrepreneurs in our study perceived capital as a barrier to their entry and were able to obtain the capital they needed to continue and expand their operations, they were judged to have overcome this barrier to entry. Venture founders who responded that the capital requirement was a barrier "to a great extent" were compared to those who responded "some extent."

The results provide limited endorsement of the consequences of barriers to entry on high technology business performance. Some support is found for the idea in the dimension of increase in employees, but not the measures of market share or ROI. High technology ventures that were able to overcome some capital barriers to entry had greater employee growth than did ventures that had great barriers to entry, but this was not statistically significant overall. Considerably more support was found as far as sales growth was concerned. The ventures with some but not great capital barriers to entry had much faster sales growth; this finding was statistically significant. The ventures also had higher average sales, but the result was not statistically significant.

We cannot say that overall results are conclusive, although they are valid for sales increases. This may be because the average sales growth criteria somewhat ameliorate large venture–large sales gain experiences in a short period of time. Several of the venture founders commented that it was easier to raise capital if the venture was launched on a large scale and showed rapid growth than if the venture were a more modest initial foray. They may have been aiming for rapid sales growth instead of a slow gain in average sales.

We next consider a particularly prominent entry barrier for the high technology business launch. Preliminary information provided by several high technology executives who were interviewed in advance of our formal survey suggested that patents and other forms of legal barriers sometimes constitute barriers for entry. Patent barriers were especially evident in biotechnology fields. Some venture representatives expressed the view that obtaining a patent was a difficult but crucial step toward commercial success. A few others noted that competitors who held important patents obstructed their entrance to the market. The legal obstacles cited were the high cost of paying attorneys for legal searches for patents, for their acquisition, and for other rights to utilize technologies. Obtaining licenses to use software was an obstacle in many other instances. Some founders cited the expense of using SAS statistical analysis software. Although there were less expensive software packages available, none had the level of support that the SAS Institute provides and this was critical for a number of the engineering analysis ventures. It was also an entry barrier because established firms could better spread the licensing fees over larger organizations.

The idea that high technology ventures experience greater market success as they manage to overcome patent and legal barriers was the next idea we tested. This was posed as the assertion that there is a difference in performance between ventures that overcome less severe patent and legal barriers to entry and those that overcome more severe barriers.

The speculation was tested in the same manner as the previous ideas. Our results were that the speculation is supported on two performance measures—average sales and sales increase. These results led to the conclusion that ventures having lower patent and legal barriers to entry enjoy higher average sales and higher sales growth. However, as far as staff increases or the other measures are concerned, there is no statistical support. Much of that may be due to the small number of ventures providing data with these other measures.

We did find strong evidence of a connection between venture performance and patent and legal barriers to entry. In this case, the two sales measures are in concert. The effect was in both sales measures, although it should be acknowledged that these are interdependent measures.

Another barrier to entry is access to technical skills. It, too, can be a powerful deterrent to entry in this skills-dependent enterprise. Established high technology companies enjoy the advantages of this barrier if they can obtain or even monopolize a technically trained labor base. Since collecting workers is not subject to antitrust action, it may even be an attractive way of securing a market advantage. There are high technology companies that scout research universities to snatch up newly minted PhDs. They may not plan to capture all the available talent, but they know a knowledge business is propelled by knowledgeable people.

With this in mind, we established the related contention that there is a difference in performance between high technology ventures that are able to overcome less severe access to technical skills barriers to entry and those that overcome more severe barriers.

The results of this test once again supported the idea that access to technical skills matters for the technology venture. It was in average sales and increase in sales where ventures that have overcome less severe barriers enjoyed the sales edge. Average sales and sales growth were much higher for young firms when these firms faced only some access to skilled personnel barriers to entry. This was statistically significant. For ROI, market share, and staff increase performance measures, this same conclusion was not reached, owing mainly to a far fewer number of respondents.

Industry Life Cycle and Venture Performance

Based on the definition offered earlier, our aim was to introduce the industry life cycle as a possible consequential factor in venture performance. What we looked at was the assertion that high technology ventures that enter the industry life cycle in introductory or growth phases experience better performance than those that enter in maturity or decline phases.

What is the basis for this assertion? There is no body of work on industry life cycle and performance of high technology ventures. Practice is another matter, though. It would seem that most high technology entrepreneurs enter the industry at the early stages of the life cycle, a conjecture proved true in the initial rounds of our interviews. Entrepreneurs constantly referred to the prospective industries they would do battle in as "growth industries" or "emerging industries." Nonetheless, there are technology ventures launched in mature industries. Of the three high technology groups examined (biotechnology, computers, and electronics), the electronics industry is the most mature and segments of it are in decline, particularly electro-mechanical devices. But because ventures get into the market at different stages of the life cycle, it is worth exploring the performance relationship here, too. There are at least a handful of new electronics firms incorporated every day, in spite of the industry's relative maturity.

To test the idea, we compared ventures that entered during introductory and growth phases to ventures that entered during maturity and decline phases. In all, over 80 percent of the ventures were in the early half, so there were few late stage ventures for comparison. The result was that there is a performance–life cycle connection—in a very limited way. Specifically, staff growth was associated with earlier entrance, and this is logical but not statistically significant. For ROI, there were only three late-stage entrants and four late-stage entrants for market share data, so although ROI was better for earlier entrants, the small number of examples prevents us from asserting an ROI and life cycle connection. No conclusion can be drawn as far as the two sales measures and market share measure were concerned. Average sales and absolute sales growth is not significantly different for early or late-stage entrants.

The consequence of all this is that there is no compelling case for industry life cycle determinants of new technology performance levels. The industry life cycle remains a possible contributor to venture performance, but in and of itself, we have not found any tight and certainly no conclusive relationship.

Technology Ventures and Environment: Similar or Different for Each Group?

At this point, let's return to one of the major questions that prompted this inquiry: Are the various businesses in what is known as high technology similar or different? In this part, we address this question as it pertains to environmental factors.

We attempted to determine if there are performance differences among the computer and computer-related group, the biotechnology group, and the electronics group. This is a further development of our effort to distinguish these three high technology groups. The specific idea we evaluated was whether venture groups exhibit performance differences among one another on the dimension of capital barriers to entry.

To get to the heart of the matter, and the results as well, we noted substantial differences among our high technology businesses. For instance, there were clear and statistically significant differences in 4 of 15 possible outcomes that represented the combinations of "great," "some," and "no" capital barriers and the five performance measures explained in Chapter 5 and the appendix.

When there were great entry barriers, there were significant average sales differences. When there were great or no differences, there were sales increase differences. There were also group differences in ROI when there were differences in the great barriers category. Why were the results not more clear-cut? The reason is that we have only reported differences that can be substantiated.[4] In almost a third of the possibilities, there was reason to believe there were differences among the groups; however, the separate findings are noteworthy. Group ventures had different average sales experiences when faced with great capital barriers to entry. The same is true for increase in sales and ROI. When there were no capital barriers to entry, the venture groups had significantly different sales increase results. Finally, ROI results differentiated the groups when great capital barriers to entry were encountered.

Again, we have confirmation that these three venture groups are different from one another in a consequential way. This suggests that lumping the three businesses together may not be appropriate. High technology venturing is more a matter of computer venturing or electronics venturing or biotechnology venturing because of the differences among these ventures and their business performance.

Continuing with the barriers to entry exploration, we next evaluated the contention that the venture groups had different performance outcomes based on the extent patent rights, government regulations, or legal restrictions were a barrier to entry. The specific idea tested was whether the venture groups would exhibit performance differences among each other on the dimension of patent rights, government regulations, and/ or legal restrictions as barriers to entry. The test of this hypothesis was done in exactly the same manner as before. Our results showed significant differences among the electronics, biotechnology, and computer ventures. These differences were significant when some barriers to entry were experienced in the measures of average sales, ROI, and market share. In 3 of 13 cases (no data was available for 2 possible outcomes), the evidence was fairly strong favoring group differences.[5]

The last of the entry barriers considered was that of access to skilled personnel. At this stage, we reviewed the barrier as a possible point of difference among the venture groups. Put more formally, we wanted to determine if the venture groups exhibited performance differences among each other on the dimension of access to skilled personnel as a barrier to entry.

Once again, there were group differences that were significant and others suggesting but not proving differences. In average sales, there were significant differences when there were some barriers. The same was true for increase in sales when there were some barriers.[6]

Our results showed reason to believe that the venture groups had different average sales and increase in sales results when there was some access to skilled personnel barriers to entry. Put another way, when ventures find some but not extensive barriers in getting employees, we can separate out biotechnology, computers, and electronics ventures on that basis. Overall, this was the weakest distinction of performance among the groups as far as barriers to entry are concerned. Only 2 of 14 outcomes showed a reason to believe there were well-supported group differences, although in most other cases differences are indicated.

The last idea in this set of environmental factors has to do with industry life cycle. In this part, our concern was with performance of the ventures across life cycle stages. Introductory and growth phases were put into the early stage while maturity and decline were considered the late stage. Specifically, the idea we evaluated was that the venture groups exhibit performance differences among each other across the early and late stages of the industry life cycle.

Our evidence strongly supported an average sales performance difference among the groups for both early- and late-entry stages. The groups have early-stage performance differences in the sales increase measure and late-stage differences in the ROI measure. This conclusion was significant. In all, four of ten outcomes suggested group differences on the industry life cycle entry dimension. There were again indications of group differences that were evident but not statistically convincing.[7]

There are differences among the electronics, biotechnology, and computer ventures with regard to industry life cycle. This was further verification of the differences among the technology ventures—now sufficiently so to advise that these groups be treated as separate businesses rather than grouped under the big tent of high technology.

Summary

This chapter had environmental factors as its focus. Many different environmental conditions impinge on the new technology venture. Some possible conditions have been discussed, but our focused look was on economic forces. Within these, it was evident that barriers to entry and industry life cycle effects could not be ignored. There were performance differences discovered in some circumstances. Ventures with fewer capital, patents, and access to skilled entry personnel had better sales results than ventures with greater barriers. However, these environmental factors did not produce uniform results on the set of performance measures. All of the ideas we developed and evaluated were partially supported and partially rejected. On measures where there was support, the results were specific on the relationships between dependent (performance) and independent variables (environmental factors).

Barriers to entry and industry life cycle entry are associated with some performance differences. This finding was consistent with what others have discovered about general ventures. What it amounts to is that technology ventures share a lineage with general ventures as far as the performance–environmental question is concerned, yet they share no links within their own family. The venture groups act and respond differently to their environment. These factors also provided some genuine distinctions among the venture groups.

It appears that the environmental dimensions considered do affect high technology venture performance to some extent, but different types of ventures are affected differently.

This concludes our review of technology ventures and their performance. We'll carry these findings on to older technology companies to see if these discoveries ring true later in the life of science and engineering-based businesses. We'll also introduce new ideas and evaluate these as well. What should result is a more fundamental understanding of what the business of high technology really is.

Notes

1. These factors are noted in L. Tyson, *Who's Bashing Whom: Trade Conflict in High Technology Industries* (Washington, DC: Institute for International Economics, 1992).

2. See K. Vesper, ed., *Entrepreneurship and National Policy* (Pittsburgh: Carnegie-Mellon, 1983).

3. Testing this hypothesis involved a comparison of aggregate performances of the three types of ventures. The ventures were compared each time separately for three categories of barriers to entry. Since there were three levels of barriers (some extent, great extent, and none), there were three tests for each performance variable. The single response category was compared with the other two in these tests. Wilcoxon scores were generated. The Kruskal-Wallis test was used for probability estimates. The number of cases in the cells or consolidated cells ranged from 6 to 23. A two-tailed test was conducted. The probability values for great barriers–average sales was .058. For great barriers–sales increase it was .073, and it was .083 for none–sales increase. For great barriers–ROI, it was .083. The 10 percent significance level was used here.

4. The null hypothesis of no group differences was rejected at the 10 percent level. Using this as a decision criteria, when some patent, legal, and regulatory barriers to entry were present, there were average sales, ROI, and market share differences among the venture groups. Such distinctions cannot be made, however, for any other level of this barrier to entry nor for any other performance measure. Overall the results provided weak endorsement of group differentiation on this particular barrier.

5. For some average sales, the probability was .019 and for increase in sales it was .075; a 10 percent significance was the rejection level. There were no electronics firms reporting some entry barriers, but in all other instances except market share great barriers, there were sufficient cases for testing.

6. For the overall evaluation of life cycle differences, the null hypothesis was that there was no difference in performance; 10 percent was the significance level. In average sales probability, the value was .084 in early life cycle and .082 in late life cycle. For increase in sales, it was .024 for early life cycle. For ROI, it was .007 for late life cycle.

References

Bain, J. (1956). *Barriers to New Competition*. Cambridge, MA: Harvard University Press.

Block, Z., and I. MacMillan (1993). *Corporate Venturing*. Boston: Harvard Business School Press.

McDougall, P. (1987). "An Analysis of Strategy, Entry Barriers, and Origin as Factors Explaining New Venture Performance." Doctoral dissertation, University of South Carolina.

Porter, M. (1980). *Competitive Strategy*. New York: The Free Press.

Roberts, E. (1991). *Entrepreneurs in High Technology: Lesson Lessons from MIT and Beyond.* New York: Oxford University Press.

Sandberg, W. (1986). *New Venture Performance.* Lexington, MA: Lexington Books.

Scherer, F. M. (1970). *Industrial Market Structure.* Chicago: Rand McNally.

Stigler, G. (1968). *The Organization of Industry.* Homewood, IL: Irwin.

III

ESTABLISHED
HIGH TECHNOLOGY
COMPANIES

9

FROM THE VENTURE
TO ESTABLISHED
HIGH TECHNOLOGY
COMPANY

Few and far between are the high technology venture founders who are satisfied with having their business remain shops, populated by a few skilled technology craftspeople who turn out only a few products for a few customers. Founders have something bigger in mind. As visionaries, they most often see themselves as long-lived inspirational fixtures and their enterprises as enduring and prominent features on the high technology landscape. As leaders, they see other businesses following their footsteps with copycat technologies, perhaps feeling flattered as much as challenged by the competition. They also see themselves as leading their employees to new heights of technical proficiency as they pass on the innovative arts to their staff. In all, it is a perspective centered on the individual as the source of organizational change, instead of one centered on external factors. In this chapter, we lay out how ventures become established companies. That requires that we consider some of the external factors, not just the power of the individual to shape a technical business.

Our focus here is on the transition from venture to established high technology company. We look at the characteristics and operations of established high technology companies or, to put it in the context of our review, companies that are no longer in their venture phase. We compare these characteristics and operations with companies that are still ventures, helping us frame that vast middle ground of business maturation.

Change Is Inevitable

Technology venture founders pride themselves on how steadfast they are in their vision. "Sticking to the knitting" was a term we heard a lot, both from them and from their venture capitalist benefactors. The phrase was also a pronouncement of cause for success in the Peters and Waterman (1982) book, *In Search of Excellence* (although we learned later that many of the fevered knitters faltered). Nonetheless, "sticking to the knitting" was taken in a positive light by the technology founders whom we spoke with. There were many tales of "other" venture founders who strayed from their initial focus and perished. Those we spoke with were determined not to have this happen to them, so they stayed the course as much as possible, perfecting a single product or providing exemplary service and support with sales and advertising.

But even if their vision remained fixed on the business at hand, their organizations did change. The new technology ages and things are done differently the older it becomes, in spite of visionary quest and rigid discipline. It may even get to the point where the founder sees little in the evolving organization that he saw in the beginning. Seymour Cray serves as an example of someone who left a company he started. Cray Research took a development path different from Cray's, so he left that firm to start Cray Computers. At such points of organizational evolution, the departures lead to abandoned progeny of disinheriting parents.

The founder's leaving is the most visible indicator of organizational change. It is an unmistakable sign that things are being done differently as the organization grows. If the technological business remains small, the entrepreneur can frequently be found at the helm. Conjure the image of a grizzled private lifelong tugboat captain and contrast that with the aircraft carrier captain who is at the helm for a brief time, then replaced by another who fits the current command profile.

Founder rocks amid environmental seas—that's the dialectic of the technology launch. Both are elemental yet both work against one another. New challenges emerge as the high technology company grows. Founder centrisim is almost always flung aside by the centrifugal forces of the venture's environment.

Some of the force components, like increased staffing, are quite predictable while others, such as being sued for copyright infringement by an unknown foreign technology company, are not. Known or unexpected, the impact can be equally profound. Founders know they will need people as they grow, but surprises lurk in other areas. Founders and those who follow these companies need to know how things will change as their high technology companies grow. In doing so, they can accommodate the environmental changes they cannot possibly control.

Change is not inevitable for all technology ventures. It is only inevitable for surviving technology ventures. These changes are not minor perturbations, either. Tweaking employee productivity or shaving a few more cents off the cost of circuit boards is not the major challenge that advancing technology companies face. The changes are deeper and last longer for the most part.

These changes affect most every aspect of the way business is done. Even if a small high technology company has a single founder who wants to remain in business on a small scale, there will be significant changes in the way things are done. At minimum, the changes will be in the technology itself. So even if the company is a part-time, single-person consulting operation, that consultant must upgrade his software and hardware or face the prospect of adding no value to his clients' business.

For software consulting firms especially, there is a need to adapt quickly to new ways of computing. Out with the COBOL, in with C + + —such are the leaps the thriving consultant is expected to take. Major companies take on software consultants to find better methods, not to act as mechanics for run-down systems, except in rare cases where antiquated programs need to be coaxed along. By and large, the software consultant is constantly challenged to change or founder, to learn and apply new applications software, operating systems, languages, and hardware. His destiny is to keep up with the technology times and change the way he does things accordingly. The most savvy software consultants have learned to share this burden. They work with clients who help their own learning along a little, giving them opportunities to improve their proficiency. Consider an Oracle consultant we interviewed who did just that, demonstrating just enough knowledge of the software to establish his superiority over internal information systems staff. That's akin to business consultants who advise new consultants to dress just one degree better than their clients. If the client is in blue jeans, wear dress slacks. If the client is in a suit, wear pinstripes and a pocket square.

The software consultant feels the urgency of change most acutely, but change is prevalent throughout high technology. Aeronautical engineers are bombarded with new knowledge about aerodynamics, materials, propulsion, and control systems. Knowledge turns into technology, which quickly changes how the best aircraft are designed. The integration of these advances is exactly what modern aircraft design is about. Similar comparisons can be made in other technology areas.

There is research support for evolution of market orientation. Edward Roberts's (1991) comprehensive study of 114 Boston area established that in the first few years after founding there was an evolution away from consulting and research and development contracting toward more product-oriented businesses. There is further evolution toward sales and marketing and less emphasis on engineering. The character of these firms shifts over time with more use of a direct sales force and representatives, sales forecasting, and analysis, as well as creation of a marketing department. Multifounder firms develop the sales orientation quicker than single founder firms.[1]

A true important and logical observation is that major technological advances are brief, especially in the high technology field. In Dante's language of the *Divine Comedy*, "O Vanity of Human Eminence. How short a time the verdure crowns the summit. . . . Worldly renown is but a breath of wind and changes name because it changes quarter." Dante comments on the passage of fame for Renaissance artists, but the thought is applicable to the more prosaic endeavor of software consulting. The winds of change are always in the air.

The changeover from cutting-edge high technology venture to seasoned high technology company that redefines the way business is done may be easy to utter, but it is not easy to explicate. Often the line between new and the old is blurred. The changeover is most fascinating when business owners themselves do not see the shift. That can be a source of trouble as venture leadership fails to give way to needed professional management, a point made by Peter Drucker and others. Maturity can sneak up on the venture founder. Other founders may depart, leaving the survivor as the sole embodiment of the quest, perhaps digging in his heels when his bare feet should be on a sandy beach.

There is no real clue to the different phases of this transition, though we see shortly that some observers have divided business growth into stages. We've found no business activity that a venture does that an established company does not do. Seasoned high technology firms search for capital, have cash-flow problems, and face risks of extinction. New ventures also act like mature firms in some respects; they may have well laid out business plans, policies, and procedures for doing things.

Still, there are obvious differences between Hewlett-Packard and a two-person, newly formed insurance software consulting firm. Size is one factor and time is another. The two go together. Most successful high technology companies get bigger over time. But there is one thing that happens to established technology firms that doesn't happen to ventures: rebirth. Many established companies have to re-create themselves when business conditions change or if they don't adapt internally. A close-up account of a particularly interesting corporate resuscitation—that of National Semiconductor—was authored by Robert Miles (1997). Then-CEO Gil Amelio helped define National Semiconductor's downward spiraling performance problem, then he guided the company to a new vision and implemented it.

In the rest of this book, we consider the established high technology company. *Established* here means companies that have been in business for more than five years. But this timeframe does not mean they are successful. Many businesses are still struggling after five years or more in business. As we just noted, some have to be brought back from the brink. But this time period does create a useful dividing line between young and old.

The companies in the group of established businesses we will explore in Chapters 9–16 have more than five years of business history. As a check on the independence of our venture group from this established business group, we cross-checked the names to sort out companies that may have appeared in both groups.

The Evolutionary Process

Those who make careers of looking at the behaviors of business organizations have divided the evolution of business organizations into more than two stages—more than just founding and maturation. These are all models with at least three stages, sometimes more.

One particularly intriguing and potentially useful way of looking at organizational evolution and its management implications was developed by Lawrence Greiner.[2] This has value in the consideration of high technology evolution. According to Greiner, as organizations grow there are predictable crises that are resolved with certain actions. For example, when a high technology organization is formed, the entrepreneur is there to solve all kinds of problems, but then more people join the company and new human relations problems occur. That is the crisis, which is solved by having a strong manager who can maintain direction and deal with human resource problems. This then leads to a crisis of autonomy, in which there is too much central management and lower level employees exert more influence, which cause conflicts with the strong manager. This crisis is solved by compromise—a delegation of authority to mid-level managers. A subsequent crisis occurs in these managers having too much autonomy, which is solved through better functional coordination. Next, the crisis is one of too much coordination and too many systems, an indication of too much bureaucracy. Collaboration and cooperation are needed to overcome the bureaucracy and solve future problems.

New technology companies experience these stages as they grow. The first-stage leadership crisis is particularly acute because many technical founders do not have human resource skills. To survive, they often need to hang up their egos and let their babies become their own people.

In our interviews, the venture founders used such terms as "jack of all trades and master of none," "Tasmanian devils," and "wearer of many hats" to depict what went on during the venture formation period. These descriptions capture the intensity and flexibility of the period. The founders had to work on a technical problem one hour and meet with their bankers the next hour. They found focus in their venture concept, not in the day-to-day life of the venture. Their daily lives were full of discontinuities and the only stability was their vision.

For established companies, with stable sales growth, the right people in place, and policies and procedures, the founders (if they are still around) are not torn by the variability and spontaneity of daily demands. It must certainly seem a far cry from the venture heydays, especially for the likes of Gates, Grove, Hewlett, Packard, and Ken Olsen. These examples are few but memorable. The larger and older high technology companies have shed their founders and operations are run by their management corps.

As we observed these emerging high technology companies in action, we saw periods of tremendous excitement and deep despair—mood swings not apparent in the more mature companies. There was also a sharing of the ups and downs in the new firms, because there were only a handful of employees and those employees were as dedicated as the founders. But not so with the mature companies.

How the Transition Is Made

Many of the descriptions that apply to general ventures also apply to high technology ventures. That includes founders who clean bathrooms, use personal credit cards to cover short-term debt, and arrange meetings with prospective customers in rented hotel rooms or restaurants because they are embarrassed to have customers see their headquarters. But the nascent technology ventures have a few special characteristics, and one of these is equipment. Since they can't afford new test equipment, they have to be resourceful. One medical electronics startup founder took advantage of his connection with a research university astronomy laboratory. The test circuitry for his medical electronics company's initial product was created from electronic components discarded by the astronomy laboratory. The founder literally went through the trash every day, scrounging for integrated circuits with clipped-off leads, used cabling, and the like. He would often be the lab in on the weekends without his tweed "university" attire and in blue jeans and Tee-shirts, a sure sign he was off-duty. Yet it was the same person, hunkering over the technical literature in the library or soldering a custom-made circuit board; but his intent was on biomedical electronics instead of programming telescopes.

The venture turned out to be successful. The company is now a successful small electronic devices firm, but few know that the instrumentation it makes started as "space waste" from the floors of a university R&D lab.

After the company grew into its own, the founder remained at the university, electing the security of tenure for the uncertainty of the market, proving that at least in this case risk taking by entrepreneurs is not absolute. This founder saw his ideas flourish, but at a distance. Professional managers took over early and helped the venture become successful.

Another way that high technology startups have made the transition to established businesses is to use acquaintances in former companies. Even after their business had been operating a few years, they still called on their old friends for favors. Although they were reluctant to talk about it, there are entrepreneurs who could not afford to buy test equipment and asked their colleagues at their old firms to do some testing as a "government" (i.e., personal favor) type of job. In other cases, government laboratories such as the Oak Ridge Lab have been made available to entrepreneurs for development testing. These two scenarios illustrate how high technology growth can be dependent on access to the resources of others, and that these resources can be obtained by hook or by crook. "Scrounging for sources" is an applicable phrase to describe what happens in such circumstances. It also helps show how the business of high technology may be different from other businesses. High technology ventures, as they transition to established firms, must make changes not only in areas such as production and marketing but in research and development and technology adaption as well. This is most often less of a demand for traditional firms.

The growth model developed by Lawrence Greiner has a recent and more technologically focused incarnation—that done by Kazanjian (who has written extensively on technology and stage-of-growth) and Drazin (1990). In an article entitled "A Stage Contingent Model of Design and Growth for Technology Based New Ventures," the authors present a growth model for a technology company. Their stages are: Conceptualization and Development; Commercialization, Growth and Stability. Kazanjian and Drazin see the essential definers of these stages as the main problems in the life of a venture. A venture's growth depends on the fit between its stage of growth and aspects of its organization, such as decision-making centralization, formalization, functional-level specialization in marketing and sales; and manufacturing and engineering technology. For example different patterns of decision making and functional specialization are needed at each stage if growth is to be achieved.

Where Change Happens

So what changes as high technology firms age? The answer is almost everything changes.

The best way of seeing just how the venture matures is to go into the bowels of its operations. Dividing the growing organization into the usual departmental components is a convenient way of coming to grips with organizational evolution, so it is the technique we have employed.

In the *personnel* area, the venture changes from having a few, multitasking individuals to having job specialization among many more employees. What can be said of the relationship between high technology employees and the founding entrepreneurs? During the first days of the venture, the founder is most often a team member, an equal among equals in the short-staffed, overworked tiny group of venture employees. But maturity brings distance. The founder may not be as accessible as he was before. He may be visiting important customers far away. He may be looking at places to set up regional operations. He may be at national trade shows or with investment bankers. In sum, the founder may not be as visible as the company ages. This also frequently means that he is less involved in the day-to-day operations of his growing company.

In *marketing*, the transition is from a single product in a single market to multiple products in multiple markets. Customers are more numerous and more diversified in their needs. There are changes in the entire promotion area as well. The new technology company relies on word of mouth advertising rather than paid advertising—that is about all that can be afforded. But as the company grows, a formal approach to advertising begins. The transition happens when the company decides to make itself known and can afford to do so. This might mean buying display advertising in a trade publication. If that generates sales leads, the next step is to organize a series of advertisements and run an advertising campaign. Perhaps other media will be used, such as radio or television, but the print medium is usually the first and least expensive way into the market. The sign of the well-established technology company is advertising done externally. An external advertising agency is hired to develop and place advertising. Since the advertising agency needs a certain amount of revenue from a client before it agrees to do advertising, an agency does not take on small technology companies as a rule. (There will be more about this in Chapter 14.)

Pricing gets more complex as the company grows. There is more price variation. The "one price for all" falls to the wayside. Ventures either try to skim a market with a relatively high price or penetrate a market with a low price. Established firms, on the other hand, use a wider variety of pricing techniques. Warranties, terms of sale, and lease-buy options all come into play with maturity, as well as pricing variants.

Sales practices and personnel change. The venture may not have a dedicated sales force—owners do the selling and product design and development. Sales for a venture is an internal function, but this too changes with time. The growing technology firm begins to look for a dedicated salesperson or possibly a manufacturer's representative who will sell the product or service for a percent of the price. The well-established firm has an extensive professional salesforce.

As far as *production* is concerned, ventures move from nonexistent production systems to actual production. In that sense, true production is a function of the established company. In the venture phase, the products tend to be customized for clients. Customers may, in fact, be very involved in the creation of the first products. But as the business grows, there is more distance—almost literally—between customers and the company. It's only recently that customers have again been drawn close to the established company, mainly through Total Quality Management (TQM) practices. TQM stresses involving the customer in defining quality through such techniques as having customers come to corporate headquarters and be involved in product design.

Rapid productivity growth in the computer industry has been associated with rapid technological change. A considerable portion of this growth comes from economies of scale. That is a conclusion reached by two researchers.[3] Norsworthy and Jang used econometric modeling to arrive at this conclusion, with the implication that there is an advantage to grow a technology company. Economics of scale become available to the larger company—this is a consistent theme of industrial organization theorists as well.

Like their nontechnology-intensive counterparts, high technology companies seem to move toward more formalization of processes as they age. The illustrative case here is the tenuous partnership between staid IBM and upstart Microsoft (Carroll, 1993). IBM programmers assiduously documented codes they wrote while the young Microsoft programmers tried codes and then recorded what they had done on paper later, when problems came up or when nagging IBMers called them on it. There were other differences as well, IBM programmers were evaluated on how many lines of code they wrote while Microsoft programmers were evaluated on whether or not the program worked. This difference between the two firms in business approach became a source of persistent conflict. It demonstrates the style and substance differences between young and old technology companies.

Another functional area that changes with time is *finance*: financial systems become more complex. The company may begin to consider making an initial public offering. The older technology companies have formalized budgeting with budget control systems in place; younger companies do not. Youthful ventures often expect to have formal budgeting, as they may have promised in their business plans, but we've seen good intentions fall apart and "seat of the pants" and "trial and error" budgeting take its place.

With regard to *structure*, the entrepreneurial high technology firm best resembles what Henry Mintzberg has described as simple. There is a direct accountability line to the head of the organization. Considerable centralization of decision making occurs in a relatively simple, dynamic environment. This is a benevolent dictatorship, with power residing in the owner-manager. That is also one of its vulnerabilities. If the owner gets a long-term illness, so does the whole organization. Yet the sense of mission is high in this young organization.

The change occurs with growth. The manager-founder cannot have as much of a direct influence. Systems begin to substitute for people. The more mature firm resembles what Mintzberg calls a professional bureaucracy. There is heavy reliance on individual skills and knowledge of other people who work in the firm. That is very much the case with technology companies, too. The organization essentially becomes bureaucratic. The researchers and product managers must go through written approval processes in order to invent. This is a far cry from when the research director stuck his head in the owner's office and said, "I'm planning on spending a few days on that new application I was talking about."

Another form the high technology company might take is a machine bureaucracy. It might assume this form if it is in the business of production. Intel microprocessor production is an example of this form.

The company might also assume a divisional form, in which functions such as marketing, finance, R&D, and production are divided up and headed by different managers. The divisional form is quite common among very large technology operations. It is a tried-and-true organizational form, though not particularly imaginative.

There are other forms of organizations that the high technology operation can take. These will be detailed in Chapter 13, when we show how our sample established firms are actually organized. What we have discussed here are the most common forms. The important point is that there is quite a bit of variation in the organizational structure of the established technology company compared to the venture firm.

The relationship between high technology growth and structure and planning processes is another area that has been researched. Robert Kazanjian reported that there was "structural specialization" in the functions of engineering/technology, manufacturing, marketing/sales, finance/administration, and employee relations. More formalization and centralization of budget objective setting and new product development also happens as high technology ventures mature. In all, Kazanjian concluded that the stage of growth of the venture explains venture structural and planning changes.[4]

The Stages of Business Evolution

High technology organizations do not move directly from being ventures to being established companies. The transition is somewhat less sudden, and there is at least one middle stage involved. Most commonly, the stages of growth can be described as: venture, growth, and maturity. That's not to say there are no further distinctions, such as a postmaturity stage, but that is not our interest here. We are following only new and established firms.

The stages of growth idea has been applied to the financial development of the high technology company as well by Edward Roberts.[5] Others have taken a different approach to organizational evolution. For example, Kazanjian (see note 4) found evidence to support for a four-stage growth scenario for high technology ventures: pre-startup, startup, growth, and maturity.

The number of stages and characteristics of these stages may be debated, but all ventures start and end at opposite poles. The initial phase of the high technology company is quite different from the established phase, however it is described. Looking at both ends greatly enhances our perspective on what high technology firms are like.

To highlight the distinctions between the new technology venture and the established company, Table 9.1 contrasts some crucial corporate functions.

Table 9.1 Differences Between Technology Ventures and Established Firms

	Technology Ventures	Established Firms
Structure	Simple, flexible, founder-centered	Formalized with many variations
Personnel	Few, team-focused	Many, specialized
Entrepreneur	Actively involved	More distant or absent
Products	Single product	Multiple products
Customers	Few and well known	Many and different
Promotion	Word of mouth	Paid advertising
Pricing	Skim or penetration	Many variations
Sales	Internal with few or no personnel	Professional sales force
Production	Small lots	Large lots, dedicated facility
Finance	Informal and flexible	Formalized budgeting and control
Corporate planning	Informal	Formal planning processes

Summary

This chapter has been our bridge from the venture to the established high technology company. What is suggested here is that the longitudinal (time) dimensions of high technology maturation could be so significant that they effectively distinguish new technology formations from their more mature counterparts, thus reinforcing the importance of the stages of growth idea in explaining how technology businesses change. A case could even be made for the convergence of very mature technology companies, much the same way that huge corporations of any type become horizontally diversified behemoths. Although that is not within our scope here, it is deserving of attention.

In this chapter, we have explored how high technology ventures mature and how they change as a result of that maturation. Although the idea of stages of growth as presented by various researchers in this chapter applies, it is imprecise particularly on the time dimension. None of the researchers mentioned put meaningful time scales on their models. There is no convention for setting the ranges for stage of growth. Is the venture startup phase one year for technology ventures? Even posing a question like this shows how difficult it is to codify an answer. The technology venture founders who were our contacts cited startup phases from a few months to ten years. There was so much variance on the issue of time that it was not possible to set a representative period. However, there was acknowledgment that the stages of growth rang true, even if some industry-specific or personally specific circumstances distorted the time dimension and obliterated generalizations about the duration of these stages.

The chief ideas in this chapter were that high technology companies change and that change involves more rather than less complexity. To understand high technology, we need understand both sides of this mutation. Having done this, we can now turn to examining those thriving entities that have gone from ventures to established companies.

Notes

1. As described by E. Roberts, in "Evolving Toward Product and Market Orientation: The Early Years of Technology Based Firms." *Journal of Product Innovation Management* 7 (1990): 4.

2. Presented first in the July-August 1972 *Harvard Business Review* article "Evolution and Revolution as Organizations Grow." Greiner's views about organizational growth have reappeared in management texts and have been popularized by business consultants.

3. See J. R. Norsworthy and S. L. Jang, *Empirical Measurement and Analysis of Productivity and Technological Change: Applications in High Technology and Service Industries* (Amsterdam: North Holland, 1992).

4. Fully described in "The Organizational Evolution of High Technology Ventures: The Impact of Stage of Growth on the Nature of Structure and Planning Processes," a 1983 University of Pennsylvania dissertation by Robert Kazanjian.

5. In *Entrepreneurs in High Technology* (Roberts, 1991), Roberts called these phases startup, initial growth, and sustained growth. Startup begins with founding and ends with the generation of sales revenue. Initial growth happens when the company develops a product line and ceases when it operates profitably with growth more or less expected. Sustained growth is marked by sales in the millions of dollars and employee numbers in the hundreds.

References

Carroll, P. (1993). *Big Blues: The Unmaking of IBM*. New York: Crown Publishers.

Kazanjian, R., and R. Drazin (1990). "A Stage Contingent Model of Design and Growth for Technology Based New Ventures." *Journal of Business Venturing* 5:1, pp. 137–150.

Miles, R. H. (1997). *Corporate Comeback: The Story of Renewal and Transformation at National Semiconductor*. San Francisco: Jossey-Bass.

Peters, T., and R. Waterman (1982). *In Search of Excellence*. New York: Harper and Row.

Roberts, E. (1991). *Entrepreneurs in High Technology: Lessons from MIT and Beyond*. New York: Oxford University Press.

10

HUMAN RESOURCES

We start our look at established high technology companies with close consideration of the most essential resource of all for high technology: people. High technology, unlike many other business activities, is a business that can be quickly reduced to the two elements of technology and people. Other businesses depend more on nonhuman production functions, and some strive to substitute capital for labor on very large scales such as manufacturing operations. But take the people away from technology, and what is left is an abyss of undeveloped potential. Some would even call it a prelude to economic stagnation, owing to the contributions of this sector. High technology—even in its most spectacular form as space exploration, where big machines dwarf human creators—is so fundamental a human knowledge business that the people dimension hardly needs proof, although it does require the occasional reminder for those who might take it for granted.

Software code does not write itself. Decoding genetic structures does not happen in nature. Deciding to curve a wing tip so as to cut down on drag is not an accident. Figuring out where to put transducers in the stream of petroleum at the refinery is not a whim. These are all the acts of humans—the workers of high technology.

People in high technology are oriented to use their minds to save labor. How quickly and how well they can do this are often part of their reward. There is no other business in which the distance between the elimination of labor and the invention is so small, yet the effort so intense to create that distance. "Put that in the code so we can save a step next time" is an instruction heard frequently among software design engineers. It's as though these workers are crossing a sea on ice floes they chip out themselves. They cut, they step, then they move on to the next floe. They make obsolete and useless what they have just done. Mind over matter is their occupation—technology workers are special if only viewed on this conceptual level, although we will see that there is more to it than that.

The subject of this chapter is high technology human resources. Our focus is on the workers themselves; executive management will be separately covered in the next chapter, when we turn to high technology CEOs. We are concerned here with *who* works in high technology, *how* they progress in this business, what their education and other characteristics are. We also discuss a subject that has not been covered extensively before: human resource policies.

This chapter complements our visits to high technology companies by answering in aggregate what we could not get though on one-on-one visits. Our accumulated answers come from exploiting the second major database created for this book—that is, a CorpTech Corporate Technology Directory–based sample of more than 200 established high technology firms. This directory is described in the resource guide at the end of the book.

As the topic of human resources is developed, we establish how the management of high technology has a special character and how the people in this field are both unique yet similar to those in more traditional businesses. Also, we bring in the views of high technology business executives in the form of CEO perspectives on managing people. The result is at least an appreciation and quite possibly an understanding of an aspect of high technology that is intrinsic to organizational performance.

Our approach is to frame the matter within the conventional treatment of human resource management and then move into the realm of high technology employment. This is a suitable approach because the field of human resource management is well developed in theory and practice. There is no separate field of high technology human resources that can be used for a framework; rather than invent our own contrivance, we make use of what is available. Because that itself is an immense body of work, we use the work of human resource management text authors to provide the key issues.[1]

There is consideration in this chapter of what is generally regarded as the special character of high technology human resource management. Our aim is quite simply and quite importantly to determine if there is anything special about these businesses. Also, we examine the similarities between human resource management in the high technology field and in nonhigh technology businesses. We rely on our interviews with high technology human resource managers to find these similarities. These informal interviews, conducted through 1995 and involving fifteen high technology personnel managers, sought to discover some general issues about high technology human resources. The interviews were personal and open ended with questions about human resource management practices being asked of the managers. The managers were mostly from small high technology firms.

Human Resource Management in Technology Businesses

The high technology employees we spoke with are universal in their opinion of human resources: the crucial resource of the business. And that's the opinion of CEOs and industry observers alike. We cannot find more uniform agreement than on the importance of people.

Although employees are seen as crucial resources, they also perceive substantial human resource management differences among companies. Compare any two high technology companies in the same line of business, and the most apparent difference you'll find is in the workforce character. "That's a very conservative company—too conservative for me" may be one impression. It is quite different from, "What a bunch of cowboys—real flakes, you'll never get me to go there." These are the categorizations that people use to make decisions, based essentially on perceptions.

While human beings are the agents of high technology performance, human resource management is the business function that concerns the flow of people into, through, and out of the high technology organization.

Human resource management is taken here to be the selection and development of people who provide labor for high technology organizations. The management of human resources is the way the technology unit employs its human capital for organizational purposes. While we cover human resource management, we also address the larger matter of workers in high technology. That's because intellectual labor is absolutely critical to any type of high technology, and the more we can understand the labor-organizational connection, the better we will know this entire business.

The function of human resource management is usually divided into a number of sequential functions. The different phases are very much like an individual's career progression, thus human resource management covers the topics of workforce planning, recruitment, selection, orientation and placement, training and development, compensation management, and performance appraisal, as well as employee separation. This sequence closely follows the typical textbook presentation, as mentioned.

High technology companies exhibit most of these functions, though the larger high technology firms have more complete and elaborate functions. In fact, in the human resource functions, high technology firms have more similarity to conventional companies than in many other dimensions. Entrepreneurial high technology firms are more concerned with recruitment than they are with job progression and separation, however, so we cannot say that entrepreneurial and established firms are the same in this sense.

Of the steps of human resource management, *workforce planning* is probably the most neglected, based on our informal interviews with executives of established companies. For the most part, no workforce planning is done—that is, the CEOs don't develop and execute plans addressing their current and future worker needs. Changes in the types of employees needed, when those changes are expected, and how the organization will change are left untouched in almost all cases. Succession planning is virtually nonexistent. We have not seen plans for substituting people in positions as incumbents leave, except at the apex of an organization or when the organization is a large one.

In *recruitment*, there is more activity. High technology firms are active recruiters. They need to be because of employee turnover. Recruitment is similar to the way it is done in traditional businesses, but in smaller firms much of the recruitment is done by the employees or executives themselves. Contract programming operations have more formalized, routine, print display advertising than informal network recruitment.

Selecting among the qualified job candidates is mostly a function of the size of the organization. Smaller firms use personnel screening panels less and usually do more direct interviewing. But no matter what the size of the company, the selection steps seem to be as formal and "legalistic" as they are anywhere else. Reference checks, formal written offers, and careful explanations of compensation and benefits are part of the process of working for high technology companies at virtually any level.

The salary and compensation plans, performance appraisals, training, and promotion systems are so varied in high technology companies that it is impossible to generalize. The variety of plans and programs make these businesses seem similar to other types of companies. There are a few ways that high technology companies are somewhat different, though. One way is chronicled in Tracy Kidder's *Soul of a New Machine*.[2] If you do well in a high technology company, you are awarded with another, perhaps better project. You are what you did and you will be what you are doing next.

The other difference is the compensation carrot of equity. Employee ownership is common in what is regarded as high technology, though it is more common in computer firms than else where. The classic illustration is Microsoft, which made millionaires out of its early engineering crews. By some accounts, every fifth employee is an equity millionaire at Microsoft. The equity package still plays a major part in luring and keeping employees. It compensates for the demands of the work. For those who sleep under their desks at Microsoft, perhaps this is their dream.

In all, the process of managing high technology human resources is very similar to traditional businesses. Next, we look at the character of the employees themselves, both as individuals and as they act together.

What Kind of People?

Raising the matter of human resource needs immediately poses the question: For what kind of high technology? There are two versions of high technology we are considering in this book and they have different human resource needs. Based on interviews and contacts, but not on survey data, we determined that the needs of a high technology venture are different from the needs of an established high technology business.

Entrepreneurial high technology companies require engineers who are immediately productive and flexible, not a commonly occurring combination. The new firm needs products or services developed quickly by people who may have to do many different functions—engineering, sales, and production. The new firm's most critical act is likely to be hiring its first technical team.

The needs of the mature firm, on the other hand, require employees who are team oriented and are patient with the systems and processes that have been put in place in established companies. The needs may not be as diverse, but they still place at the forefront. Established high technology CEOs talk about the shortage of qualified personnel. "We can't get the people we need. We are moving so fast. The people we need we don't have. Our best hope is for somebody who is willing to learn while working," stated high technology entrepreneur William Maasberg at the St. Louis University–BYU joint entrepreneur and academic participant conference in the fall of 1995. He had a message for the academic participants: "You can change their minds first, then we can do the training. We can train them easily, but we need the right people."

Paul Sybrowsky, a proud father of the very successful Ameritech Library Services, a database service for libraries, put it as, "We have to teach what commitment is. Even if we don't know how, we still have to do it." Sybrowsky's concern is that high technology employees sometimes do not share the drive and commitment to customers that founding entrepreneurs have.

The creation and survival of high technology businesses depend on both entrepreneurial personalities and technically proficient employees. Yet there is often a pronounced difference between the two, and in many cases there is an uneasy coexistence between individuals in different categories.

On first observation, the heart of the entrepreneur is inimical to the head of the high technology employee. The technology entrepreneur, either alone or as a team member, seeks to create a new business based on a new idea. The technology employee seeks to effect the vision of that entrepreneur. Yet we have heard from both technology founders and employees that the dichotomy is not absolute. The presidents of several high technology companies told us of their wish that more employees would act as entrepreneurs, and take risks and assume responsibility for pushing good ideas. We have also heard employees long for an environment where risk taking, when it turns bad, is not the occasion for punitive action. Employees of high technology organizations have expressed a sense of isolation, a feeling that the founder will not listen to them. When there is this distance, it is apparently is a great distance. And that separation can disrupt a functioning organization. The geographical concentration of high technology companies we noted at the outset of this book means easy transferability of employees. Sometimes it is simply a matter of walking down the street to the next software company. The result is that established firms are brood mares for yearling entrepreneurs. Little wonder that the founders have concerns. Little wonder, too, that the employees have temptation.

The wanderlust of software workers caught the attention of Kanter (1995), who studied over 2,600 companies and found that many "software gypsies" are constantly on the move from one company to another, from one project to another, without long-term employment in any single company.

But there is something not quite right about the statements from founders who seek entrepreneurs as employees. It cannot be an unqualified truth because the entrepreneur both takes advantage of opportunities to work independently and seeks these opportunities. The true technological entrepreneur looks for a way out, not a way further in. It must be that they seek partial, not pure, entrepreneurs as employees. Still, the motive must be recognized as an effort to close the gulf between entrepreneur and employee.

Characteristics of the Workforce

Because a high proportion of R&D expenditures defines high technology, we would expect the workforce to be populated with R&D professionals. This is not entirely satisfactory, though, because R&D work—usually confined to scientists and engineers—does not capture the work of some related professionals, such as the manufacturing quality assurance managers and laboratory technicians found in many operations. For the R&D professionals, there are some distinctions between their approach to work and that of others.[3]

There are many R&D professionals who do not work in high technology companies. One can easily envision the university scientist or the industrial engineer at a papermill as exceptions. Nevertheless, there are many scientists and engineers in the high technology industries we have identified, and they make a deep imprint. Part of that imprint is their role as founders. David Packard and Bill Hewlett's audio oscillator invention was the basis of the whole Hewlett-Packard enterprise, and the two engineers shaped their company for years. High technology firms are usually founded by scientists and engineers, but more recently also by teams that include experts in business areas. The impact of these R&D founders, either in teams or as individuals, is nevertheless a lasting one. No matter who founds the high technology firm, its core remains new technological invention, even for production-dominant firms like Intel.

What kinds of companies employ non-R&D technology workers?— the computer industry. What is curious is the inverse relationship between industry job and firm size. While 85 percent of all computer firms employ fewer than 100 workers, the firms with over 500 workers employ 91 percent of the total number of computer business employees (Harrison, 1994). Harrison describes a growing division of the workforce into an internal permanent workforce and a subcontracted outside labor base, an observation shared by Kanter (1995). Harrison believes this division has emerged from the flexible manufacturing practices of large production technology companies and hence represents its dark side. An employee survival tactic was offered by Winner (1995), who suggests that career stability is a fading dream and that such employees should see themselves contractors prepared to change jobs frequently. Thus, if a high technology worker can be typified, that worker is in the computer and computer-related field, more than likely a programmer for a large company.

Motivation

The motivation of employees, or for that matter human motivation as a whole, is a subject applicable to high technology. It is an immense topic, but one that has been untouched as it pertains to high technology. We do not intend embark on that research now. What we can do is describe some of the methods used by high technology companies to motivate their workers. These methods are varied, and varied results also happen.

The standard mode of motivating employees—that of *incentive compensation* (bonuses, profit sharing, and the like)—was reported to be very "tough to do on the technical side" by a high technology CEO whom we talked with. It requires accurate and impartial measurement of performance and, in a sense, either complete technical knowledge by a single evaluator or a fair appraisal by peers or technical managers. Disproportionate awards demotivate the parsimoniously recognized, while uniform awards become bland gestures of corporate largesse. Some CEOs are reluctant to have incentive compensation plans because they are not comfortable doing performance appraisals that lead to different compensation levels. They also express dissatisfaction with incentive compensation not based on performance as being an insufficient motivator. They are thus troubled by the whole question of compensation. All the CEOs we interviewed were, to one extent or another, relatively dissatisfied with their incentive compensation systems, regardless of type of system used. They plunged ahead, though, mimicking the compensation systems of their more traditional predecessors with a certain angst.

Even if employees achieve success from their individual effort, it is not systematic or predictable. Technical performance has no guarantee of success and also shows irregular progress. The early days of Data General were marked by times of no progress, according to the engineers who worked on the computer (Kidder, 1981). Engineers would spend days sitting alone, without coming up with anything new. And if this happened with the successful Data General computers, it happens far more frequently with less successful technology systems.

The motivation of high technology workers has perplexed managers as has motivation in most other lines of business. People respond differently to different kinds of motivation. The measurement of performance can introduce bias problems, depending on the person charged with evaluating. The amount of reward and frequency of reward often need to vary, and it is difficult to pinpoint when is best and by how much.

Our contacts led us to conclude that there have been no breakthroughs in achieving motivation within high technology. There are variations, though: working on better projects and socialization perks.

Socialization is a specially evident aspect of high technology. Having people interact with one another in activities other than work is one way some high technology companies have distanced themselves from other companies. The extent of this is almost the stuff of legend. And both image and policy reinforce this: the barefooted Jobs, the open offices of Data General, the campus atmosphere of Hewlett-Packard where engineers often take classes at nearby universities, and flexible hours. It has been suggested that socialization may even be more important for motivation than specific technical skills. The opportunity to socialize with computer people can be a more important motivator than the opportunity to display technical accumen. That may be hyperbole, but if the big successes of high technology are reviewed, they have emerged from dedicated, socialized teams.

Managing People: From the Mouths of the Illustrious

The personalities of high technology—the founders and managers of the mighty leading firms—are among the best sources of information about large-firm human resource practices. What they have to say about managing people is consequential because they represent the premier technology firms. And because they all have something different to say, it's worth exploring their thoughts on the matter.

David Packard's (1995) views about managing people in a technology environment are standard-bearer views. Hewlett-Packard almost invariably occupies a high position in the best American companies to work for. Until his death in 1995, Packard himself spent quite a bit of time on the human resources management issue. He wrote about and practiced model human relations within Hewlett-Packard. The company invests in its personnel and their development—that is exemplified in supporting continuing education, promotion from within, and stock option plans.

Then there is Andrew S. Grove (1983), founder and president of Intel. Grove was sufficiently stirred about managing human resources to summarize his ideas in his book *High Output Management*. He concentrated on productivity improvement and aimed the book at mid-level managers. The productivity focus is understandable because of Intel's manufacturing bent, but much of Grove's thrust is not on the Intel production machine but on the Intel personnel machine. His subjects are team building, getting more from meetings, and effective decision making and planning. Grove, like other technology counterparts, believes that effective teams are the way the business of innovation gets done.

While Grove's engineering background is evident in his emphasis on processes, a quite different human management approach was manifested in the behavior of Steven Jobs during his Apple days (Young, 1984). In the early days, it was reported that there was a "no gloves attitude" with a freewheeling give and take among employees at all levels. As the company grew, this was less prevalent except in the case of Steven Jobs, who often managed by provocation. To call Jobs abrasive is to utter only the obvious. He questioned managers under him very aggressively but to good effect. The Apple people learned to anticipate the argument and make their case while covering the prodding questions. What was ironic was that, as Jobs pressed people with direct and brutal questions, the culture of Apple was becoming less formal and more relaxed. By available accounts, a more polished, perhaps even subdued Jobs has returned to Apple to revitalize the company.

Jobs's successor at Apple, John Sculley (1987), was a stark contrast to Jobs in his conception of workplace relations. "The new heroic style—the lone cowboy on horseback—is not the figure we worship anymore at Apple. In the new corporation, heros won't personify any single set of achievements. Instead, they personify the process. They might be thought of as gatekeepers, information carriers and teams. Originally, heros at Apple were the hackers and engineers who created the products. Now, more teams are heros" (p. 321). Sculley's endorsement of teams is accompanied by a view of corporate structure as malleable, changing as new circumstances require. He rejects the traditional, hierarchical corporate form. That is a resounding theme of high technology leadership.

These leaders have been identified with a growth sector that itself has been heralded as a "new economy" by *Fortune*. The way people work in high technology has shifted from machine centered to team centered, a move away from the wizardry of products to the dynamics of people working with people and the provision of services. The shift to services is due to the microchip, which the *Fortune* described as being as transformational now as the internal combustion engine was in its time.[4] The new economy was instigated by technology but carried out by employees. *Fortune* has identified the power of invention to transform modern work.

Categories of High Technology Workers

Having sampled the views of the esteemed of this business group, we next address the matter of workforce composition. Who are technology workers? What kind of jobs do they have? How much do they make?

We used two ways used to gather information about high technology human resource practices. The first was our survey of high technology companies and the second was an additional sample of five firms where we simply observed human resource management practices, read available material, and talked with human resource managers. Two of these firms were large high technology companies. The other three were small entrepreneurial firms. Our aim was to collect in-depth information about the management of people. Questions followed the sequence of human resource management issues: workforce planning, recruitment, selection, career development, and separation. The human resource managers were also asked how different high technology firms were from general firms in these functions.

The broad-based survey of human resource characteristics was aimed at discovering the number and composition of the workforce, extent of unionization, and salaries. There was a pointed need to better understand the composition of the workforce. The usual way of doing this is a simple dichotomy between technical and clerical workers. Particularly in the small high technology firms, human resource managers tend to separate the technical workers, or engineers, from the nontechnical side—administrators, managers, and clerical help. Our approach was to establish ten different job categories, which more completely describe the range of high technology jobs. The categories were scientists, chemical engineers, mechanical engineers, software engineers, other engineers, electronic technicians, other technicians, clerical, management, and other administrative. These categories were created by reviewing job descriptions in approximately 20 high technology firms.

The respondents told us their average number of employees over a three-year period and what percentage of their workers were in each of the ten categories. The overall distribution is shown in Figure 10.1.

Our research showed an even split between purely technical jobs and nontechnical jobs. Grouping the various scientists, engineers, and technicians as the "technical" side and the administrative, management, and clerical jobs as the "nontechnical" side, we found the split to be 51 percent nontechnical and 49 percent technical.

This can be taken as showing some balance between technical and nontechnical functions (the human resource managers are not totally off base by splitting most companies into two groups). But what is more striking is the prevalence of technical positions. Nearly half the workforce of our respondents are technical workers. This far exceeds the percent of R&D workers (usually less than 20 percent of the total workforce) that

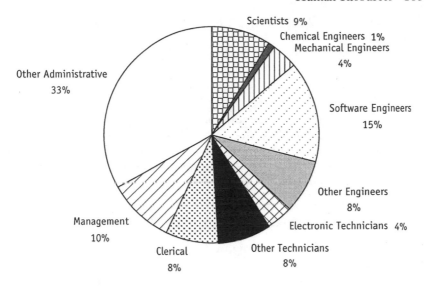

Figure 10.1   Composition of workforce

customarily defines the threshold for a high technology business. So our group is very high technology by this account; that's expected because staff follows science as companies grow.

There are many more engineers than scientists in established technology companies and more technicians than scientists in typical workforce compositions. More scientists are found in high technology firms engaging in basic research than those in applied research. Biotechnology is our example in the category where scientists are extensively employed.

Among engineers, software engineers are the predominant group, followed by mechanical engineers, chemical engineers, and other engineers. The category of "other engineers" includes civil engineers, aeronautical engineers, and nuclear engineers. Software engineers have the largest single imprint, however, because software engineering is an area of growth, especially within high technology itself. The need for software engineering crosses all kinds of high technology fields, while the same is not true of other engineering. Aeronautical engineers do not work on civil engineering projects—but maybe they should have in some cases. It was the aeronautical engineer Theodore von Karjian who diagnosed the aerodynamic forces behind the collapse of Washington's Tacoma Narrows Bridge—after the wind tore it apart.

Software engineering is a broad occupation that accounts for many positions. There are the engineers themselves, electrical engineers, and software engineers, as well as the programmers, developers, and other specialists who are collectively, and perhaps a bit too loosely, called software engineers.

What we see from these results is a triad of functions in high technology composed of technical design and experimentation (scientists and engineers), production and development (technicians), and administrative functions.

Other administrative functions take up the largest share of the workforce, with a third of workers. These positions are not management nor clerical, so this group could include administrative support personnel, business planners, finance and accounting positions, production support, sales and marketing, buyers, and other related positions.

Education and Technology Workers

What educational levels have high technology workers attained? Our information shows that they are highly educated. This was no surprise, given the character of the business—one that snatches up virtually all privately employed scientists and engineers. It also collects an increasing number of MBAs as the technology firms realize they must have a talented business side as well.

Our high technology businesses provided information on education levels of employees in five categories: had a high school diploma, had some college education, had an undergraduate degree, did graduate work, and had a postgraduate degree. The findings are summarized in Figure 10.2.

Our research showed that 62 percent have graduated from college, a figure that makes the workers in this field stand out from general businesses, where it is closer to the reciprocal percentage. Almost a quarter of those in high technology have postgraduate degrees, an exceptionally high figure. Businesses grouped under the banner of high technology have a highly educated workforce overall. Indeed, the mark of a well-educated workforce is found in all sizes of these technology-intensive companies. On the technical side of the business, it means that one is likely to encounter people with technical degrees and on the administrative side, managers are likely to have business degrees.

Labor Organization in High Technology

Unionization of high technology workers is a subject that has escaped attention in other reviews, probably owing to the impression that these businesses are small, entrepreneurial, and filled with young employees. Nonetheless, it is worth knowing how many high technology workers are union members. We addressed the organization of labor in high technology by asking what percentage of the workforce was unionized. The answer was that unionized workers are a small element. Companies that were not unionized made up 93 percent of the total. A handful of com-

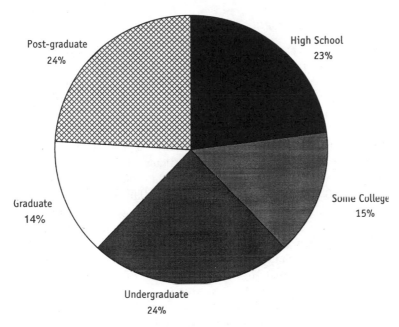

Figure 10.2   Education level of employees

panies (3 percent) were union shops, and these were larger, historically unionized firms. A fewer number of companies were partially unionized. This information does not suggest that the number of unionized workers is only 3 percent. The larger historically unionized companies have larger workforces. It does say that unionization is a sparse commodity when viewed in terms of number of companies.

This scant unionization situation could be looked at as a potential area for unionization. High technology employees without equity interests do not own the "means of production" in the classical sense. They are younger and the businesses they work for are fragmented—conditions that favor unionization. What also has to be taken into account, though, is employee turnover—a feature entrepreneurs such as William Maasberg, whom we introduced earlier in this chapter, told us about. Mobility abounds and there is no disgrace in it. Young engineers have been advised to take chances and leave companies if they have dead-end jobs.[5]

What also has to be considered is the size of many companies that have passed their entrepreneurial stage but remain small. The median number of employees for these companies was 15. These factors could make unionization more difficult as socialization factors overshadow economic considerations.

Salaries in High Technology

How much people are paid is one of the most fundamental definers of high technology industry. There are quite a few factors that influence salaries—length of service, competition, location, and work qualifications being among them. There is much variance in high technology salary classifications as well. Still, salaries among our reporting group appear higher than what might be expected in more traditional employment. Recent annual salary information is presented in Figure 10.3.

The salary distribution showed no particular skew one way or another. There is a lumpen "middle class" of high technology workers who earn between $25,000 and $75,000 (data for 1995). This range takes in 60 percent of reported salaries. Only 13 percent earned more than $75,000, which again partially reflects the small size of high technology firms.

While the salary distribution is somewhat better than it is for most jobs in the United States—particularly in the above $50,000 range—overall these businesses are not the occupations of millionaires. We found few people at the top end of the salary range.

In order to make a comparison between high technology occupations and more traditional occupations, we used Bureau of Labor Statistics 1991–1995 weekly median salaries for selected occupations. Table 10.1 shows the comparison.

The traditional positions we selected showed a greater range of salaries and the high-end positions earned more than the high technology samples, but the information helps indicate that high technology positions paid better than most traditional occupations.

Returning to our specific results, we probed some high technology managers about the relatively low salaries for the apex—the business-owning jobs. Founders told us they avoided paying themselves well, and instead wanted to build equity in their firms for later enrichment. We know exceptions, of course, including a team who paid themselves such high salaries that their income tax obligations were questioned. Year after year they were audited, and year after year they had to justify their salaries to the IRS.

Productivity

It's been stated many times that technical employees are productive. They consume few raw materials. (Caffeine-riddled beverages don't count, although their consumption rate comes close to qualifying them as a raw material.) They add considerable value to their organizations, for salaries that are only moderate.

Between 1993 and 1995, high technology companies considerably increased their productivity, measured in sales per employee. As shown in Figure 10.4, productivity was due to an increase in sales and a slight decrease in total number of employees.

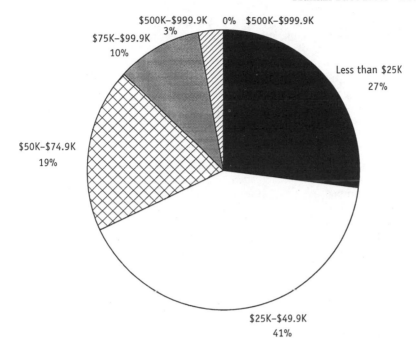

Figure 10.3   Salary composition

Between 1993–1995 inclusively, these firms managed to increase productivity while many other sectors declined in productivity. Had the total workforce remained the same, the productivity of the high technology worker would have risen anyway.

Productivity can be measured in many different ways—sales per employee is just one. A much more complete productivity picture emerges by examining a broader range of measures over a longer period among more companies. But we found it striking that there was actually a decline in employment in what most analysts would consider a constantly growing business endeavor.

General Observations on Worker Characteristics. What can then be said about high technology human resource characteristics compared to other kinds of firms? The employees are better educated, somewhat better paid, nonunion, and highly productive. High technology workers are younger in the emerging companies where we did open-ended interviews. In all the companies, we saw younger people than one might find in more traditional businesses. This age characteristic may depress earnings for a group as a whole and better age corrected salary comparisons. From these general characteristics we'll move to the subject of managing human resources in the high technology operation.

Table 10.1   Weekly Salaries for Selected High Technology and
Traditional Occupations

Type	Occupation	Weekly Wage
High technology	Programmer	$ 743
High technology	Operations/systems analyst	$ 758
High technology	Mechanical engineer	$ 925
Traditional	Hairdresser	$ 286
Traditional	Machinist	$ 368
Traditional	Bookkeeping clerk	$ 386
Traditional	Cashier	$ 466
Traditional	Auto mechanic	$ 466
Traditional	Elementary school teacher	$ 640
Traditional	Nurse	$ 695
Traditional	Lawyer	$1,125
Traditional	Physician	$1,140

Managing Human Resources

Recruitment is the most active area of high technology human resource
management. Managers of the smaller 10-to 50-person companies
had employee turnovers of approximately 20 percent per year. Turnover
was greatest in the software consulting companies. Freshly schooled engineers typically work in the companies for two years, then they go on
to another consulting firm or are hired by the client customer. In other
cases, they decide they don't like software consulting and they change
fields.

In the area of workforce composition and retention, one fact observed was the internationalization of the technical workforce. According
to a *Business Week* article, non–U.S. citizen students who trained in U.S.
schools and worked for U.S. companies have returned to their home
countries.[6] In Russia, South Korea, India, and Japan, technical positions
are occupied by former employees of Wang or Hewlett-Packard. It's a
brain drain with a twist: U.S. companies provide a postcollege training
ground for departing engineers.

More anecdotal information from the same source seems to contradict the effects of this technical emigration. Another *Business Week* article
reported that technically trained foreigners were replacing domestic workers.[7] These people come in as temporary workers, students, or new hires.
Since there is no proof offered for the assertion, it can't be concluded
that such displacement is significant, though it may be a problem in companies that pay U.S.-level consulting fees and compete with firms using
much lower cost imported labor.

How people work in technology organizations has received attention—whether as individuals or in teams. What has been noticed is a

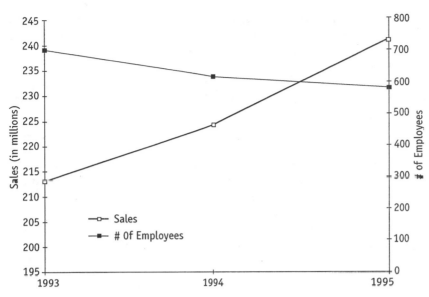

Figure 10.4   Salary versus number of employees

change in work habits. Monsanto, Hoechst Celanese, and Xerox moved to the team structure by establishing formal, permanent teams. These firms have allowed innovation teams to act in an entrepreneurial fashion, unleashing the power of the creative group in developing new products.[8]

As a final point on human resource management, we've noticed that human resource managers in high technology have a tendency toward informality rather than formality. We reported in Chapter 6 that technology ventures seldom have written policies and procedures. This applies to established firms as well with respect to human resource management. We saw no employee manuals and very few forms. What forms there were seemed limited to employee reimbursements. Educational policies were not codified but were handled informally. Even though these companies were small, their written processes were scant.

The Work in the Main

Recognizing that the case is building for the high technology businesses to be more different from each other than they are similar, they share a few elements.

High technology work is oriented toward solving science-based problems for some commercial purpose. The work is dependent on using the tools of technology to accomplish this. Indeed high technology workers constantly use instrumentation and computers to do their work. The field is rigid in its use of the science disciplines to develop new products and services, but loose in its demands on individual creativity. Though the work requires creativity, there do not appear to be intractable schedules for individual creative production.

There are constant peaks and valleys of progress and stagnation; steady, routine work is not common. Periods of broad conceptual thinking can be followed by solving focused technical problems. People tend to work in teams and on projects rather than as individuals on unending quests.

High technology is more a project than a process business. The engineering activities require that new decisions be made in unstructured settings and those decisions have cumulative effects. Successful projects must be proven by the tests of science and must be accepted in the market. Thus, the outcomes of high technology work are rendered successful or not, in the dimensions of both science and business. The kinds of problems faced by those in the engineering part of high technology have been detailed.[9] This theme of science-based solutions through team interaction comes through quite plainly.

These are several of the unique aspects of high technology work. But what has not been detected in our research is perhaps the field's most telling mark: the dedication of high technology professionals. This was seen in our on-site visits and in our open interviews. The stories about Microsoft engineers sleeping under their desks at night, the competition among aeronautical engineers to come up with the best design for the F-22 fighter, the molecular biologists' giving up weekends and vacations to get a patent application ready—these instances were repeated so frequently that dedication became a mode of work in our observations. It is a characteristic that really distinguishes the high technology endeavor.

Future Workers in the Future Business

Near the top of the list of what the Bureau of Labor Statistics projects to be the fastest growing occupations between 1994 and 2005 are the high technology occupations. Positions for systems analysts will grow by 92 percent and for computer engineers by 90 percent. Interestingly, positions for computer operators will decline, reflecting a reduced reliance on mainframes.

In addition to career changes or displacements within the field (as PCs displace mainframe operations), an impending development among high technology workers is the changing nature of the workbase. This sector is not immune from the demographic imperatives affecting other businesses; the diminution of the traditional and the elevation of the new will occur in this business as well.[10] For example, in the software engineering firms of tomorrow, nearly half the engineers will be women. This fact is being addressed by companies such as Hewlett-Packard, as shown in Chapter 3. A much more sizable percentage of employees will be immigrants. The white male will remain, but he will be far less prevalent. Indeed, by 2005, the percent of white workers will dip to 73 percent, from 78.5 percent in 1990. The growth will come from Black, Hispanic, and Asian workers. Technology firms will become much more populated by non-white workers. This will be a major change from the largely white, male-dominated business that high technology was.

Our expectation is that this greater workforce diversity will be experienced more slowly in the high technology area than in other businesses. One reason is that the technical skills needed for high technology work will initially block significant nonwhite representation. However, the top ranks of high technology are more likely to change as U.S.-educated, foreign-born workers choose to open businesses in the United States. Another influence could be the immigration of scientists and engineers into American firms.

These sources will change the face of high technology from a white, male-dominant activity (An Wang was an exception) to something much different, although the transformation will be slow. The ranks of senior engineers awaiting the summons of corporate leadership are still mainly male Caucasians. Women head companies that retail used computer equipment, but these companies remain on the fringe; they are not the giants of high technology.

High technology labor will emerge as a competitive advantage in international trade. In fact, the importance of a technologically trained worker base has been described as essential to U.S. competitiveness.[11] Rapid commercialization of high technology products has been viewed as antidote for lagging American international competitiveness by Reich (1989). Presumably, it will be the new workforce that will unleash the international potential of domestic technology. *Internationalism* and *workforce composition* will affect how high technology work will be done for at least the next half-century. These factors are top-of-the-mind considerations for human resource managers in the high technology field.

Summary

Human resources are critical to high technology businesses. That fact makes this industry similar to others. But individual high technology companies have different marks or human resource reputations. In most respects, the tasks of planning, recruitment, and retention resemble traditional businesses, though there are important variations, particularly in motivation. Young entrepreneurial firms have difficulty finding employees; these CEOs seem to want entrepreneurial employees, though they may not realize the implications of that.

Prominent technology CEOs speak out quite clearly about the importance of human resource management, though their emphasis on what needs to be done differs. The businesses in our group had well-educated, well-paid workforces with little union representation and high productivity. The work of high technology was characterized as science based and in many ways different from conventional businesses. Finally, inevitable changes in demographics will reshape the labor base of high technology in the near and long term, and will further internationalize market activities.

In the next chapter, we will complete our survey of the human resources dimension of high technology by considering the leaders and their impacts on their organizations. We will also consider the matter of whether leaders really make a difference.

Notes

1. Texts such as *Human Resource Management* by Cynthia Fisher, Lyle Schoenfeldy, and James Shaw (Boston: Houghton Mifflin, 1996) provide a framework with topics such as planning for jobs, acquiring human resources, building individual and team performance, rewarding employees, and maintaining human resources.

2. Kidder's book is a celebrated account of life inside Data General. It describes the challenges of engineering leading-edge computers, especially the daily struggles of computer engineers. T. Kidder; *Soul of a New Machine* (Boston; Little Brown, 1981).

3. See Donald Miller, *Managing Professionals in Research and Development* (San Francisco: Jossey-Bass, 1986), as described by Mary Ann Von Glinow in *The New Professionals* (New York: Ballinger Publishing, 1988). Miller observed that people who are attracted to the R&D environment seek intellectual and technical challenges, see the potential to improve the quality of their lives, aim for making breakthroughs and obtaining recognition, follow technology cycles, relate to management beliefs and politics, and are rewarded with emotional and financial connections to their firms.

4. See J. Huey, "Waking Up to the New Economy," *Fortune*, June 27, 1994.

5. A comment made by Samuel Florman, "The Humane Engineer," *Technology Review*, July 1995, p. 65.

6. This is the theme of "Have Skills, Will Travel-Homeward," *Business Week*, November 18, 1994.

7. In "Give Me Your Huddled High Tech PhDs," *Business Week*, November 6, 1995.

8. Reported in "Innovation Congregations" by Tom Kiely, *Technology Review*, April 1994, pp. 54–60.

9. Described in H. Petroski, *Invention by Design* (Cambridge, MA: Harvard University Press, 1996).

10. Much more detail on the changing nature of the American workforce is found in A. Carnevale and S. Stone, *American Mosaic* (New York: McGraw-Hill, 1995). The implications for high technology workers in this chapter come from this book.

11. As an example, Richard Cyert, and David Mowery, argue that new technology adaption is a remedy for underemployment and an enhancement to international competitiveness. In "Technology, Employment and U.S. Competitiveness," *Scientific American*, May 1989, pp. 54–62.

References

Grove, A. (1983). *High Output Management*. New York: Vintage Books.
Harrison, B. (1994). "The Dark Side of Flexible Production." *Technology Review*, May-June, pp. 38–45.
Kanter, R. (1995). "Nice Work if You Can Get It: The Software Industry as a Model for Tomorrow's Jobs." *American Prospect* 23 (Fall), pp. 52–58.
Kidder, T. (1981). *The Soul of a New Machine*. Boston: Little, Brown.
Packard, D. (1995). *The HP way*. New York: Harper Collins.
Reich, R. (1989). "The Quiet Path to Technological Preeminence." *Scientific American*, October, pp. 41–47.
Sculley, J., with J. Byrne (1987). *Oddysey*. New York: Harper and Row.
Winner, L. (1995). "The Culture of Technology." *Technology Review*, August–September, p. 82.
Young, J. (1984). *Steve Jobs: The Journey is the Reward*. New York: Lynx Books.

11

HIGH TECHNOLOGY CEOS

In Chapter 10, we explored the high technology workforce and discovered that these workers are well educated, fairly well paid, and productive. In the last three years, their productivity grew rapidly, providing evidence of just how vibrant this enterprise really is. In this chapter, we continue our exploration of the human side of technology, but we narrow the beam to the extreme. We look at a single, critical worker—the leader of the business, the CEO. We examine, as we did the technology venturers, the prevalence of education. But we also look at prior experience, how related that experience is to the present position, and the leadership style of the CEO. Very importantly, we look at the consequences of these factors on the organization's success. This brings a new intensity to our overall examination of high technology by exposing the possible connections between these characteristics and business performance. How the CEO affects what technology companies do, the decisions they make, and their business success will be explored in the remaining chapters.

The CEO—Yesterday and Today

Think about prominent high technology CEOs, and you have an immediate study in contrasts: the conservative church deacon Ken Olsen of DEC and the free-spirited and volatile Steven Jobs of Apple, the computer nebbish Mitch Kapor of Lotus and the urbane John Sculley of Apple, the methodical and analytical James "Mr. Mac" McDonnell of McDonnell Douglas and the California beach-surfer-looking visionary

Scott McNealy of Sun Microsystems. All are different, but all are or were successful high technology CEOs.

For some—Olsen and Sculley—success is in the past; they have not regained their leadership glory, but this doesn't diminish what they did and when they did it. A particularly useful research project would be to explore why formerly successful high technology people were not able to succeed a second time.

Taken together, these technology CEOs are the *Drosophila melanogaster* of management science; they are found in many mutations, making it easy for management theorists to concoct virtually any taxonomy. (We hope readers will pardon our fruit fly analogy for the sake of making our point.) This also illustrates that success in the field—something all of these CEOs have enjoyed at some stage—does not come from a golden-edged cookie cutter that outlines a uniform type of CEO. There is no cookie cutter. Efforts to be prescriptive about who should run these companies defy research proof. Despite a considerable number of studies on leadership, we are far from able to say who makes a good leader and why they succeed. Hard taskmasters run successful technology companies and easy-going managers run successful technology companies.

The essence of what we recognize as corporate leadership has also changed over time. Not only are CEOs a varied lot, but what we have come to expect from the CEOs has changed as well. There are two vectors—individual difference and historic change.

Compare a contemporary technology CEO to a business baron of the Gilded Age, and you'll observe drastically different behaviors. The leadership function is done differently now. Being a CEO involves more human relations skills. Decisions are not as unilateral, nor are they as centered at the top of the organization. The CEO is not as removed from employees as he was in the "old days" of management. Much of the authoritarian quality of the CEO position has given way to participative means of doing things. Our high technology CEOs are modernity in spades. There are old-style corporate leaders in high technology to be sure, but they are the exceptions.

If we took a CEO from an earlier era—maybe only as far back as the 1950s—and transplanted him to a 1990s technology company, we would expect him to have a short corporate lifespan. His style would not be tolerated. If there is any doubt about this, just compare the letters and communications to employees of the *ancien regime* of corporate leadership to the pronouncements and messages from today's presidents. The older style CEO—probably authoritarian, probably elitist, probably an older white male—would likely not fit in to the contemporary technology organization. The technology company heads we talked with saw themselves as modern managers who recognize their dependence on committed, able employees. To a person, they were interested in and practiced management techniques that stress cooperative goal setting.

There is one thing the modern technology CEO has in common with his precursor, however. That is an intimacy with his technology. Just as the head of a foundry had to know his casting techniques a century ago, the modern technology CEO has to know something about the technology base of his enterprise. Within high technology, a professional class of nontechnically oriented managers has not emerged, as there has in other fields. High technology has not been turned over to professional management because the people who run high technology companies emerge mainly from the technical professions, as far as we have seen. Even when there are exceptions, such as John Sculley from Pepsi, there is a base of science knowledge. Sculley engineered color television technology when he was quite young. Tom Watson Jr. was more a businessperson than an engineer, but he maneuvered IBM into the swift current of high technology that swept along IBM and other adaptive firms. He also acquired the engineering talent as IBM needed it.

What Technology CEOs Do

When we refer to the CEO, we mean the individual in the high technology organization who functions in this capacity. We do not expect that all high technology organizations have CEOs as titled positions, yet almost all have people who do the CEO functions: set corporate strategy, develop and execute budgets, hire key managers, make major business decisions, and otherwise lead the organization. Sometimes, especially in smaller technology businesses, those with one or two employees find it pretentious to call themselves CEOs—the default title is president in these circumstances. That title is usually a requirement of legal incorporation, but the small-firm heads in our sample did not use the title flippantly. More often, they simply called themselves "owner."

Let's consider some of the major CEO functions in businesses that are driven by science and engineering. In doing this, we base our comments on the direct observation of seventeen technology CEOs, rather than our survey data.

In setting corporate-level strategy, the role of technology CEO is inclusive. The CEO is expected to provide a vision for the firm, act as a technology seer by interpreting the larger environment, and plot a course to the future. And in this activity, a curious thing happens. It used to be that technology's main intrusion in the modern corporation—the information technology division, department, or group—made its own decisions about what kinds of computers, networks, and software the company would use. Now we see these decisions being moved up to the CEO's office because these information technology decisions can be important for the whole company. Stated another way, these have become strategic decisions because the costs are high, the impact is extensive within the firm, and the duration is long term. So information technology decisions complement the other major decisions made by CEOs for most major nontechnology firms.

For high technology CEOs, this involvement is nothing new. They have been making technology decisions all along. But it does raise the ante. Technology CEOs must be even more foresighted than their conventional counterparts or else they are not truly in technology companies. All of the high technology companies we profiled depend greatly on information technologies. The computer is the lifeline that holds together the infrastructure of all high technology businesses—biotechnology, electronics, chemicals, and of course the computer business itself. Information technology keeps high technology ahead of traditional business. It also enables technology businesses to make technology leaps themselves. CEOs in this field know that progress can be made if the advantages of information technology—processing masses of information, decision support, and artificial intelligence—are realized, thus they elevate information technology acquisition and usage decisions as strategic in nature.

High technology CEOs do some of the same kind of strategic planning operations as other CEOs do, such as arbitrating disagreements on corporate direction between top managers and exhorting their company to carry out its strategic plan. This can best be described as managing strategy development, and more will be said about this in the next chapter.

Turning again to the special character of technology leaders, the CEOs of technology companies provide leadership in an area that is unique to the field—intellectual leadership. Technology CEOs are frequently the most technically proficient members of their organizations—in some circumstances. This is certainly the situation if the CEOs are founders of new companies. It is much less the case for CEOs of mature organizations because, while the CEOs have worked themselves up through the ranks, they often have lost much of their professional currency. New technical expertise has to be brought in, mainly at the junior level.

In the entrepreneurial technology organization, the founder-CEO is the innovation leader—that is the basis for building the new company. However, a particular hubris occurs. A software engineering consulting firm had a CEO who felt competent with their technical skills. The CEO talked frequently about how "entrepreneurial" and "innovative" the firm was even though it had been in business for nearly ten years. Yet when we talked with the software engineers, we sensed an informal conspiracy to keep the CEO away from clients. The engineers believed the CEO was well behind both the clients and them in any knowledge of programming languages.

This CEO did not realize that their knowledge was out of date. The software engineers "hid" the CEO by arranging meetings with clients when the CEO was unavailable and by otherwise conducting business without the knowledge of the CEO. Quite possibly, this perceived lack of currency would have surprised the CEO, since the CEO had been out of school only five years. Established company CEOs can succumb to the glorification of the past.

Another way that CEOs show leadership is by coaching their staff. A few have learned along the way, saying that coaching sessions with newly hired engineers was their way of keeping up with new technologies. Some companies, such as Microsoft, are renown for their coaching practices.

We have seen the technology CEO act as a kind of firefighter, reengaging in the business by taking a hands-on approach when a major problem occurs. We know of a medical electronics firm CEO who turned over day-to-day operations to his son, yet needed to reassert his stewardship when the company didn't "stick to its knitting" and embarked on a series of unpromising businesses. This CEO was still officially the CEO, but the "firefighter" was brought back from his semi-retirement.

In this review, our references to the CEO have been to individuals acting in this capacity. But that's not the only way of running the organization. There are some interesting and successful exceptions to this individual-centered idea of CEO. Isthmus Engineering in Madison, Wisconsin, is a headless but successful high technology organization. (We profiled Isthmus in Chapter 3.) This organization builds customized, automated machinery. It is organized as a cooperative rather than a corporation. There is no president or CEO. The cooperative makes decisions through employees, who are represented by director-like members of a planning group. In this solitary example, there is a technology company that runs without a CEO. Having a CEO is a convention, but not a necessity, for technology organizations.

Characteristics of Technology CEOs

We now turn to information provided by our survey of established high technology companies. One of our first questions was if the current CEO of the company is its founder. Since the companies in the sample were well established, having been in business for more than five years, it would seem that the founder's initial influence may have waned and that they were no longer leading the company. Though we can easily recall many computer and software companies whose founders are still at the helm, these examples are memorable because they are exceptional, their CEOs heading hugely successful companies. Gates is memorable because he is still in charge, but his founding partner Paul Allen has not achieved equal fame and retired from Microsoft for health reasons. More recently, Allen has reemerged in the computing community.

Some technology companies founders in the past stayed with their companies just as their names do: McDonnell, Douglas, Boeing, Hewlett, Packard, Picker, Cray who had his name on two companies, and duPont. But more recent history has been marked by a departure from namesmanship to functionality—companies with names that say what they do. The result is that when the founders leave, their names do, too.

Our findings showed that founders still play the major role in technology companies. In 67 percent of surveyed businesses, the CEO is the founder (or one of the founders) of the technology company. Another 28 percent reported that the current CEO is not a founder.

The prevalence of the CEO-founder may be due in part to the fact that these are small companies and the founders may not be able to abandon their fledglings. For instance, they may not have recruited staff with sufficient technical skills to run the company. Nonetheless, it is confirmation that founders are not short-time business leaders. These venture founders kept leading their organizations, through the startup period and well beyond. If you consider that there are a few years devoted to conceiving the venture, you are led to the conclusion that the technology venture could possibly be a fifteen-year experience for these entrepreneurs—and that is just so for one business. The habituation of entrepreneurship causes founders to start other companies—until founding ventures becomes a truly lifelong experience. Silicon Valley legend Gene Amdahl, for example, founded three high technology companies. Kamran Elahian founded six.

What can be said about the backgrounds of the CEOs in our sample that might further characterize them? A way of capturing the characteristics of a CEO is to find out something about their experience. Do CEOs have more technical experience or nontechnical experience? The answer is very much on the side of technical experience. Two-thirds of the sampled company heads had no nontechnical experience. On average, the CEOs had 2.8 years of nontechnical experience and averaged 14.7 years of technical experience.

Based on this, technical experience appears to be a strong characteristic of technology CEOs. It also may suggest that these CEOs may be less acquainted with the nontechnical side of their business operations— finance, marketing, sales, and production. This relative lack of experience in nontechnology areas could be the source of problems later if the company does not acquire people with these skills at some point.

There is a distinct slant toward science and engineering as producers of these technology leaders. Among the companies we surveyed, the top executives were software engineers, electrical engineers, chemical engineers, and scientists. We asked if the CEO's experience was related to the industry they were currently in. The result was that 76 percent had industry-related experience and 19 percent did not. The rest did not reply. By and large, these CEOs did not depart from a major industry group, though that industry might be broadly defined.

Of this group of surveyed businesses, it could be said that talent was mainly home grown. The CEOs knew the industry as a whole, which no doubt partly reflects entrepreneurship within the field. The entrepreneurs among these CEOs stayed in their broad industries. We saw little to

Table 11.1   High Technology
CEO Education

High school diploma	1.7%
Some college	7.7%
Undergraduate degree	21.5%
Graduate work	31.3%
Postgraduate degree	33.5%

suggest that technical CEOs have crossed industries to take control of different companies. We left it to the respondents to define the bounds of their industry, but our conversations with many CEOs left us satisfied that they knew the parameters that separate industries.

The next issue we looked at was the extent of the CEO's experience. The CEO might have all his experience in the same industry, perhaps even in the same business, but it means little if there is not much of it. So how experienced were the CEOs? The answer is very experienced. On average, they had 17.4 years of prior business experience. That experience is not normally distributed, however. It had multimodes, with peaks at ten, fifteen, twenty, twenty-five, and thirty years, perhaps reflecting a rounding off by the respondents. There were four CEOs of 233 companies who had no prior business experience, and one who had fifty years. The range of experience, like the amount, was considerable.

It seems that the high technology CEO is an experienced CEO, at least judging from the information reported to us. This puts the average age at just under 40, if we assume they worked continuously since college graduation.

We previously reviewed the educational background of technology workers and found them better educated than most other workers. Does the same hold up for CEOs? The answer is definitely yes. The level of education for the CEOs is shown in Table 11.1.

In our sample, nine of ten CEOs have a college degree and a third of all CEOs have a graduate degree. This indicates a very high level of education, perhaps more so than in any other type of business activity. This level of education is somewhat higher than for technology employees, even though the CEOs may be slightly older and come from a generation not as well educated.

Leadership Styles

How high technology chief executives get things done in their organizations is the matter of leadership style. Leadership is an extraordinarily popular subject, is yet it is devoid of experimental verification as far as understanding why leadership works. And high technology leadership is no different.

Studies of leadership have mainly concerned examples of great leaders, with the implication that emulation means replication—that leadership characteristics might be copied by others. The character of the leader is dissected and explained. As laudable as this is for the sake of at least historic record, it doesn't explain many things, such as how to predict who will lead and who will follow, the transitory nature of organizational leadership, or why some leaders succeed and other don't.

The response of those who study organizational leadership has been to open up the realm of leadership and accommodate many "styles" of the art. That way, it is no longer necessary to be precise about leadership. What results is a taxonomy of leadership, a classification scheme in which an A leader type fits in one classification and a B leader type fits in another.

Leadership in high technology bears a resemblance to conventional business leadership, but the science base makes a difference. Companies driven by scientific breakthroughs need technical and business leadership that engages the workforce or facilitates leadership within the organization. High technology is a human knowledge business that is only as good as its people, and its people are only as good as they propel themselves or are propelled to achieve.

Although there are myriad theories about leadership style, one depiction separates entrepreneurial high technology leaders from leaders of proven companies. The work of Bernard Bass separates transformational leaders from transactional leaders. The transformational leader leads through vision and personal energy—very fitting for entrepreneurial stages—while the transactional leader decides what the organization needs to do and then enables employees to do those jobs.

As we encountered this issue, our choice was to make use of a classification scheme that differentiates among leadership behaviors and also translates into some meaningful and practical leadership categories. This scheme is richer than the simpler two-category schemes. There are six behaviors of leadership that can used to separate high technology leaders. These categories are an amalgam of what organizational behavior theorists have proposed about leadership styles.

The first is a *give-and-take* management style. This is a nonauthoritarian approach in which decisions are made and tasks done through negotiation with those involved. It is akin to a political style of managing where there is trading of obligations and collection of debts. "You owe me one" is the parlance of the give-and-take management style.

The second style is one of the *inspirational leader*. This technology CEO gets people to do things because of their personal attributes as a leader. They lead by example and they engender the admiration of others. It may be that these CEOs inspire by taking on excessive work, by creative genius, or simply by their personality. The inspirational CEO doesn't coerce, but pulls people up to their capabilities by setting a sterling example. "Come by and visit" may be a comment issued by this type of leader, knowing that the mere invitation to associate may inspire employees.

A third style is *people focused*. This CEO is involved with more than the task itself. They are concerned about the beliefs, attitudes, and behaviors of employees they seek to achieve through them. "How are you doing today?" is the earnest utterance of this focus on the people in the organization.

A fourth style is *laissez-faire*. The CEO gives employees free rein to come up with solutions. There is no constant supervision of work. The sign of the laissez-faire CEO is the expression, "Let's see what you and your group come up with. Let me know when you want me to look at it."

The fifth style is *team management*. The CEO believes that work teams are important for getting the job done. The CEO may be an active participant on a team, but even if not, they have high expectations from work teams. "We need to get together on this one" is a common expression of the team-oriented CEO.

Lastly there is the *task-oriented* management style. The exhibitor of this style fits the description of the typical business manager quite well. The focus is on the accomplishment of tasks, how to do them on time, how to get the job done better, more efficiency, and the like. The task-oriented manager is likely to utter such things as, "We're getting all our projects through."

These six styles are not mutually exclusive. Adept CEOs know that a contingency approach is often warranted, and they vary their style in accordance with the circumstances, perhaps changing from being people oriented to task oriented as the need arises. In spite of such shifts, almost all organizational leaders are able to identify their anchor style, their own zone of comfort in which they manage.

For high technology businesses, we can easily construct situations in which any of these six styles might work best, but our immediate interest is in discovering how technology CEOs actually place themselves. The results of our survey are shown in Table 11.2.

Although an autocratic management style is one way of managing, it was not a response choice because respondents were not likely to place themselves in this category. In our study 6.4 percent of the 233 respondents did not reply to this question.

It is quite interesting that over half of the respondents decided they were team-oriented managers. This was by far the most popular choice. There is a definite skew toward the team-oriented style, although by observing behaviors rather than relying on self-reported assessments, we might have seen more variations in style. There is, after all, a contemporary emphasis on team building as a mode of management; the responses could very well reflect this. The management approaches of notable high technology leaders (e.g., Bill Gates, Scott McNealy, Lew Platt, Eckhard Pfeiffer of Compaq, Michael Dell of Dell Computers) have

Table 11.2 Technology CEO
Leadership Styles

Give-and-take style	0.4%
Inspirational leader	18.9
People focused	5.2
Laissez-faire	4.3
Team management style	51.1
Task-oriented style	13.7

been profiled by James (1996). In their own words, these leaders emphasize team dynamics. Gates is such a luminary in this category that profiles of him have been done by Stross (1996), Wallace and Erickson (1993), and Manes and Andrews (1994). These may all be different books, but Gates emerges as someone who has established and enforced hiring directives calling for fresh-from-school, highly intelligent employees with lots of drive.

The CEOs of technology business now face new realities, which partly they have created and partly have inherited. This affects their working relationships with their staffs and the management style they use. The common thread is the change in the way work will get done in the future a theme discussed in Chapter 11. Kanter (1995) surveyed companies of all kinds and reported that they were experiencing more competition, greater use of technology, greater cost pressures, higher quality needs, and the need for greater speed. Most companies favored flatter structures, more outsourcing, and—quite to the point—greater use of problem-solving teams. So the team-style management mode of our respondents may reflect this new reality.

As noted in Chapter 11, Kanter looked at the software industry as a model for future workforce planning. She described the "software gypsies"—contract software specialists who move from job to job yet accept this migration. For the high technology CEO concerned with team effectiveness, the issue is how to make cohesive and productive teams of contract workers whose organizational allegiance may end at the end of their contracts. Those who run technology companies will be first to recognize how the nature of work is changing. This situation is of particular interest for relatively new companies, which have to make the transition from an intimate entrepreneurial style to perhaps a more distant contract-worker arrangement.

Does the CEO Make a Difference?

Do the characteristics of technology CEOs have performance consequences? The firms responding to our survey were placed in one of seven technology groups, based on two-digit SIC codes: Group 28—chemicals and allied products, Group 34—fabricated metal products (except machinery and transportation equipment), Group 35—industrial and commercial machinery and computer equipment; Group 36—electronic and other electrical equipment and components except computer equipment; Group 38—measuring and analyzing and controlling instruments, photographic, medical, and optical goods, watches and clocks; Group 73—business services; Group 87—engineering, accounting, research, management, and related services. The categories were also combined into a single group, which was a composite of respondents.

We used these seven groups because there were a sufficient number of firms in each group to facilitate analysis. Other groups had insufficient numbers for us to analyze results.

Sales growth over the past three years was the measure of organizational performance. Our reasons for using this measure are cited in Chapter 5.[1] Sales growth was split into categories of more and less than average for each major group. This let us compare two performance measures against two different classifications of independent variables.

The first look at the performance-CEO connection was in the area of educational background. High technology CEOs are very well educated, as we have discovered. They need to be because science and engineering are taught at the undergraduate and graduate-school level. This is where students acquire knowledge about the latest discoveries and most complex concepts. It's true that maturity in the sciences starts in grade school, as it does in mathematics, but a career in the sciences is decided in college. The phrase "Differential equations separate the men from the boys" is something at least one of us heard as the way true prospective scientists get separated from the pretenders.

Our interest was in the level of education as a discriminating factor in business performance. We compared relatively higher levels of CEO education to relatively lower levels of education.

Our results were surprising, in a way. Splitting the responses into two groups, one including high school and college-educated respondents and the other including graduate and postgraduate CEOs, we found that CEOs who had more education did not have better sales performance. In fact, CEOs with less education were associated with firms that had greater sales growth. The difference was statistically significant. This carried through in the largest technology subgroup—the computer industry—where more education was negatively associated with sales growth. This was also statistically significant. It was also true for the engineering field in general.[2]

Next, we considered the question of whether past CEO experience was associated with sales growth. The respondents were split between those who had 17 or more years of prior experience and those who had less than 17 years of experience. (Seventeen years of experience was the average CEO past experience.) The results showed no significant difference between the less experienced and the more experienced CEOs. The exception here was in the fabricated metals industry (a category that was also explored as a major group for comparison purposes), where there was a statistically significant difference, with more experienced CEOs doing slightly better (although there were only a few companies represented). In the measuring and analyzing instrumentation sector, there were more examples and statistically significant results indicated more experience was connected to better sales growth.

Does it matter if the current CEO was a venture founder or not? The respondents were split into the two groups, and it was discovered that there was no significant difference between the groups. Founders as CEOs don't seem to have companies with better sales growth as a whole. For chemicals, though, there was a statistically significant association: founder CEOs had better sales growth than nonfounder CEOs. The chemicals industry had a definite composition difference from other high technology businesses. The larger businesses were in this group, compared to the other high technology groups.

As a final consideration of CEO associations with sales performance, we investigated the matter of leadership style. For all of the technology companies, there was no significant association between sales growth and particular style of leadership. For computer firms, however, all of the high sales growth firms had inspirational leaders. This could be significant, but it is not conclusive because of the small number (fewer than ten) firms that fit into the different style categories.

Portraits of CEOs

Can we typify the high technology CEO? In some ways we can, but we must also add that there is considerable variation within this group. With this in mind, our depiction of typical high technology top management is generally but not absolutely true. This section is based on our interviews working with high technology CEOs.

Acting in the function of a CEO involves both traits and behaviors. There are certain traits we observed no matter the size organization. One is, to put it bluntly, their energy level. For some CEOs, high energy was immediately detectable in their mannerisms and speech, but for others it lay behind a laid-back exterior and was revealed only when they described how they used their time. These CEOs were in continuous communication with their organizations and were frequently double-booked for meetings.

Another trait we noticed is that of holding opinions and a willingness to share those opinions with others. Even those CEOs who looked quiet—in other words, did not have a commanding physical presence— were not in fact quiet. In a room of talkative academics and CEOs, the CEOs held their own, frequently challenging the academic view of things.

For information on the behaviors of CEOs, we looked at how they handled their organizations. We relied on their descriptions of past actions and found what seemed to be a high level of involvement with employees. The CEOs made efforts to know their people well, using the tried-and-true management by walking around technique.

Are high technology CEOs for established companies more like high technology entrepreneurs or more like professional managers? While we did not interview CEOs of large technology firms who may have pushed our impressions another way, our inclination is to say that CEOs are more entrepreneurial than managerial. Our perspective for assessing this is based on the work of researcher and management theorist Henry Mintzberg (1973), who classified managerial roles as interpersonal, informational, and decisional. The manager as entrepreneur is in the decisional role. We saw CEOs who wanted to improve their organizations by changing them, by making new innovations, and by shaking up the current structures and processes. This led us to consider these CEOs as "entrepreneurial" in behavior.

What we also found were differences in results for CEOs in different high technology businesses. Most of the experimentation and innovation in CEO management techniques happen in the business consulting fields while there is less in fabricated metals, where CEOs are more traditional in their human resource management practices.

A Perspective on Technology Leadership

There has always been a dichotomy within philosophy between the person of action and the person of thought. It has been subject of long-standing debate. Even Plato in his *Republic* sought to create a meritocracy in which pure thinkers could become the leaders of state. Yet the separation has endured. People of action are different from people of thought. One is one or another, not both. Certain historical figures, such as T. E. Lawrence, are historic because they are exceptions to this rule—they are the rare combination of action and thought, able to leap into action at any time.

As a group, though, high technology CEOs come closest to being a fusion of thought and action. They can take basic science and shape it into useful goods and services. In their own way, they bridge the gap, not only as individuals but as a group.

Summary

High technology CEOs are individuals who are both obvious and inexplicable. They are prominent as leaders of their organizations, yet they defy efforts to lay out a path for ascendancy. There are no prescriptions for their success.

This chapter has shown that technology CEOs can be quite different from one another and still enjoy success. They exercise many different styles of leadership, but there are no clear-cut performance consequences. Responding to different situations, effective CEOs know how to mix leadership styles for best effect.

High technology executives do many of the same things that other CEOs do—set corporate strategy, prepare and execute budgets, hire key managers. But they also provide intellectual leadership over the research and development that is at the heart of high technology. In this environment, they are frequently seen coaching employees, perhaps making a second generation of technical innovators. They also act as firefighters to help solve business problems or provide goal reorientation.

Our review of CEOs in going concerns demonstrated that the impact of entrepreneurship is high. About two-thirds of the CEOs sampled were founders of their organizations.

The CEOs were very experienced, averaging over 17 years of experience, in a technical field. Over 80 percent have college degrees and many had higher degrees.

A team management style is the most evident self-reported way they manage employees. This may reflect the popularity of techniques in management literature today; both academics and practitioners have sounded this theme lately.

When it comes to assessing the results of the CEO on business performance, we found that there is no demonstrated connection. Better education didn't mean better sales growth. There was no clear connection between background of the CEO and performance of the technology company. Even the relationship that seemed reasonable—more education and better performance—was inverted in our research: less education was associated with better performance.

In the other dimension we explored—length of experience—there was no apparent relationship between performance and this factor.

Lastly, simply because the CEO is a founder does not mean the business does any better. Companies that still have their founders on the board as CEO derived no detectable sales growth advantage, though there are certainly other advantages and disadvantages of keeping the founder at the top.

All this suggests that no single type of CEO has the keys to the kingdom of organizational success. While education may be a requirement, possession of a degree does not translate to a successful business. Those without degrees can be cheered by this, since it indicates openness in high technology leadership. This openness seems to be especially apparent in the software business, where many self-taught programmers thrive in a fragmented and highly competitive arena.

In the next chapter, we'll look at the primary instrument of the CEO in guiding the organization—the business strategy. And we'll look again at possible relationships between strategy and business results, this time in terms of the established high technology company.

Notes

1. There were a few possible ways of measuring performance, but sales growth was used a factor widely reported and could be verified. Other performance data collected, such as ROI and market share, were determined to be less reliably reported. Employee numbers growth was not taken to be a good performance measure, either.

2. Chi-square approximations were done. The null hypotheses of no differences were rejected at the .05 or below confidence interval.

References

Bass, B. (1990). "From Transactional to Transformational Leadership: Learning to Share the Vision." *Organizational Dynamics*, Winter, pp. 19–31.

James, G. (1996). *Business Wisdom of the Electronics Elite*. New York: Times Books.

Kanter, R. M. (1995). "Nice Work if You Can Get it: The Software Industry as a Model for Tomorrow's Job." *American* Prospect 23 (Fall), pp. 52–58.

Manes, S., and P. Andrews (1994). *Gates*. New York: Touchstone.

Mintzberg, H. (1973). *The Nature of Managerial Work*. New York: Harper & Row.

Stross, R. (1996). *The Microsoft Way*. Reading, MA: Addison-Wesley.

Wallace, J., and J. Erickson, (1993). *Hard Drive*. New York: HarperBusiness.

12

STRATEGIES FOR ESTABLISHED COMPANIES

Our exploration of the loose collection of businesses known as high technology has shown that they are distinguished from other businesses yet also are distinctive among themselves. In this chapter, we take on the issue of how CEOs and employees alike align themselves to a course of action—a core business strategy.

We are concerned here with strategy in the high technology business. We look at the term *strategy* and its many crucial dimensions. But beyond the definition, we need to discover what the meaning of strategy is, not for the musing theoretician but for the technology organization whose stake is its own fate. We see if there is a connection between strategy and what happens to the high technology organization. Does strategy have a performance consequence? Our aim is also to discover if certain strategies are better or worse than others. And determine if particular strategies are more appropriate at certain times during a firm's evolution and in different competitive circumstances.

This chapter builds on the work reported in Chapter 7, but our interest here is in technology companies that have already passed muster—those established firms (at least to the degree they are still in business) that are continuing to grow or have stabilized in a profitable market. Our earlier ideas of strategy are intact, but their application here is at a different point of development.

There are questions we can ask that will help us determine the nature and consequences of business strategy. How important is strategy for the high technology company? At what level of the organization should it be done? Should it be done at the very top or should it percolate from below à la MBO style. Is it an important task for the CEO or should it be parceled out through the business? Ken Olsen of DEC had answered that last question by saying that his job was to make sure the company has a strategy everyone followed. The other questions will be answered by looking at the connection between strategy and business performance in our sample of high technology companies. It will also be answered by looking at how strategies play a part in the day-to-day activities of established companies.

The Meaning of Strategy

The term *strategy* as it applies the business organizations has been defined a number of ways. "An organizational plan of action intended to move an organization toward its goals to achieve its mission" is one definition.[1] This definition is just a start, however. "An integrated and coordinated set of actions taken to exploit core competencies and gain a competitive advantage" is the definition offered by another group of strategy authors.[2] Another definition puts some emphasis on the outcomes of strategy: "An organization's strategy consists of the moves and approaches devised by management to produce successful organizational performance."[3]

Other strategy texts contain other definitions. These very different definitions show that the word's meaning seems to be a function of individual authorship—there is no commonly accepted definition of *strategy*. And these divergent definitions were all produced by teachers of strategy, who might be presumed to be moving close to a common definition, since these authors are all in the same field.

This is a complication, not a fatality, however. If we consider a business strategy in relation to the organization's goals or objectives, strategy is the path to those goals. That is our building point. It is a simple one, but it distinguishes objectives from methods. An educational software company might want to be the leader in the industry—that's its objective. Its strategy is to develop and commercialize a grade-school-level integrated English-Math instructional program for PCs and Macs.

Most often, goals and objectives are quite visible and articulated, but strategies are less so. There is a sense of pride in announcing what your company wants to do, as Bill Gates often does. Sagacious speechifying is a function of success. But strategies are more elusive, partly because they may change and partly because true strategies may be disguised from competitive purview.

But the path, the strategy, is one of the crucial determinants of high technology success. A bad match between strategy and business environment dooms the incipient technology venture and wounds the prosperous established company. For high technology businesses, strategy is especially important because of the fluidity of the environment and the necessity of refining strategic responses to that environment.

As we consider the subject of high technology strategy, we need to recognize that strategy is not an extricable function in the business of high technology. It is not an isolated happening, divorced from normal business operations. It is also not something tangible that comes into or leaves the business. It is an internalized process reflective of management assumptions, information, and judgment.

We have looked at strategies in high technology companies that have been recognized as "best of class." The process these companies use to write their strategies are models for other technology companies. Owners of these companies do not devote themselves to a singular, hermetic approach to the crafting of "strategy" for a business. They do not do strategy the way they might do taxes: fill out, file, and forget until next year. Strategy permeates the company on a continuous basis, done by many people. From what we have found in our look at exemplary high technology business strategists, they involve different functional units and have frequent checkpoints (often quarterly) to consider revisions to their strategy.

The ubiquitous nature of strategy, especially regarding the good practice examples we cite shortly, itself poses a conceptual problem. Strategy is everywhere in the organization. There is a strategy for marketing, a strategy for production, a strategy for financial investments. Taken loosely as "the way objectives are met" or a path to objectives, it's understandable that strategy has a puttylike quality, seeping into all business functions, performed by many players.

There is a routine to strategy making. That routine leads to the problem of not staying at the pinnacle when technologies or markets change—that is the premise of Bower and Christensen (1995). Companies use an ingrained rational, analytical investment process for known markets. It's hard to divert resources to unknown markets and firms. To accomplish the change, firms should look for a disruptive technology (provides for a major change in the business environment) and exploit it. To spot these technologies, the firm should determine if the technology is disruptive or sustaining; establish the significance of the disruptive technology; locate a market for it, place the responsibility for marketing the technology outside the organization; and keep the disruptive organization outside.

If strategy is done at all levels and in all functions, can it really have any special meaning?

While it may seem true that the term *strategy* is so diluted and pervasive it is without meaning, that is true only to the extent that the definitions themselves are feeble and the task of developing strategy enormous. To simply say it is how a business accomplishes its objectives offers no consolation for the preparers of corporate strategy and provides scant guidance.

More can be found in the doing of high technology strategy than in the describing of it. If we pursue the idea of strategy from this different slant, especially as applied in the business world, we find that there is more than putty to it. There is something substantial, and as we look at it, we see that there is reason for strategy making to be very broad in the "good example" firms that craft and execute strategies as though their corporate lives depended on them.

There has been sufficient historical development of the subject of business strategy to considerate it separately, as we do in this chapter. The subject has developed from military history, economics, finance, and the social sciences. While we merely allude to these origins, the essential point is that strategy has preceded the high technology organization, as well as the business organization. It has been around as long as there have been human communities. High technology business is the mere inheritor of the legacy. Modern high technology companies have adapted strategies from their traditional business precursors, rather than invented new forms of strategy. That's not to say there are no variations on the theme. There certainly are.

The most telling aspect of the special nature of high technology strategy is in its distinction from older technologies. This distinction was made by H. Igor Ansoff (1987), a corporate strategy classicist who wrote about the uniqueness of high technology strategy. Ansoff distinguished between the technology of the past and that of now. He considered modern high technology firms as part of the second industrial revolution instead of the first:

> The Second Industrial Revolution is science based. The great inventions, such as penicillin or the transistor, are still products of individual genius. But today, science guides the invention process, provides explanations of why the inventions work, and shows the way to further inventions. As a result, every major technological breakthrough triggers a competitive race in which technology is continuously improved and applied to uses other than those for which it was originally intended. (p. 29)

In the first industrial revolution, individual entrepreneurs had inventions and built large factories with long production runs of goods based on those invention. More recently, technology firms cannot be based on production but must be effective at understanding and working in their environment. Ansoff put it as "a predicted trend from technology orientation to the strategic orientation in an increasing number of high tech industries" (p. 29).

We can add that the second industrial revolution has been one of applied science, not basic science. It is also an era of competition in the uses of technology, which can be interpreted as the emergence of a true business of technology. Technologies are commercially successful if they are developed by companies that are market driven rather than product driven.

All this drives home the importance of strategy for the high technology company because it is strategy that makes the match between the company's internal operations and its external environment.

A particularly incisive aspect of Ansoff's observation has to do with the computer business. Progress in the early stages of this business was guided by technological advancement, and the needs of the market were not considered. It became harder to differentiate products once the initial industry growth wave passed. When growth stopped, there was a steep downturn and even bankruptcies. In diagnosing this condition, Ansoff described high technology business as one where there are many products to choose from and the consumer can substitute one type of technology for another. This distinction separates high technology from medium technology, in which there are many products, and low technology, in which there are few products. More will be said later about this view of high technology. It will, in fact, be part of a framework for looking at high technology strategies.

Strategy at Its Core

All strategies have scope, resource deployment, competitive advantages, and synergies.[4] In the high technology arena, strategic *scope* is the extent the company acts within its environment. For example, scope might be how broad a market a computer firm chooses to compete in—the broad scope of all mainframe computers or the narrow scope of mainframe supercomputers.

Resource deployment is where and how resources should be applied to meet goals and objectives. A high technology firm makes resource deployments when it sets budgets for new product development or assigns engineers to a project. A pattern of deployments can be referred to as its distinctive competencies.

Competitive advantages are the positions the company arrives at in relation to its competitors as a result of its resource deployments. How might competitive advantage for a high technology firm be exhibited? An electronics company may be able to stave off competing firms because it was the first to manufacture a light-emitting diode, even though other firms have produced superior technology. The advantage is in the knowledge of the market that newer entrants do not have.

Synergy is the combined effects of resource deployments and scope decisions. But it is more than the simple addition of resource deployments and scope. It may be that an extremely productive resource deployment in a narrow market scope caused much greater market potential to exist without new resource requirements. For example, a high technology company may have used its resources to have its "industry standard" adopted in a relatively narrow telecommunications arena. If that standard is accepted in a larger arena and the firm can sell to that arena, then synergy has occurred.

There is one other observation that must be made to present high technology strategy as a definable and real concept. That is to show the three levels of strategy. Strategies fit within hierarchies, depending on the objectives. By considering the level of the strategy, we can put some needed clarification on the concept and practice of strategy.

The highest level is the *corporate* strategy level, which concerns what types of activities the business should be engaged in. This is an issue for a complex and diversified business such as General Electric, which must manage different and separate businesses, some high technology and some not.

Next, there is the *business* strategy level. The main issue here is how to compete in a specific industry or market. Many very large high technology companies have only a business-level strategy focus. Cray Computers operated as a single business—that of supercomputers—with only a business level strategy. It was not in multiple businesses, which are addressed by corporate-level strategies. (Note: The hierarchy names are by no means ideal, but they have stuck. We might be confused about the use of "corporate" strategy, since it would seem to apply to corporations even if they are single businesses. However, we stick to convention for the sake of remaining consistent with the field of strategic management.)

The third level of strategy is the *functional* level. It is the concern of functional units of the organization, often the domain of different departments. In the high technology organization, an example is the marketing department's strategy to introduce a new spreadsheet product.

All three levels of strategy must fit with each other and be bounded by each other. That's easy enough to say, but the congruence of strategies at different levels is one of the most troublesome areas of effectiveness for high technology organizations. We show how this can be a problem in advanced technology firms later, as we review product vs. market orientation.

The levels of strategy that we deal with here are the business level and the functional level. The exceptions of corporate level will be noted as needed.

Part of our task was to find out if there are differences in strategy between high technology and traditional firms. One important study has shown that there are major differences. Covin, Prescott, and Slevin compared 52 high-tech and 61 low-tech firms. They found major strategy differences. The high-tech firms tended to have entrepreneurial management styles, had freer-flowing organic organizations, a mix of what the authors call microstrategy variables that emphasize advertising, innovative marketing, patents and copyrights, new product development, operating efficiencies, industry awareness, external financing, extensive warranties, high quality, and high price.[5]

Strategy Choice

A high technology company makes a choice of strategies even if it is an unarticulated choice. There is no escaping strategy choice—all technology firms follow some path to some objective. But it can also be said that strategy choice happens in dramatic and widely separated points. If strategy happens at different levels and is done by different functions, choosing a strategy is often so continuous an action that it cannot be confined to a single point in time. Instead, for the established company it is a process of frequent readjustment—but it is strategy choice nonetheless.

So what does a high technology company or any company do about strategy choice? It invents it and executes it. To be hoped, it also evaluates it. These three actions capture the way strategy is done by technology organizations. We will concentrate on strategy invention (or formulation) and execution.

Strategy Formulation

Strategy formulation is posed as strategy choice for the new firm and as strategy revision for the more mature company. One distinct advantage of being a new high technology company is that there is a choice of strategies that can be pursued. The primary business-level strategy can be conceived and put into the business plan well before business operations begin. In fact, venture capitalists often insist on a strategy statement before considering any funding. A further advantage for the new technology firm is that the founders have the luxury of planning a strategy before they enact it. Planned in advance, the strategy can be critiqued by allies of the new organization before it is implemented.

Strategy is derived from the Greek meaning "art of the general" and the art referred to here is the task of creating a plan of action that ensures the survival of the firm. This is to say that the war is planned by the technology general and the battle is fought by the technology sergeants. So the captains of industry are really its strategy generals.

For the established technology firm, the essential strategy choice is the maintenance of the business level strategy or a departure from it. Abandoning the initial strategy choice is warranted when the initial choice proves not correct or if external circumstances require a change for the firm to survive. It is also possible and desirable that both types of departures may be required—that is, correcting the initial choice of strategy to accommodate new realities in the organization's environment.

Consideration of the kinds of strategies that high technology firms pursue would logically start with a search for strategies within the field itself. Unfortunately, there is no *Baedeker* for the explorer. The practice of high technology strategy formulation remains uncodified and ambiguous. Ideas for initial strategies may be copied from successful firms or concocted by particularly adventurous technology entrepreneurs. Suggestions for strategy are certainly passed by word of mouth among the founders, venture capitalists, potential customers, suppliers, and others with an interest in building a technology organization. But the practice has not been well recorded. Furthermore, it has not related particular strategies to particular predicted performance outcomes.

Instead of trying to uncover an oral history of high technology strategy advice, we need to step into the academic field of strategic management to discover usable generic strategy schemes that can be applied to high technology strategy formulation.

Without examining specific situations for new technology companies, we see that the number of strategy choices is not unlimited. It may also be apparent that the optimal strategy in most situations is a focus strategy. This has a number of advantages. It builds on the new firm's rather narrow initial range of skills. It also helps avoid direct competition with major competitors although as a practical matter large competitors can simply wait out the highly probable extinction of a new technology firm. A focus strategy also provides a good base from which to expand to undifferentiated strategies or other variations. Put another way, a focus strategy is flexible enough to be abandoned if new opportunities arise for the company.

Having introduced the general reasons for a focus strategy, we now offer a cautionary note. Strict adherence to a focus strategy over a long time, however, can hamper the ability to see emerging competitive forces.

Should Technology Drive Strategy?

In developing their strategies, high technology firms face a question that most other firms merely flirt with. What is the role of technology in formulating and executing a business strategy? The degree to which high technology companies incorporate technology into their strategy making can have an impact on organizational performance. Technology is a powerful force for change, in and of itself.

It could be said that all strategies should be hinged to technological development. But a firm that bases its corporate strategy on its technological abilities faces the danger of tunnel vision, gaining a distorted impression of the superiority of technology in the market. It may also fail to see opportunities because of what it falsely interprets as the superior technology of its competitors. Either of these perceptions produces something akin to Theodore Levitt's "marketing myopia"—in this case technology marketing myopia, a shortsightness that comes from making assumptions about market conditions, not relying on facts.

The opposite extreme is when a high technology company does the utterly incongruous: eliminates the consideration of technology in formulating and implementing strategy. Such might be the case in larger firms, where corporate strategy creation is distant from the technology-driven business units, where corporate strategic goals are aimed totally toward profitability, or where there are conflicting strategies that general managers fail to resolve.

Strategies and technologies have interactions and some effects of these interactions have been reported. Itami and Numagami (1992) reviewed the effect of technology on current strategy, the effect of current strategy on future technology, and the effect of current technology on future strategy. They found that strategy capitalizes on technology, strategy cultivates technology, and technology develops an understanding of strategy. As the interactions progress from first to last of those mentioned, those interactions become more inventive, development, process, and organization oriented. The findings also depart from classical economic theory with their progression through the interactions.

High technology firms often follow the first path—technological dominance. This is especially true of entrepreneurial technology firms, whose original impetus is often a particular and specific device or software. The founder remains close to the invention and they organize the company around it. The innovation is their baby and they view the baby's world through the eyes of a proud parent, not as an impassive corporate manager. Quite naturally, the technology is the driver for strategy. What market shall I be in with this? How should I enter the market? These are the strategic scope questions of the newly formed technology firm. These questions often linger for established companies especially as entrepreneur-founders are inclined to persist with initially successful strategies.

The connection between strategy and technology has not been studied extensively yet for the high technology firm. But there can be no doubt about the potential importance of this connection. Suitable business strategies are the better past decisions of successful companies as well as unsuccessful ones. Boeing pursued a commercial aircraft focus market while North American Aviation went unsuccessfully into many different aircraft markets.

A reasonable objective is to develop a framework for the analysis of the strategy-technology connection for both entrepreneurial firms and mature high technology firms. A dual approach recognizes the differences between the two types of firms in separate stages of development. As such, in this book we consider the strategy issue from the two perspectives.

Corporate Strategies in Established Companies

Earlier we mentioned that some technology companies serve as good examples of how strategy is done. Our first such example is the General Electric Company. What does this company do to get the attention it receives concerning strategy formulation?

In a well-established, technology-driven company such as G.E., the choice of corporate-level strategy is a function of its size. Size alone determines part of the strategy development. With 14 separate businesses, the company is the largest diversified business in the United States. It shows the challenge of fitting technology strategies within the context of a diversified corporation. This is high technology strategy development at its most complex level.

How did G.E. corral its 14 separate companies? There are three "circles" that circumscribe the 14 business groups: technology, services, and core manufacturing. The three circles are sufficiently different to merit three strategy orientations.

For example, one strategy is to continue with sufficient sales revenue to fund ongoing research and development. The research and development effort is itself focused and does not simply conduct basic research because it has a sales revenue–secured budget. This applied to all circles. Research is directed into several prioritized areas. And its own research dollars are combined with external sources to produce the directed research. This can be read as an example of synergy. The researchers at G.E. have an enforced flexibility. They are often reassigned to new areas.

The task of meshing technology and strategy is not neglected at G.E. Before this strategic business unit approach, planning was not as integrated with the business units. G.E. now considers technology from the point of view of how it helped the entire business and how it would shape new business. All corporate business functions were linked through a "R&D Center" funnel. However, technology strategies are not their own end. They are part of overall corporate strategy. That strategy reflects both the 14 strategic business units and the 3 major circles.

G.E. is not a typical high technology company. Much more typically, there is far less formalization or complication to strategy formulation. In the typical small high technology company, the CEO decides on a particular strategy and executes it. G.E. has since varied its circles approach, but the strategy still demonstrates the kind of strategy innovation that originates in a complex corporation.

Strategic Management, Planning, and Analysis

If strategy is a part of the whole organization, how is it done? In the technology companies we interviewed, it is accomplished through strategic management.

The management of strategy is part of the technology corporation's strategic plan. This strategic plan is the official register for corporate, business, and functional strategies for a set period of time. But the strategic plan is not immutable simply because it is written. In fact, most high technology businesses stray from their documented plans as new realities impinge from the outside or sales forecasts fall short of expectations. For new technology ventures, the strategic plan is most often the business plan. And that business plan evolves into the strategic plan for mature high technology companys.

One of the most useful devices for the analysis of strategy, and one we employ here, was devised by Porter (1980), and is known as the Five Forces model.

The Five Forces model puts strategy into a dynamic realm. Prior strategy analysis took place in a much more rigid and formalized way. The Porter model is a breakthrough in strategy analysis. Prior analysis would examine business functions and policies. Business-level strategy came from the interactions of functional units such as marketing, finance, and operations. Thus, strategy had an internal bias. But Porter depicted the strategy-making process as dynamic. He also brought in the effects of the external corporate environment.

The Porter model is useful because it has been popularized—more so than any other theory of strategy formulation and implementation. Many high technology managers are familiar with it and it is applicable as a practical model as well. It is useful because it collects the mix of forces that shape high technology businesses without being unduly pre-scriptive about their application. Another advantage is that it is somewhat nonsequential, meaning that various forces can have impacts on the business at various times. It also has the strength of focusing at the industry level (which may be a weakness if the industry is not well defined). It develops the idea of strategic groups that influence all business compet-itors. Lastly, it emphasizes the proactive, political nature of the five competitive driving forces. It does not, however, give any weight to management values and social responsibility. It should be read as an econ-omist's model.

Our look at the real world of high technology incorporates all factors of the Five Forces model: suppliers, customers, rivalry among high technology firms, new firms coming in, and substitutes for products and serv-ices. Among high technology executives we talked to, a majority were familiar with the basics of the model and said that it was helpful in giving a large view of their businesses. They used it as a checklist when reas-sessing their strategies. What was especially important for them was the entry of new technology firms into their field. They had their antennae up for new companies constantly. On the other hand, these executives were not very aware of the threat of substitutes. This may be a symptom of the myopia that technology innovators develop, thinking that their technology is the best that will happen. Even so, they conceded many examples of customer-preferred substitute products that came from out-side their own industry. An example is composite materials, which fun-damentally changed the performance of small private aircraft. For this and other reasons, the Porter strategy model has use in the high tech-nology sector.

There is a classification system for business strategies that incorpo-rates the range of choices high technology companies face (Hofer and Schendel, 1978). The first is a *share-increasing* strategy aimed at increas-ing the market share of the business. The second is a *growth* strategy aimed at maintaining the company's current competitive position in a rapidly growing market. The third is a *profit* strategy. While all companies seek profit, as products and markets evolve, profitability becomes a determinant of survival. Still another strategy is *market concentration and asset reduction*. Market concentration can be evident in having a focus on market segments, particular groups of customers.

The last two strategies are for high technology businesses not doing well. The first is a *turnaround* strategy, designed to reverse poor performance. Very simply put, tries to save a high technology company in performance trouble.

At the most dire extreme is a *liquidation and divesture* strategy. Its purpose is to make cash as the company leaves the industry. For all the growth in the high technology sector, there are companies that cannot be saved—Osborne Computers comes to mind. In high technology, successful liquidations enable persistent entrepreneurs to start again.

These all have elements of strategy types, but together they depict the range of strategies used by high technology businesses. There are other strategy classifications, but none represents the wide range of choices nor are as sensitive to the ways markets evolve in the high technology sector.

An additional strategy management issue is the lifetime of a technology, which relates directly to the lifetime of a product or service. Our general sense is that technology lifetimes are shorter than in the retail and industrial sectors. It's true that fashion may change, but only the color and lay of the fabric changes, not the loom. Technology shifts change the fabric itself and cast away the loom.

The S-curve of Foster (1986) applies here. It can be described as a situation in the industry in which there is slow technological progress, then rapid improvement, followed by slow advancement as a product ends its useful life and a new technology leads to a different and more innovative product. Returning to our survey respondents, we asked if this this description of product lifetime reflected conditions in the respondent's industry (without our indicating the curve as an S). The result was lukewarm endorsement of the idea that S-curve elements apply. In fact, 53 percent found the elements evident in their experience while 47 percent did not.

Overall, the S-curve was not a well-observed occurrence in this group, even if it is demonstrable by other means or even if, in fact, it is accurate. There is probably limited application of the concept, since technologies do not decelerate evenly—their last stage may be quite protracted. It has also been difficult to define *innovation* in such a way that the mature invention resembles the early invention.

When it is applicable, however, the S-curve can explain when a substitution might occur: when much effort is made to improve something but it gets no better—say, the way vacuum tubes were tinkered with in the 1950s. Substitution may also happen for nontechnical reasons—for example, consumer whim.

Strategy Types and Selection

What strategies do established high technology businesses employ? Table 12.1 shows the different types.

The popularity of a strategy for increasing market share is sensible, since most of our sample companies were in an intensely competitive arena—software services. Growth and profitability strategies also generally fit high technology businesses. It is quite interesting, though, that a sizable number of high technology companies (11 percent) recognized that a turnaround strategy was appropriate. While we may think of high technology as a fast-growing and successful line of operation, for about one in ten firms it is a struggle.

These results show the usefulness of a strategy classification scheme. All but 5 percent of the surveyed companies were able to assign themselves one of the strategy choices provided.

Because high technology businesses are especially sensitive to their environments, their strategy orientations reflect this sensitivity. We also noted that not all technology companies are in industries exploding with growth. For instance, in the electronics components industry, firms like Allen Bradley of Milwaukee have produced the same, simple carbon resistors for sixty years. Industry sales are not expanding and are quite stable, even though other sectors of electronics have grown. The same circuit board may have many generations of microprocessors but a single generation of resistors.

We used a strategy classification scheme that covered industries at different evolutionary stages. As explained earlier, this was captured in Hofer and Schendel's (1978) description of generic strategies. This classification scheme was the one utilized in our study. The scheme fit a few critical criteria: it made sense to a few technology CEOs with whom we reviewed the scheme and we saw how well it captured almost all of the respondent's strategy choices. The categories were distinctive from one another, and it factored in time, a very important aspect of technology firm development.

Other Views on Technology Strategy

One perspective that counters the view that strategy making by the technology business is homogeneous, without separate effects of technology impacts on strategy has been studied. The impact was found to lead to more success among defenders rather than aggressive, prospecting firms, according to Dov Dvir, Eli Segev, and Aaron Shenhar (1993). There is another cut at the technology strategy, done by Feeser and Willard (1990). The authors looked at high- and low-growth firms in a single industry, and discovered that initial strategies differed substantially among the companies. Strategies were important in explaining growth, at either degree of growth.

Table 12.1 Strategies Used by High
Technology Firms

Strategy	Percent Using
Market share increasing	34.8
Growth strategy	21.7
Profitability	21.7
Market concentration	4.5
Turnaround	10.9
Liquidation/divesture	0.9
Other	5.4

Competitive advantage—the ability to keep superior skill, technology, or production efficiency over competitors—is a common corporate strategy. One question frequently raised is whether that advantage can be sustained or if it fades away. This has not been extensively investigated; however, in an article by Doutriaux (1992), the subject is illuminated somewhat. The author considered 72 Canadian high technology firms and found that initial sales levels and initial government orientation were apparently two longstanding startup effects. These factors endured longer than initial capital, past experience in marketing and finance, founder age, or similarity with market served by the previous employer.

There have been attempts to explain why some companies respond well to technological change and others don't. One example is Nicholls-Nixon's (1995) results. The author acknowledges the turbulence of the technology marketplace, partly owing to obsolescence of technical skills. "Absorptive capacity"—the ability to get new technology information—is seen as critical for reacting to change. What determines absorptive capacity is the firm's research and development investment, allocations of research and development resources, and interaction with outside sources through joint ventures, contracts, and the like.

Burgelman and Grove (1996) (Grove famous for starting and leading Intel and Burgelman prominent in the academic field of business policy and strategy) have introduced the concept of strategic inflection points (SIPS), which are caused by changes in industry dynamics, winning strategies, and dominant technologies. SIPs generate strategic dissonance in the firm because they are associated with the divergence between the basis of competition and the firm's distinctive competence, and between top management's strategic intent and strategic action. Top management can take advantage of the information generated by strategic dissonance to develop new strategies and lead the organization through the uncertainty associated with SIPs—but this requires a capacity for strategic recognition by top management, which is accomplished through an unbiased internal selection process. Strategic recognition is the foundation for exerting strategic leadership; it is encouraging and concluding debate at a point that establishes a basis for competition and distinctive competence and strategy and action.

Strategy and Performance

Returning to the Hofer and Schendel (1978) generic strategy classification model, we assessed the idea that high technology firms that use strategies that fit their position in the industry life cycle do better than those that do not have such a fit.

The first step was to see if our sample firms had a good match between business strategy and stage of industry life cycle. When the share-increasing, growth, profitability, concentration, turnaround, and liquidation/divesture strategies were compared to the life-cycle stages of introduction, growth, maturity, and decline, there was a highly statistically significant match. Thus, growth strategies were found in growth stages and so on.

The next step was to find out if these fits made a difference in sales performance. In fact, the respondent businesses that used the appropriate strategy-life cycle match had better sales growth than those that didn't have such matches. This was found to be statistically significant (though at the .066 level). This does indicate that business strategy is important for the established high technology firm.

Summary

This chapter concerned high technology business strategy. A definition of strategy was developed, and the elements of strategy were detailed, along with the levels of strategy. We included the thoughts of high technology CEOs about strategy and also reviewed the issue of technology and strategy. The Porter Five Forces model was described and the Hofer-Schendel strategy classification scheme was detailed and applied to our survey responses.

The S-curve of technology innovation was described, and established businesses provided only mild agreement with the S-curve concept as they observed it. In examining the connection between the type of strategy and sales performance, it was found that high technology firms that matched strategy to their industry situation did better. Strategy counts for the established high technology company.

Strategy is how the technology firm interacts with its environment. In the next chapter, we will discuss the business environment for established high technology companies.

Notes

1. This definition is contained in Jeffrey Harrison and Caron St. John, *Strategic Management of Organizations and Stakeholders: Theory and Cases* (Minneapolis-St. Paul: West Publishing, 1994).

2. As presented by Michael Hitt, R. Duane Ireland, and Robert Hoskisson, *Strategic Management: Competitiveness and Globalization* (Minneapolis-St. Paul: West Publishing, 1995).

3. This is the definition of Arthur Thompson and A. J. Strickland III, in the seventh edition of *Strategic Management: Concepts and Cases* (Homewood, IL: Irwin, 1993).

4. A more complete discussion of these characteristics, as well as a conceptual development of strategy for organizations, is found in Charles W. Hofer and Dan Schendel, *Strategy Formulation: Analytical Concepts* (St. Paul-Minneapolis: West Publishing, 1978). This work is the basis for our depiction of strategy in this chapter.

5. J. Covin, J. Prescott, and D. Slevin, "The Effects of Technological Sophistication on Strategic Profiles, Structure and Firm Performance, *Journal of Management Studies* 27 (1990): 5. Note the number of marketing-related microstrategy variables considered by the authors.

References

Ansoff, H. I. (1987). "Strategic Management of Technology." *Journal of Business Strategy*. 7: 28–39.

Bower, J., and C. Christensen (1995). "Disruptive Technologies: Catching the Wave." *Harvard Business Review* 73: 1, pp. 43–53.

Burgelman, R., and A. Grove (1996). "Strategic Dissonance." *California Management Review* 38, no. 2 (Winter), pp. 8–28.

Doutriaux, J. (1992). "Emerging High-Tech Firms: How Durable Are Their Comparative Start-up Advantages." *Journal of Business Venturing* 7: 4, pp. 8–28.

Dvir, D., E. Segev, and A. Shenhar (1993). "Technology's Varying Impact on the Success of Strategic Business Units Within the Miles and Snow Typology." *Strategic Management Journal* 14, no. 2 (February): 155–162.

Feeser, H., and G. Willard (1990). "Founding Strategy and Performance: A Comparison of High and Low Growth High Tech Firms." *Strategic Management Journal*, February, pp. 87–98.

Itami, H., and T. Numagami (1992). "Dynamic Interaction between Strategy and Technology." *Strategic Management Journal Special Issue*, Winter, pp. 119–135.

Foster, R. (1986). *Innovation: The Attacker's Advantage*. New York. Summit Books.

Hofer, C., and D. Schendel (1978). *Strategy Formulation: Analytical Concepts*. St. Paul-Minneapolis: West Publishing.

Levitt, T. (1975) "Marketing Myopia" *Harvard Business Review* September-October, pp. 26–44, 173–181.

Nicholls-Nixon, C. (1995). "Responding to Technological Change: Why Some Firms Do and Others Die." *Journal of High Technology Management Research: Special Issue on Entrepreneurship and Technology* 6: 1, pp. 1–16.

Porter, M. (1980). *Competitive Strategy*. New York: Free Press.

13

ORGANIZATION AND ENVIRONMENT

This exploration of established science and engineering-based companies has thus far dealt with human resources, the CEOs, and strategy. While certainly critical subjects, they are not the only determinants of what technology companies do, how they do it, and why. In this chapter, we move much further along in that direction by considering factors that can be assembled under the broad categories of organization and environment. Part of our aim is to discover if the environment has an influence on organization; thus, there is a link here between the immediate environment and organization, even though they may seem to be independent factors.

By immediate environment, we mean the world in which the company dwells. That is not just a matter of where it is domiciled, though that is important, but also conditions that the company cannot control in a major way, such as the cost and availability of labor, community life, attitudes of people in the community, governmental regulation, legislation and taxation, the practices and disposition of the courts, and abundance and quality of natural resources to name several. A key point is that these factors change, and that change affects the prosperity of the business of technology without equal reciprocal consequences.

The effects of the environment are contained in the industry life cycle model, that powerful wave of introduction, growth, maturity, and decline that frames our overview of these business. Environmental factors combine with economic factors to drive the overall industry and individual businesses, one way or another, toward more growth or less, toward profit or loss. These various forces are known as *driving forces* in the field of strategic management. As described by Porter (1980), these drivers are industry growth rate, customer composition and usage, innovation, technological change, technology dissemination, marketing innovation, entry and exit barriers, globalization of an industry, cost and efficiency changes, regulation, socioeconomic changes, and business risk reduction. Notice that three of these drivers—innovation, technological change, and technology dissemination—are outputs of the very businesses we are looking at. To an extent, then, technology businesses can shape their own drivers, though the effects of single business are marginal.

How are these companies organized in their environments and does that make any difference in their business performance? By looking at the immediate environment, we examine the connection to proximate factors such as the firm's suppliers and economic conditions and other strategic drivers. We will not explore larger environmental conditions, such as macroeconomic issues of fiscal and monetary policy, political and international trade issues, national socioeconomic trends, or cultural factors that affect high technology. It is little concession to say that these are important—which the are; it is more to admit that they have been excluded owing to practical necessity, of putting an outer limit on what constitutes a high technology firm's environment.

One of our projected outcomes is a characterization of the relative sensitivity of these organizations to their environment. How do high technology firms compare to other companies in either reflecting or shaping what happens in their environment? Since this was not specifically appraised in our survey research, the conclusions are based on observed tendencies in the 17 firms we visited and among the managers we spoke with.

At the outset, we learned that these companies in our broader survey research are independent operations rather than units of other companies. At least this was the case for 82.4 percent of them, while 14.2 percent said they were units of other companies. By and large, these are organizations making their own destinies, at least as far as accountability is concerned.

These companies are privately held in 86.7 percent of cases. Their shares are not publicly traded on a stock exchange. This further establishes the companies as their own organizations. With these two factors in mind, we studied companies that are not bound to the vagaries of public trading nor the hot breath of a looming parent firm.

Business Structure

The organizational structure of a technology business can be thought of as being the way the organization puts it people together to mesh with its environment. Marketing deals with customers, Legal with government, Purchasing with suppliers, and so on.

The structure of the high technology company is truly a cloak of many layers. It is often visible, like a cloak, because the structure is usually well publicized—organizational charts are often available on request. The "many layers" refer to the changes that occur in organizational forms, from venture to established firm. Young high technology companies have simple structures, while mature companies take on complex forms. Structure evolves as one form is shed, replaced with another more suitable for the present business climate.

We found earlier that technology ventures say they take on organic structures—flat, horizontal structures with little hierarchy, job duties are flexible. Since the established firms in our sample tended to be small companies, there was a similar expected result. Small firms in any industry do not build hierarchies; they don't have the problems that require top-heavy structures.

Casting back to our findings on high technology ventures, there are some points on structure that can be carried over to established firms. The venture founders provided diagrams of their organizations during their first five years of operations. Some founders went to considerable effort to draw organizational charts. The changes were quite noticeable. One started with a conventional fork-type chart with functional management, but within a few years the organization was depicted as a series of concentric circles with the CEO in the nucleus, a free-form effort at displaying an organization where people support one another—or at least that was what was intended to be conveyed to outsiders.

It might be expected that the growing technology company leaves that organic structure to adopt another that better coordinates the different functions. If so, the established high technology company would resemble its more conventional counterpart. The results of our research were solidly on the side of flat structures rather than hierarchial structures. Companies that had flat structures constituted 77.3 percent of the total, while companies with hierarchical structures made up 11.2 percent. The rest had either other structural forms or did not respond.

While the companies in our database are fairly small, and thus more likely to have organic structures, it was nonetheless striking that the organic form is still so dominant. Most likely it was preserved because it was the base on which the company was built.

It is not really enough to say that these companies are mostly organically structured with flat reporting arrangements rather than hierarchial structures. While comparison to ventures was helpful, in practical terms a simple dichotomy is not useful. CEOs would have a difficult time classifying their companies using such simple terms. High technology companies, and most other companies for that matter, use a different way to organize themselves. Put in its most basic way, organization can be by customer, product, geography, function, or some combinations of these. Thus for our research we grouped customers and products together. The results of asking how the companies primarily organized themselves are

Table 13.1   High Technology
Organizational Forms

Organization Type	Percent of Companies
Customer/product	47.6
Geographical	2.1
Function/division	41.6

shown in Table 13.1. No responses and inapplicability of the question accounted for 8.5 percent of the replies.

The conclusion is that almost all of these established companies are organized on the basis of customer, product, or function. Geographic organization was not expected to account for so little; this may again reflect the relatively small size of these companies.

The results suggest that established high technology companies use conventional ways of organizing themselves, which removes the organizational mode as a way of differentiating these businesses. As we looked at high technology organizational structures apart from our group, there was confirmation of traditional structure. There were only a few exceptions and these were a matrix-managed companies or very small software consulting businesses with one customer and one product. Thus, something can be said for the convergence of high technology and general business as far as organizational form is concerned.

The next concern was whether any organizational structures have performance consequences. By splitting sales growth into the classifications of greater growth and lesser growth, and further dividing the structure into customer/product, geographic, and function/division groups, we found that geographic organization had lower sales growth than did either of the other forms; 75 percent of geographic firms had low sales, while 61 percent of the other organizational types had low sales. However, there was no statistical significance to the differences, and we cannot conclude one form is related to better sales performance.

When the database was sorted into the two-digit SIC classifications, in order to see if the results from the entire group also applied to the business subgroups, there were an insufficient number of cases to draw any conclusions for all but one of the categories. In computer programming, there was confirmation that the form of the organization did not matter in terms of sales results.

Competitors

The factor of competition is integrated into almost every framework for analyzing an industry, either explicitly or implicitly. In Chapter 7, we discussed this from the perspective of strategic management. Porter

Table 13.2 Percentage Market Share of Firms and Competitors

	1993	1994	1995
Your business	22.7%	24.6%	26.0%
Primary competitor	28.4%	28.2%	not reportable
Secondary competitor	13.3%	not reportable	13.2%

(1980) considers competitors a major force in his Five Forces model of strategic management, which we referred to in Chapter 12.

We let the companies in our survey group decide who their competitors are, instead of relying on indirect sources. Accurate source information in this field is really nonexistent. Besides, if the firms themselves don't know who they are competing with, other ways of obtaining this information are problematical.

How much of the market do these surveyed companies command in relation their competitors? The percentage estimates for their narrowly defined markets is shown in Table 13.2. Data were not reportable when there were problems with the reported information (e.g., market share means above 100 percent).

These sample firms see themselves as smaller but more effective than their primary competitors. They believe they have less market share, but that they are gaining in relation to a primary competitor. The secondary competitors are judged to have significant but unchanged market shares.

This information needs to be taken, not as a definitive depiction of market share, but as an assessment of market positioning. It is doubtful that these firms accurately measure their market share or can access an outside party that does. But the positional measurement—their market share versus ours—is useful information. It does not have to be absolutely accurate to be relatively helpful, and that is the way we take it.

To help complete this picture of competition, we asked our sample firms what kind of a threat competitors were to their businesses. This time we assigned adjectives rather than numbers to the categories: serious threat, major threat, significant threat, and not a significant threat. The companies responded to the degree of primary competitors and secondary competitors as shown in Table 13.3.

Fifty-seven percent of respondents rated their primary competitors as at least a significant threat. Only 42 percent of respondents considered their secondary competitors as at least than a significant threat. In a way, that is why they are secondary competitors, but the responses still gave us an indication of the degree of concern these companies have about their competitors.

From the market share and the competitor rating information, it was clear that high technology companies are concerned with—and should be concerned with—their primary competitors. Secondary competitors

Table 13.3 Degree of Threat by Competitors

	Primary Competitor	*Secondary Competitor*
Serious threat	14.6%	4.7%
Major threat	10.7%	9.0%
Significant threat	31.3%	29.2%
Not significant threat	28.3%	40.3%
Not applicable/missing	16%	16.8%

are also important, but they do not pose as immediate a threat. These firms are very aware of their competitive position, as well as that of their closest rivals. In this respect, they are sensitive to this aspect of their environment.

Yet another way of gauging the depth of competitive sensitivity was to ask the companies about reaction times to consequential market events. The question was, How long does it take for competitors to react to the new products of a technology company? The companies reported as shown in Table 13.4. Twenty-three percent of the companies did not respond to this question.

It was between one and three years of reaction time that accounted for the largest percentage of replies. It is especially consequential that almost 20 percent of the firms discovered a competitive reaction within a year; about two-thirds evoked competitive reaction to the new product within three years. This is an indication of just how competitor-sensitive these high technology companies are.

While the nature of reactions to competitors was not cited, it could have ranged from very mild to very serious. A mild reaction could be a competitor's pricing change while a serous reaction could be the introduction of a head-on competing product or service. Whatever the case, these companies reported some type of competitive reaction, so these companies are crediting their competitors with quick reflexes.

The reverse side of the competitor reaction to new products is the reaction time of the company itself to what a competitor does. Do these companies react quicker or more slowly? When asked about their reaction time to new competitive product introductions, the responses were varied

Table 13.4 Competitor Reaction Time

	Percent of Firms
Less than 1 year	19.3
1 to 3 years	44.2
More than 3 years	13.7

Table 13.5 Reaction Time of the Firms to
Competitors

	Percentage of companies
Less than one year	39.1
1–3 years	29.6
More than 3 years	5.2

and the results are summarized in Table 13.5. Missing and nonapplicable responses were 26.1 percent of the total.

These companies have depicted themselves as very fast responders to competitor product introductions. With 68.7 percent saying they respond within three years, these firms believe they are considerably faster than their competitors. The nature of the reaction could be the introduction of a similar product, or it may be a change in the way an existing product is marketed. We believe these are substantive reactions, however, because these companies took competitive moves very seriously. When competitive reaction was observed first-hand, the immediate reaction was to screen for a new product rather than change a pricing arrangement.

What has emerged is the distinct impression that competitive actions and competitive reactions matter in the area of new product development. They certainly matter to managers we interviewed. So many of these companies survive because they effectively serve a market niche, and if competitors pour in with superior products, that niche advantage becomes imperiled.

New product offerings are the manifestation of the competitive character of this industry. The dynamics of introduction and responses, because they are rapid, give it further distinction. It is notable that an industry that presumably is research and development driven can respond so quickly to competitors. There maybe a resonance of technological knowledge or companies may simply believe it is essential to respond and are willing to make the necessary investment to do that.

A curious aberration of the commonly accepted idea that high technology companies compete without equivocation was provided by David Packard, in *The HP Way* (1995). Packard told of working out a cooperative relationship with competitors on product introductions. Hewlett-Packard and a competitor would both offer similar products to the same customer. The rationale was that some customers needed the reassurance that at least two companies believed enough in the offering to market it. That was new product collusion for the risk-averse customer. Packard reported it worked.

While we have provided a broad overview of high technology competition, we must acknowledge that actual competition can be much more complex. Take the semiconductor business. There are at least three

levels of competition: a captive market in which the microprocessor manufacturers resell to computer manufacturers who are vertically integrated
with one another; a merchant market in which large microprocessor manufacturers produce industry-standard circuitry as a commodity; and a
merchant market for specific applications.[1] Competition is mainly within
the different levels and not among the levels. As an example, The merchant market for specific applications competes within itself more than
with the other levels.

Getting Into the Arena

We addressed the matter of barriers to entry in Chapter 8 regarding new
ventures. The amount of capital needed to start a technology business
could be a barrier for a new firm. Technologically superior products may
not be enough. If the present players in the market are there because
they have met high initial capital requirements, then they enjoy a market
advantage when other companies cannot meet this requirement—for example, if very expensive equipment must be purchased. As we found,
ventures that overcome less severe capital barriers to entry have better
sales performance than ventures that have to overcome more severe barriers to entry. In Chapter 4, we approached the issue of capital from the
perspective of venture capitalists, and we established that the process of
funding ventures was done informally but carefully by venture capitalists.

Now, we want to determine what access to capital means for established technology companies. Since these companies have already entered
the market, our concern is not with capital as a barrier to entry but with
it as a more or less important ingredient for the ongoing firm. In pursu-
ing this idea, we asked our survey sample how important access to capital
was for them. The results are reported in Table 13.6. Missing replies
amounted to 6 percent of the total surveyed.

The responses were unevenly distributed. While the mode was that
capital access was very important, the second most popular response was
that it was neither important nor unimportant. It is thus not possible to
derive an overall conclusion about capital access among these technology
firms.

This consideration of capital needs leads to the question of whether
access to capital yields any difference in sales growth. For this, respondents who stated that capital access was very important or somewhat
important were compared to respondents who stated it was very unimportant or somewhat unimportant. The results showed no significant difference between the two response categories for the total of technology
firms surveyed.

The story was different for computer firms, where there was statistically significant difference between the two response categories. Companies that regarded capital access as less important had better sales

Table 13.6   Importance of Access to Capital
for Established Firms

	Percent of companies
Very unimportant	9.9
Somewhat unimportant	16.7
Neither	24.5
Somewhat important	17.2
Very important	25.8

results. In the other subsets of technology firms, there were not enough
example firms to confirm or deny this association.

Riding the Industry Wave

Where do these companies see themselves in relation to the industries
they inhabit? How does a computer manufacturer see itself in relation to
the computer industry?

When the sample companies were asked if their businesses were
growing faster, at the same rate as the industry, or slower than the industry, there was a descending order of response led by growing faster
than the industry (39.1 percent), growing at the same rate (33.5 percent)
and growing slower than the industry (22.7 percent).

This information is intriguing in several respects. First, more companies of our sample believed they were growing faster than the industry
as a whole. There is already extraordinary growth in one of the industries—computers—of which we have many companies in our sample. In
fact, just over a third of *Fortune's* October 14, 1996, list of fastest growing firms were technology firms. This was, in fact, the single largest intact
industry category in the *Fortune* report on fastest growing companies. To
grow faster in an already fast-growing industry is indeed an accomplishment.

Second, for as much as these firms reported themselves as healthy,
robust, and fast-growing organizations, some made quite an honest concession by describing themselves as a business that was growing slower
than the industry. This is a telling observation. Almost a quarter of the
companies in our sample were not keeping up with their industry. The
companies reporting this were in some of the electronic components businesses.

This introduces an interesting sidelight to high technology in general—how some high technology businesses have made other high technology business outdated. For example, in the semiconductor business,
large-scale circuit integration forced resistor, transistor, and diode component manufacturers, and wire cable and connector manufacturers, into

Table 13.7 Where High Technology Companies Place Themselves in the Industry Life Cycle

Phase	1993 % of firms	1994 % of firms	1995 % of firms
Introduction	17.2	11.2	9.0
Growth	36.5	39.3	44.6
Maturity	24.0	29.2	23.6
Decline	13.3	11.6	14.6
No response	19.1	8.1	8.1

new lines of work. High technology is a ruthless displacer of the slow to adapt.

Irrespective of how individual businesses may be growing, another consideration is industry life cycle. To review the life cycle effect, we carry forward the conceptualization of industry life cycle that was applied to technology ventures in Chapter 7. Table 13.7 shows where our sample companies place their business in the industry life cycle.

For the most part, these companies believed that their businesses were in the growth stage for the three years pertaining to our study. When we put the introduction and growth stages in one category and the maturity and decline stages in another, we find that companies are in the first half of the industry life cycle.

Industry life cycle information can be put together with individual company growth rate reports to depict various combinations of industry and business conditions. For example, a growing business in a growing industry may be only keeping pace in relation to how it could be doing— market share needs to be attended to in this situation. Erosion of market share in a growing market equally would be a concern.

One facet of the industry has to do with the tempo of change. We would expect that in high technology, this tempo would be evident. Thus, how fast the industry changes and in what ways it changes is a defining dimension for high technology. That prompted us to look at aspects of industry change.

There were two dimensions of change that explored: substitution of new technologies by customers and capacity of the firm to adapt to technological change. These factors help distinguish this business category from others. Commodity businesses, for example, have little concern with technological substitution.

When we asked how important the customers substitution of technologies was for our sample high technology companies, our results were as shown in Table 13.8. Missing replies amounted to 7.7 percent of the total surveyed.

Customer technology substitution is important for these companies, with close to 60 percent placing this factor on the "important" side.

Table 13.8 Importance of Substitution of Technologies

	Percent of companies
Very unimportant	6.0
Somewhat unimportant	11.6
Neither unimportant/important	19.3
Somewhat important	28.3
Very important	27.0

Since technology substitution was important, we could expect that the customers of these firms might have ongoing searches for new technologies to apply to their businesses. This is illustrated by a bottling plant, which is constantly searching for more automated machines. A clear implication for the technology company is to keep up with customer technology substitutions.

We pushed the question directly with the responding companies, we found even more importance on the technological change issue. When they were asked how important their capacity was to adapt to new technological change, we discovered a profound pattern. The firms provided the replies shown in Table 13.9. We had 3.9 percent of the total who did not reply to this question.

There is an pronounced skew in this data toward the "very important" side. In fact, this is by far the most dramatic attachment to one side of the scale reported in this chapter. Very clearly, these companies believe that they need to adapt to new technology changes. This need to change does not solely apply to their R&D efforts but to their business operations as a whole. Thus, this represents a defining character of these companies and confirmation that they are truly high technology firms. If they were not, the skew would not be as evident toward the importance of technological adaptation.

These results provide even more reason to look at how high technology companies take in new technologies themselves. Though not part

Table 13.9 How Important Is Adaptation to Technological Change?

	Percent of companies
Very unimportant	3.0
Somewhat unimportant	5.6
Neither unimportant or important	10.7
Somewhat important	24.9
Very important	51.9

of our scope, the matter of what new technical systems and ideas are acquired and where they are implanted is a topic with considerable merit.[2]

Suppliers

The suppliers of high technology businesses range from raw materials providers, to silicon dioxide for semiconductor manufacturers, to other high technology companies such as avionics firms for finished commercial aircraft. So it is with this thought in mind that we discuss a generic concept for a broad business endeavor. The approach may be general, yet we knew that there was something useful here if we could grasp the topic and find the nature of technology businesses.

Our particular device was to pose the question of how many suppliers are internal or external. That answer separates suppliers into two different classes. It may seem that all suppliers are external because that is what they do—supply a firm from the outside or shovel coal into its firebox. But suppliers may, in fact, be internal if *internal* means another unit of the same company. When technology firms buy from their own subsidiaries, they have internal suppliers.

We found that 72.7 percent of our responding technology companies had external rather than internal suppliers. Much of this has to do with the small size of these companies (they are not likely to be such a market force initially that they can vertically integrate), but the response marks a pronounced disposition toward independent outside suppliers rather than internal suppliers.

Once we knew the suppliers resided outside the companies, we asked where they were located. In our earlier inspection of high technology ventures, we considered supplier proximity. At this point, we wanted to determine if this carried over to established companies.

The established firms we surveyed reported the distribution of supplier locations as shown in Table 13.10. About 14 percent of the companies surveyed did not reply to this question.

What clearly emerges is the conclusion that suppliers are not in the immediate vicinities of these companies. However, there is a sizable cluster of nearby suppliers—nearly 9 percent of suppliers were located within ten miles of the companies. And, as distant as the remaining suppliers are, they are located in the United States in 92 percent of the cases, according to our respondents.

The nature of suppliers needs to be discussed to some degree. It is not enough to say there are suppliers for high technology firms. Suppliers need to be characterized, at least broadly.

To begin, the kinds of suppliers could be material or natural. In considering the latter, we saw that an especially prevalent characteristic of high technology companies is their use of human resources in the supply role. This reference is to contract employees, especially in the com-

Table 13.10   Location of Suppliers

	Percent of companies
0–10 miles	8.6
11–20 miles	5.6
21–30 miles	6.0
31–40 miles	4.7
More than 41 miles	60.9

puter field. Contract programmers provide their expertise on an as-needed basis for these companies. Contract programming supplier firms have flourished, and so have supplier firms that install networks and hardware. So suppliers include human labor in our consideration.

Another way to be more specific about suppliers is to consider their number. What was discovered about the number of high technology suppliers? In this instance, we considered the number of suppliers for the industry itself, not for particular businesses. There was so much switching of individual suppliers that stability was found only at the industry level.

Our results showed quite a dispersion in the number of industry suppliers, as shown in Table 13.11. Missing responses were 11.6 percent of the total.

There is a curious feast or famine among suppliers, as reported by these technology companies. There are either fewer than 50 or more than 100 suppliers. The famine is in the midrange percentages, for the 50 to 100 supplier range.

Using the number of suppliers, and considering that supplier power derives from fewer suppliers, we cannot really say that supplier power exists. It does for some of the firms, but not for many others. As a result, these high technology companies in aggregate do not seem to be beholding to suppliers.

In our discussion of this topic with a few company heads, we noticed that computer businesses have more suppliers than some of the other business classifications. The number of suppliers is tied to the complexity of the product or the number of production inputs. Thus, the aerospace business has a large number of suppliers while biotechnology has fewer. In all, supplier numbers seem to be a way of distinguishing among technologies rather than a way of unifying them.

A substantial majority of the supplies needed by the high technology companies are provided by a few suppliers. Ten or fewer suppliers account for slightly over 67 percent of the supplies, so although the suppliers are many, they are also concentrated.

Turning to the matter of whether supplier arrangements make a difference, we divided the respondents into categories of more than 50 suppliers and fewer than 50 suppliers. Then the sales growth for each category was obtained, as done previously. The results showed no appre-

Table 13.11   Number of Suppliers

	Percent of companies
Less than 10	19.7
10–49	24.0
50–99	6.9
More than 100	36.1

ciable nor statistically significant differences between the groups based on the number of suppliers. Supplier number was not connected to sales performance.

Other Teaming: Teeming Possibilities

For most high technology companies, there are major risks in developing and offering new products. The consumers may not be ready. The product itself is untested in the market. The risks may not be all in marketing, either. They may also be in production, where it might be extremely risky to shut down a working line and set up for something that is complex to make and requires very tight tolerances.

These are some of the reasons why there has been a move to set up high technology colonies where small firms can share one roof, often literally, and operate as mall-like businesses. These colonies are found in Madison, Wisconsin, and, in Silicon Valley, and at other geographic technology concentrations. The firms may share advertising agencies and development laboratories.

The subject of resource sharing or joint ventures has recently provoked a host of articles, so much so that we group them by theme.

On *obtaining alliances* Anderson and Narus (1991), suggest that through partnerships, suppliers can leverage their resources through joint efforts with customers, can gain the benefits of customer ideas, and can obtain higher profits. On the other hand, transactional relationships let suppliers prune the relationships that customers see as undesirable. To keep the different types of relationships, a coherent strategic approach is required. The segmentation, targeting, and relationship-building approach has customer value as its cornerstone. But not all firms want the same working relationship. Managers have to decide which customers want collaborative relationships.

Dubini and Aldrich (1991) propose a set of navigating strategies for entrepreneurs. They recommend aggregating personal networks into extended networks. These networks can be analyzed within as well as between firms. Two principles that apply to personal networks are to systematically plan and monitor networking activities, and attempt to increase network diversity. The most effective firms are those where entrepreneurs do these.

On *changing and improving alliances*, Hamilton and Singh (1991), investigated the determinants of change in strategic alliances involving technological innovation. Changes in strategic alliances are categorized as either redefinitions of original purpose (in either application or scope) or dissolution. The alliance's original purposes influence, the likelihood of its change through redefinition or dissolution. Koh and Venkatraman (1991) considered the impact of joint-venture formation strategies on the market value of parent firms in the information technology sector. Their analysis found that announcements of joint venture formations generally increase market value. After that, joint ventures were found to be superior to other cooperative mechanisms. The magnitude and significance of market valuations differ across types of joint-venture strategies. The announcements and managerial assessments of the joint venture's contributions have a positive significant relationship for a subgroup of firms. Their analysis supported the use of stock referents for assessing performance, but highlighted the differences between intended and realized strategies. The results also showed the importance of technology competence.

What do technology companies say about sharing their resources? We asked the firms how important joint ventures and resource sharing were for their firm. The responses are shown in Table 13.12. For the question, 8.2 percent of the total did not reply.

The responses demonstrated that the opportunity to conduct joint ventures or to share resources in other ways is somewhat important or very important to only about a quarter of the companies. This didn't establish actual usage, but may suggest attitudes toward these opportunities. In all, there appears to be little utilization of cooperative arrangements with other technology firms, which leaves us with the observation there are teaming opportunities but nothing more for most of these companies.

Transfer of Technology

The transfer of technology is one factor that makes high technology unique. The industry as a whole was created because of technology transfers, as federal laboratories and universities disseminated their government-supported research to the private sector at the end of World War II, something we reported in Chapter 2.

Moving basic research into applied research and engineering is still what technology transfer is about, but it has become more complex. Federal laboratories now package their transfer operations. A technology transfer agent may be present to discover and publicize technologies that can be made available to private firms. A typical process might involve identification of a promising technical process or invention. The labora-

Table 13.12 Importance of Joint Ventures and
Resource Sharing

	Percent of companies
Very unimportant	20.6
Somewhat unimportant	19.6
Neither unimportant or important	26.2
Somewhat important	11.2
Very important	14.6

tory or university then patents the technology. Contact is made with outside firms, possibly through publication of technical notes or one-on-one meetings with companies. And sometimes the companies have their own people looking for promising technologies. Known as industry ferrets, they are common fixtures at the larger federal laboratories and research universities.

It is not just high technology firms that are the beneficiaries of transfers. For example, mining companies have made use of PID weighing systems, first conceptualized in universities. But high technology companies have a longstanding connection with these programs, so they are the most eager.

The path to commercialization from the technology transfer to the public sector to the private is not an easy street. Extinction of the transfer can be quite sudden and final. The recipient company may not be able to find a way to adapt the technology. There may not be a way of profitably incorporating it, even if it can be adapted. Then, of course, there may not be customer acceptance. Such perils have led some of the public agencies to lend assistance in these aspects by doing some market research themselves.

Based on our research survey, we can say that established companies are driven by technological advances, some of which come internally and some of which originate outside. Our interest was in how significant technology transfers are judged to be by these companies. In answer to this, we obtained the distribution of replies as shown in Table 13.13.

The very high percentage (54 percent) of "inapplicable" or "missing" responses prevented our making conclusions about the group as a whole, but for those who did reply, technology transfers were deemed to be significant.

Have these companies transferred technologies outside their companies? In 73.9 percent of the cases, the answer was negative. Technologies developed by these companies were largely held by the same companies. That is certainly expected, since the release of closely held technologies may be a competitive disadvantage, or it may be the technology was simply not worth transferring outside.

Table 13.13   Significance of Technology Transfers

	Percent of companies
Very significant	10.7
Somewhat significant	18.5
Neither significant or insignificant	8.2
Somewhat insignificant	2.1
Insignificant	5.6

The information resulted in a finding that technology transfers are significant but not an overwhelming factor for these businesses. Keeping up with developments in the world of technology is a far different matter though, and that was considered very important.

R&D Intensity

Research and development intensity—how much activity is devoted to research and development in relation to all other activities—is often measured as a percent of research and development costs to total sales. And that is how it was done for this effort. The relative intensity of these research and development-related activities was a source of speculation about corporate performance. Does spending in this area mean a company does better? To address this question, needed to determine what the level of intensity was.

The responding companies reported that, on average, 16.5 percent of sales was invested in corporate research and development. That is a fairly high intensity, especially considering that this spending is well below 10 percent for most companies.

When we asked whether R&D intensity was connected with better performance, the result was that there was no statistically significant relationship between greater or lesser intensity and sales growth for the combined group of technology firms. This finding is confirmed among the computer programming businesses, although the other businesses did not have enough cases to test the connection. So, although these companies are research intense, that does not mean more intensity is associated with better sales growth.

Summary

Studying the relationship between high technology business structure and the larger environment produced some notable findings. These companies are independent businesses that are privately held. In describing their organizations as either hierarchical or flat, the respondents indicated a flat structure in the great majority of cases. The business are mainly organized by customer or product line, with functional-type organizations being the next most popular choice. Far fewer companies were organized on the basis of geography. There was no significant sales performance differences among these organizational forms.

Our survey companies believed their primary competitors are a significant threat, having greater market shares than they do, but they are not overwhelmed by the competition because they avoided placing primary competitors in the upper half of the threat scale. Secondary competitors were also judged to be significant contenders. Even though their market shares were lower than the responding firm, they still had a sizable (greater than 10 percent) market share, but were mainly seen as not significant competitive players.

The technology firms believe they react faster than their competitors do in new product developments. Typically, they react in less than a year while they asserted their competitors react in one to three years.

On the question of access to capital, there was no clear indication that it was important to these firms. This was not found to be consequential for sales performance. Access to capital really did not emerge as a great concern among these companies overall. While some considered capital access very significant, this was balanced by others who considered capital access as not significant. Quite a few companies occupied the middle ground and did not commit themselves.

For the most part, the surveyed companies believed they are growing faster than their more broadly defined industry, which itself was thought to be on the upside of the industry life cycle rather than the downside. Put in other words, these companies thought they are in a better position than their industry because they are growing faster than the industry.

The results were also enlightening when they considered the capacity of their customers to do technology substitutions—to replace old MIS systems, to adopt new polymer extrusion machines—as being either important or very important. What was especially telling was when this question was applied to the firms themselves. These technology companies felt very strongly that adaption to new technological changes was very important for them. The sentiment was the most clear and dramatic agreement point among these companies.

When we examined the relationship the companies had with their suppliers, we discovered a few notable patterns. High technology companies use many suppliers, but ten or fewer suppliers account for most sales. The number of suppliers for the industry itself revealed a somewhat different story: there were either less than 50 or more than 100 industry suppliers. On the very important question of whether the number of suppliers made a performance difference, we found that the number of suppliers made no difference in the sales performance of this group.

Joint ventures among technology companies or between a technology company and a general firm were viewed as only marginally important. The extent that these companies themselves transfer technologies outside was also very minimal.

One of the most pronounced findings was that these firms strongly endorsed the need to adapt to technological changes themselves. They appeared to exhibit a high recognition of Social Darwinism applied to technology firm survival. The technology companies reported a fairly high degree of R&D intensity—R&D expenses to sales—something expected from technology companies and well above the definition threshold for most governmental definitions of high technology. But there were no provable sales growth consequences from this.

Thus, we can say that high technology companies are organized in conventional ways. To an extent, the environment does have some influence on structure. These organizations have built research and development arms reflecting the environmental pressures for constant innovation. We can also say that they are very sensitive to their environment, given the importance they have placed on this factor in the questions they answered. These companies are in complex environments where things are frequently changing. It then makes sense that the companies have a high degree of environmental sensitivity.

In all the areas we explored—competitors, suppliers, customers, and the greater sociopolitical environment—we found many interdependencies. These high technology businesses are not islands of individual activity but active trade partners with their outer world. In fact, an isolated high technology company is sure to stagnate.

In finding that there are no consequential performance differences linked to structure or environmental issues, we can say that there are no organizational or environmental prescriptions that high technology firms should follow to organize themselves or behave in their environment to achieve better performance. Performance connections to structural or environmental factors may exist, but not as shown in our sample group.

A final but important note on environment and the technology company is in order. On one hand, the more interactive a company becomes, the less distinctive it is. Some contract software firms that quite plainly reflect workforce availability in a local area do not seem separate from the local workforce—they are merely channels to it. On the other hand, in spite of an enclave or campuslike setting, there are technology companies quite connected to their environments. They send their employees to continuing education events, constantly bring in new talent, put customers in their TQM programs, and otherwise interact with their environment.

This concludes our consideration of structure and environment. In the next chapter, we will turn to the most public manifestation of high technology: marketing efforts. We will carry forward some of the subjects discussed here, such as customers, products, and competitors, but will do so in the context of the corporate marketing function. It is in this area that the dynamic nature of this business can be found.

Notes

1. As described in Stephen R. Hill, *Fostering Competitiveness in the High Technology Industries: Firm Behavior, Industry Structure and National Policy* (New York & London: Garland Publishing, 1993).

2. This was posed as a question meriting research by G. Dale Meyer, of the University of Colorado, at the October 1995 St. Louis University–BYU conference on high technology entrepreneurship.

References

America's Fastest Growing Companies: The Top 100 (Oct. 14 1996), *Fortune*, pp. 90–100.

Anderson, J., and J. Narus (Spring, 1991). "Partnering as a Focused Market Strategy." *California Management Review* 33: 3, pp. 95–113.

Dubini, Paoli, and Howard Aldrich (1991). "Personal and Extended Networks are Central to the Entrepreneurial Process." *Journal of Business Venturing* 6: 5. pp. 305–313.

Hamilton, W., and H. Singh (1991). "Strategic Alliances in Technological Innovation." *Journal of High Technology Management Research* 2, pp. 211–221.

Koh, J., and N. Venkatraman (1991). "Joint Venture Formations and Stock Market Reactions: An Assessment in in the Information Technology Sector." *Academy of Management Journal* 34: 4, pp. 869–892.

Packard, D. (1995). *The HP Way.* New York: Harper Collins.

Porter, M. (1980). *Competitive Strategy.* New York: Free Press.

14

MARKETING OPERATIONS

As we near completion of our review of established high technology companies, we note that we have covered the subjects of personnel, leadership, and strategy, as well as environment. These certainly are critical matters, but they are not the only fate makers for technology firms. We look now at markets, customers, products, and promotion. We gather and discuss these under the umbrella of marketing operations. In so doing, we turn to where the company connects with its customers: in its markets with its products. This is an intersection that can be devastating or highly profitable, but never placid, because marketing is the most dynamic business activity in a field that is far more dynamic than traditional businesses.

The Importance of Marketing

If there is an area where science and engineering-based companies exhibit a gap between what they do and what they need to do better, it is the area of marketing. Much of this has to do with the fact that achieving an optimal marketing program is such an ideal that it is almost out of reach. After all, effective marketing is a total understanding of customer wants and needs, and the total satisfaction of these wants and need with proven products and services. It is tremendously demanding for any business organization to be really good at marketing. There are always more potential customers and more present customers who could be served better. Compound that with the fact that high technology marketers encounter new application technologies from outside their businesses that may obliterate their customer base in the wink of the proverbial eye, and you can appreciate the challenge of high technology marketing.

The downside of bad marketing can be quite severe. Take John Sculley when he was with Apple Computers. He came in as a marketing problem solver, yet he may have been seduced by the technology, the cuteness of the Newton. The Newton computer was a marketing failure and contributed to his departure from Apple. If Sculley had maintained his marketing view and looked at new things by the measure of customer acceptance and not technical wizardry, the Newton might have found a market and Scully longer tenure.[1]

The aerospace industry has its own version of precipitous failure. When fighter aircraft used to be chosen on the basis of fly-offs, General Dynamics pitted the F-16 against Northrop's F-20. The fly-off was not between two paper project plans, nor two computer simulation, but between two actual air-breathing, kerosene-burning, supersonic metal vessels. Northrop lost the contract and many million of dollars in investment.

There is certainly no single way of doing marketing, no unique success formula for high technology companies. Some firms like Microsoft take a very rational approach by forecasting demand, while Apple aims at Microsoft by emphasizing product superiority, a campaign so thorough it has bumper sticker repercussions: When we visited Apple headquarters, we saw a bumper sticker: Windows 95=Apple 84. The animosity may have been alleviated by Gates's recent investment in Apple.

Economists point out that the consumer demand curve shifts with changing tastes and preferences. Customer tastes and preferences are just that—tastes and preferences for fixed periods of time. High technology marketers need to respond to these vagaries, often by anticipating and accommodating customer preference—say, Graphical User Interfaces (GUI) on software programs that may be technically inefficient but user-friendly. The marketing manager in high technology may have to counter the internal biases of engineers for the sake of external customers. The challenge is then twofold—internal and external.

This conundrum was mentioned by the high technology marketers we interviewed. At times they felt they had to push customer simplification while facing the derision of the company engineering colony. Engineers felt they would have to needlessly "dumb down" application instructions for customers who were not quite at their level. Such are the dynamics of customer satisfaction in technology marketing units—dynamics that can mean a failure to communicate for marketers and engineers, and the possible isolation of one or both.

Given that most of the established technology organizations we contacted were headed by people who climbed up through the technical ranks, not the business professions, there is little wonder why there is such a gap. The marketing function is usually grafted onto the organization by the CEO once the product engineering has been worked out.

It is not part of the root structure. The CEO's inclination is more than likely to be toward a product or production rather than toward a marketing.

As difficult as marketing may be, it can be a great compensatory factor for mediocre technology, whereas mediocre marketing cannot be compensated for by superior technology. That's an opinion held by those who advise entrepreneurs, and it's something stated by technology founders as well. In the course of writing this book, we heard several pleas for marketing help from technology heads, reinforcing our proposition that marketing is not an easy task for technology companies. These pleas came from owners who were adept technically but who felt their lack of marketing knowledge had finally caught up with them. They weren't quite desperate but nearly so, and they wanted help with the most basic aspects of marketing, such as how to write an advertisement or produce a product brochure.

At the outset we must arrive at a definition of marketing for high technology operations. There are many definitions available; it is almost as though defining *marketing* has become a cottage industry for text book authors. Each author has his or her own definition and there is no single prevailing construction (much as in defining *business strategy*). We have pieced together some of these definitions for our own operating definition of high technology marketing.

Marketing in high technology is the practice, function, or task of identifying, classifying, and evaluating end user and reseller customer wants and needs, and of developing and promoting products to achieve organizational objectives. This definition incorporates that aspect of high technology marketing—intermediate and end user customers—that is prevalent among these organizations. It also carries forward elements of the conventional definitions of marketing—namely customer want and need identification. Since this definition also places the most emphasis on a customer orientation for the entire marketing process, it avoids the common problem of making marketing synonymous with sales. One high technology entrepreneur who operated internationally complained that many of his customers within high technology (he sold to value-added resellers) did not know the difference between sales and marketing. Our definition addresses this situation.

The other terms we use, *products* and *market*, refer to the offering to the customer (which includes services) and what broad industry market it is offered in.

Distribution Channels

The distribution of a high technology firm's products and services is a critical component of marketing. It is the "Place" of the four Ps of marketing (product, promotion, price, and place). Distribution is how the company's products and services are delivered to the end-use customer.

It is not the sale itself but the route to the sale. (We deal with sales in our later in our consideration of promotion.) How long that route is, how direct or indirect the route is, how many and what kind of middle-men are used—all these are issues of distribution.

When we consider distribution, we are concerned with physical goods—say, computers or airplanes rather than services. In service-oriented technology firms such as software consulting companies, dis-tribution is less relevant. Software consulting is not stored and shipped. It is supplied and consumed at the same time by the end user. You may notice here that distribution is way that high technology firms differ. Distribution applies to high technology products, not services.

The first question we raised in our survey was whether distribution was done by the high technology company itself. Company-owned dis-tribution consists of distributors who are part of the corporation itself or an organization owned by the company. The alternative is an indepen-dent distributor, such as a manufacturer's representative. As in the case with ventures, these individuals work for themselves, often representing several companies. They earn a commission from sales. Some actually take possession and resell while others simply make sales without handling products.

As a simplified example of distribution in the electronics compo-nents business, a manufacturing company would sell a large lot of power supplies to a national wholesaler. The wholesaler would take de-livery and send power supplies to regional distribution centers. The wholesaler then makes sales calls to local electronic shops and takes power supply orders. The orders are filled as shipments are made from the distribution center.

When we asked the question of whether distributors are within or outside the company, our results were solidly on the side of company-owned distribution. A very sizable 81 percent of the companies have their own distributors.

The dominance of company-owned distribution is quite pronounced. It may seem that because these companies are small and entrepreneurial, and as a result cannot afford an internal distribution staff, so we would argue that most small technology companies have only the recourse of using outside distributors. But that introduces a difficulty for small firms. The percent of sales done through independent distributors was fairly low, at 19.5 percent. The fact of life for many small technology compa-nies is that they cannot acquire a distributor because they do not have enough volume to make it worthwhile for a distributor to work with them.

The subsequent issue was how effective were the various distribution channels were. In this instance, we looked at direct sales, independent distributors, and company-owned channels. The respondents were asked to rank effectiveness of each of these on a five-point scale with 5 being the most effective. Direct sales was found to be the most effective, earning

Table 14.1   Number of Distribution
Channels

Number of Channels	Percent of Firms
None	2.6
1	31.8
2–3	31.3
4–5	6.9
6 or more	12.0

a score of 4.1, followed by company-owned channels at 3.0 and independent distributors at 2.7.

These scores were logical expectations. The technology firm has more control over distribution if it owns the distribution channel all the way through to direct sales. What was noteworthy is how close the satisfaction level is with independent distributors. This indicates that independent distributors are a viable means of distribution for the high technology firm.

These established companies appear to be on par with general business companies in their use of distribution, even if they are smaller firms. That result itself was somewhat unexpected, because our impression was that those who know these businesses only casually are likely to view them as technologically advanced but marketing deficient.

The rankings are those reported by the companies themselves, and this fact is important to bear in mind. In other words, the results are not from a truly neutral source. However, they do give a preliminary indication of satisfaction with distributors.

The number of distribution channels is another factor that we wanted to review. This factor may affect satisfaction. For instance, a few well-managed channels may work better than a multitude of channels that cover more of the market but do so in a less effective way.

In asking the companies to categorize how many distribution channels they use, we ended up with some widely scattered results, as shown in Table 14.1.

While nearly two-thirds of these technology companies had three or fewer distribution channels, a significant percent had six or more. An observation here is that our question asked for number of channels, not number of distributors. Thus, if two different independent manufacturers representatives were used, that would count only as a single channel.

Another aspect of distribution is how complex the distribution channels are. We wanted to know how many "interruptions" or "hand-offs" there are from the producing company to the customer. We asked how many middlemen there were in the distribution process. The result was that, in a majority of cases, there was only a single middleman. The results are shown in Table 14.2. In 33 percent of the cases, there was either no response or data was missing.

Use of a single middleman was by far the most popular answer, indicating that these companies use fairly simple distribution paths, much like the electronics company example we described. The single middleman channel is consistent with small firms be they technology companies or traditional companies.

The most crucial question we would ask about distribution is what the results are on business performance. As we have before, we used sales growth over a three-year period as the performance variable and split our firms into those with relatively more channels and those with relatively fewer. Our results fell clearly to one side and led to one conclusion: there was no statistically significant difference between high or low sales and the number of distribution channels. The number of distribution channels was not a determinant of sales growth for these firms. The implication is that the high technology firm has some discretion in adopting distribution strategies, since the number of channels may not affect performance.

In all, we found that distribution is treated seriously by technology companies. These companies have distribution systems in place, and their performance is judged satisfactory. Usually, distribution is the weakest element of the four Ps in terms of how organizations engage in marketing practices. The other aspects of marketing receive more attention; but within this group, it more than holds its own.

Our final word about distribution for these companies is that it appears to be a necessity, again a kind of membership card for entering the high technology club, yet not demonstrable enough to make a performance difference. Distributors are used by technology companies, but there is no single path to distribution effectiveness. We did not encounter a single high technology company that did not have some kind of working distribution system.

In relation to our findings on distribution among ventures, the established small technology firms may get more sophisticated in using distributors but do not seem to greatly expand the number of distribution channels they use. As ventures do, the established firms stick with single or very few channels. The core fact remains, though, that technology companies use distribution channels.

Customers of Technology Companies

The next consideration is of the customers themselves. We were interested in discovering who they are, how many there are, and where they are located. We also wanted to know if customer location made a difference in performance.

We could get no closer to the heart of high technology than by looking at customers. It is however, with apologies to Joseph Conrad, a

Table 14.2 Number of Distribution Middlemen

Number of Middlemen	Percent of Responses
None	5.2
Single middleman	51.5
2–3 middlemen	8.6
4 or more middlemen	1.3

heart of darkness. There is little very obvious about high technology customers. They come and go, for various reasons. There is also little common to them. For the defense businesses, the customer was only one—the federal government—but for a PC microprocessor manufacturer, the customers are in the millions. They are distant compared to the intimacy between the government procurement officer and the aircraft project manager. Despite this, we can and should say something about the customers of high technology firms. What we do is to paint them with broad strokes and then add the details later.

The broad strokes show where customers are located, how many there are, and how important they are. Turning to our sample of established firms, the median number of customer reported was 200. That is a fairly large number considering that these are relatively small companies. It may very well be a clue to their robustness, their ability to survive at least half a decade.

Are all the customers equally important for the company? Apparently not if measured in terms of contributions to sales. The skew is toward the smaller number of customers as bigger sales contributors, as shown in Table 14.3.

The wedge of sales is buttressed on the flank by very few customers. This may indicate vulnerability on the part of these companies, or a least a high degree of dependence on these customers. Put together with the fact that these companies are largely providing services, the results reinforce the importance of customer retention and satisfaction.

As to the geographic scope of high technology customers, we first examine how global the customers are of U.S. high technology. There has been a fair amount of advocacy here, posing high technology as a resuscitator of international trade stagnation. Much of has come from the Clinton administration, but it has also been a theme of former Labor Secretary Robert Reich, who wrote about it before his cabinet appointment. In this view, the exportation of high technology products and services is one path to a healthier trading status. It is an area of domestic comparative advantage with international demand for U.S. technology.

Our results show that much more needs to be done if the healthy global technology market is to be achieved via the high technology route. Among the companies in our study, only 15.3 percent of their sales are

Table 14.3 Customers and Sales Amounts

Number of Customers	Percent of Sales
0–10	47.5
11–50	22.5
51–100	11.2
101–1,000	11.4
1,001–2,000	3.0
More than 2,000	3.5

generated outside the United States, indicating a decidedly domestic slant of the customer base. Our expectation is that many larger technology companies, which were not part of the study, do more international sales and their inclusion would have increased this percentage.

Our interviews with technology companies not part of our survey research gave us some reasons for this low figure. Owners of small, single-employee firms told us about the difficulties they had processing orders from overseas. They said that selling was not the problem—getting payment was. Sometimes the party they dealt with initially was gone and they had to start the sales cycle all over. Sometimes the payment clearing methods got bogged down for months.

These were conditions experienced by company owners who knew how to operate internationally. Those who did not seemed to view any excursion into the area with fear and trepidation. We often heard comments to the effect "Yes, I would like to sell internationally but there are so many obstacles." One special concern was technology protection. Owners were concerned that their patents would not be honored by Pacific Rim countries and were full of stories about unlicensed software being sold overseas.

The relatively low level of international operations should signal opportunities for technology firms in spite of difficulties. A source for such advice may come from technology companies that have been successful, which includes many small, one-person operations that prosper in international markets.

If the high technology customers of American firms are not outside the country, where are they? The technology firms in our survey provided information if the majority of their customers were within 100 miles, within 1,000 miles, or outside 1,000 miles. The distribution of customers is shown in Table 14.4.

The results represent a substantial departure from conventional wisdom, which would propose that these small companies would have close customers. That happens with startups that locate near customers. It might also be recalled that ventures with closer customers had better performance.

Table 14.4 Location of
Customers

Within 100 miles	10.7%
Within 1,000 miles	25.8%
Outside 1,000 miles	60.5%

The distribution of customers for the mature technology company is a different matter entirely. The great majority of customers are outside 1,000 miles, leading to the view that, because international operations are limited, these companies have most of their customers within North America and probably within the United States.

This is a tribute to the effectiveness of their marketing if these companies sell from one coast to another or sell from the middle of the country to either coast or to any one of several other distant customer markets. To the extent that they are surviving companies, it is some endorsement of the idea of wide geographic operations. That itself makes sense because the products of high technology are high value per weight and volume, and transportation costs to faraway customers are not a major cost load. These companies are not in the business of hauling sand, after all. Customer distance is not an obstacle for high technology companies.

It is also sensible because, and this is speculation, there may be greater facility in making use of electronic channels to customers, as well as more sophisticated use of computer database marketing. Both make it easier for small technology firms to seek and find customers across the United States.

The fact is that many entrepreneurial technology firms have Internet home pages. One simply has to briefly surf to discover hundreds of such home pages. Indeed, these companies emerge from the pack as the most prevalent with home pages. They show a responsiveness to opportunities for reaching customers through the most technologically modern channels available.

A primary means of reaching customers in any business is the sequence of segmentation of different customer groups, target marketing, and product positioning. This sequence applies in high technology, though there are permutations. Poor marketing can have quite conventional explanations in technology firms. For example, case study of Prodigy's on-line services demonstrated that its initial market failure was due to a poor match between target market needs and Progidy services, something that can happen in any business.[2] But again, the hubris of technological superiority as a consumer magnet rears its head; it was simply assumed that consumers could detect and would select Prodigy for its technical prowess.

As a last point on the subject of customer location, we explored a possible connection between the location of customers and their business performance. The particular question was whether there was a significant difference in sales growth between companies that sell mainly on a local or regional basis and companies that sell in the United States and other countries. The results were that there was no statistically significant difference. On this rough cut, there is no evidence that sales growth is better achieved on either a more local or a global basis.

The question of customer location was also asked a slightly different way to see if there was a sales growth impact. When asked whether a majority of customers are within 100 miles, within 1,000 miles, or outside 1,000 miles, the respondents' answers were compared to relatively high or low sales growth. As happened in the prior test, there was no significant difference between the sales growth categories based on where the majority of customers are.

Those who have followed high technology, and particularly those who have prepared directories of high technology companies, have wondered how many high technology customers are themselves high technology firms. The answer to this would give an indication of how stacked these businesses are—how many layers there are to the technology business.

The customers of our surveyed high technology companies are high technology companies in 47.4 percent of cases. Therefore, there is a very high degree of technology firms passing goods and services to other high technology firms. Cast in this light, high technology has a self-perpetuating internal dynamic, with supplier firms feeding other value-adding firms. This can be contrasted with the 28.5 percent of companies that do not sell exclusively to high technology firms. The remaining percent sold to both or did not report on this question. When this information is put together with the facts we discovered on distribution middlemen, we can see that there may be some four layers between a supplier firm and the final customer.

Products and Innovation

Lynn and Heintz (1992) provide a good summary of the challenge of product management in the high technology sector. They observed that firms try to capitalize on technology to gain sustainable competitive advantage. As the rate of technological change increases, products and services must be more advanced, faster, and easier to use than those of competitors. A problem faced by these firms is how to convert new products or technologies into businesses. There is a need to find where the technology fits in the market, to size the market, to determine how it will evolve, and to implement the most effective market penetration strategy. The authors used a three-level technology screen (needs, economics, and time) to identify realistic market opportunities and build timetables.

One way that the very different products and services of high technology can be classified is how standardized or customized they are. This lets us put some parameters on what could be a very broad subject. Standardized products are inherently undifferentiated and completely substitutable. In this business field, the products tend to be components and raw materials—simple products more or less. A 14-pin cable connector is an example of a product given to standardization. The market is massive and there is little reason to physically differentiate the product from others. At the other end, customized products may be single items with vast complexity made for a single customer—the Boeing 777 is an example. Technical services are, of course, completely customized, for reasons stated earlier.

When we asked if the technology firm sold products that were standardized for most customers or custom designed for particular customers, we found that a slight majority (51.9 percent) in the group customized for their customers. Firms that standardized were 45.7 percent, and the rest did not reply or the question was not applicable to them.

This result did show a fairly even split among the two different types of products. What came as a surprise was the extent of customization, a far greater percent than it for businesses as a whole. That could mean that product lines closely matched customer requirements and that there was little cross-selling between types of customers. Because of the large number of customers these companies have, customization of products must be especially challenging. It is very likely that there is not total customization but rather customization for certain customer groupings.

In all, though, this slice of the respondent base emphasized the prevalence of customization within high technology. That is a point we enlarge upon in our summary comments about the nature of this business, but it does seem a distinguishing characteristic of this business field.[3]

An issue for high technology managers is whether the products should be standardized or customized. Among our surveyed firms, customized products were associated with higher sales growth. The result was statistically significant and the Figure 14.1 shows the results.

Although our exploration of technology company products has just touched on its many dimensions, it has given us a portrait of the products as being customized. The products are delivered to end users through simple distribution channels as well. Next, we wanted to discover more about the sources for new products, not just existing products.

What can be said about the sources for new products? We can cut through to an answer directly by asking if the source for new products is internal innovation or imitation. Innovation would be shown as products and services coming from the firm's research and development laboratory. Imitation happens when a technology company seizes a product offered by another company, either a rival or not, and duplicates it under its own name.

Among our surveyed companies, imitation accounted for 20.4 percent of new product origins and internal innovation was responsible for

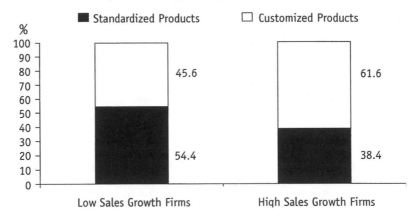

Figure 14.1   Growth versus product type

77.3 percent of new product origins. This very high percentage of internal idea creation shows that innovation is very much alive in these science and engineering-based companies. It could also be that replication of the ideas of others is very difficult. This is probably true as well, but that really does not diminish the huge dependence on internal creativity.

It's true that *innovation* is not precisely defined here, and it is subjective to say how much innovation is true innovation. Marginal innovation could come from something as prosaic as adding another logic gate to a microprocessor—that could still count as a liberally defined innovation. We left it to our responding companies to define this term and there is some evidence that they have not defined it frivolously. Instead they took a conservative approach, as evidenced in the small number of yearly innovations reported.

When the companies were asked how many innovations they had per year over the past three years, the average went from 1 in 1993 to 1.14 in 1994 and 1.4 in 1995. About a third of our firms reported no innovations in each year. The majority had two or fewer innovations per year.

A notable feature of the innovation information was the strong growth in average number of innovations, indicating a strengthening of creativity in the firms.

One measure of the vibrancy of any market is the percentage of total sales derived from new products and innovations. The greater the percent, the more efficient the organization is in getting its ideas tested and brought to market. In some industries, such as utilities and commodities, this is virtually nonexistent; new products are 1 percent or less of sales.

What is the situation with high technology? We can say that it is much more robust. What's more, there has been a recent increase in the percent of sales derived from new products and innovations, as shown in Table 14.5.

Table 14.5 Sales from New Products and
Innovations

Year	% Sales from New Products and Innovations
1993	18.7
1994	20.8
1995	23.4

When you consider that about one in every five sales dollars originates from something new in these organizations, there is a clear message that an innovation-driven industry passes innovation on to its customers. This level of new product development is extraordinary when compared to other industries.

What affects new product development time? Schoonhoven, Eisenhardt, and Lyman (1990) looked at the speed to market in a longitudinal study of young semiconductor businesses. They concluded that substantial technological innovation lengthens development times and reduces the speed with which first products reach the marketplace. There is a temptation (that we have succumbed to) to name semiconductor venturing as entrepreneurchip.

The upside and downside of the time factor in new product development was developed by Baumol (1993), who studied the matter of optimal timing of the introduction of an innovation. Baumol stated that the longer the delay in transferring the innovation from research and development to manufacturing, the more the product is likely to be improved. Yet the longer the delay, the more likely competitors can deliver the product first. Therefore, there are tradeoffs in new product development.

It is understandable for new technology ventures to derive sales from new products, but our surveyed firms were more mature companies. Perhaps this is a way for them to preserve entrepreneurship as they grow.

One area of new product development—cycle time reduction—appears to be a subject technology firms should pay more attention to. Carmel (1995) described research results involving 15 software developers in the Baltimore-Washington area. The developers were generally unaware of cycle time reduction (new product development time) as a management concept. They tended to focus on rapid development with a last-minute rush to meet deadlines. Speed of development can help with cycle time. Successful teams are small, cohesive, and entrepreneurial.[4]

Further research has been done on the cycle time concept, particularly by Ali, Krapfel, and LaBahn (1995). They stated that cycle time has become a critical competitive variable, especially for small high technology firms. The authors investigated the impact of product innovation and entry strategy cycle time on the initial market performance of small firms. They found that faster product development is associated with shorter breakeven time. They also found that firms that accomplish this do so not by sacrificing product quality but by keeping the technical content of the product simple.

Karakaya and Kobu (1993) attempted to address the issue of new product performance across high and low technology firms. They compared the food processing industry and the medical instrument technology industry and found differences in variables relating to new product success. In the food industry, the most important success variables were a firm's adherence to new product plans, competition, price, access to distribution channels, and customer cost of switching. For the medical instrument technology industry, it was product performance, poor marketing, product obsolescence, access to distribution channels, and consumer switching costs that were the most important variables.

In addition to new product development, way of vitalizing the product line is to modify existing products. Determining the difference between new product development and product enhancement might be difficult in some cases. It is not always obvious, so an illustration may help. The hand-held Global Position System used by some campers, pilots, and luxury car drivers represents a new product development. Increasing the capabilities of that system by increasing the database or improving the resolution of the screen are product modifications.

When we asked our surveyed firms how frequently product modification goes on in their companies, we found that the answer was quite frequently, as shown in Table 14.6. Other answers (not applicable or missing) amounted to 16 percent of the total.

About 60 percent of the firms do product modifications within five years. Coupled with new product development, this indicated an extremely high level of attention to products and services by these technology firms. It may also indicate that these companies show some balance between research and development intensity and product focus, a sure sign of corporate effectiveness. If a technology firm can balance its research and development efforts in the new product development area with just enough focus on maintaining existing profitable products, it is managing its product mix effectively.

Two studies of product growth strategies shed some light on high technology company practices. First, M. H. Meyer and E. B. Roberts (Roberts has written extensively on high technology ventures) reviewed ten firms in terms of their growth. The most successful companies concentrated on one key technology area and introduced product enhancements related to that area. The poorest performers tackled unrelated new technologies and entered new product-market areas.[5]

Later, a larger study of 68 small, young high technology firms provided more detail on effective product strategies. Pavia asserts that neither a technological growth path, where new products go to existing customers, nor a market expansion growth path, where existing products to go

Table 14.6 Product
Modification Frequency

Annually	24.9%
1–5 years	36.5%
More than 5 years	21.9%

new customers, is superior. Practices that minimize strategic dependencies and produce a high-quality product that suits customer needs directly (needing no further modifications after the sale) are associated with success. Firms that have used market expansion to grow show more systematic planning and describe their industry as rapidly changing. They also describe themselves as technologically innovative and their customers as well informed about the products they buy.[6]

Patents

Yet another way to look at the question of what kind of products and services are available is to examine the nature of these products. There are certainly close to a million high technology products and services, so it is difficult to generalize about them. Not all innovations lead to new products; most innovations do not pass new product feasibility tests. Other innovations are internal process improvements—the progeny of TQM, for example—and are not commercialized. But some innovations are certified and made available to customers. We turn now to one of the most pronounced manifestations of innovation: patents.

Different technology businesses approach patents in different ways. For biotechnology companies, the chase for patents is a modus operandi of the field. Indeed, patent acquisition is the game for biotechnology more so than tangible product sales, although there is a sector of biotechnology that sells equipment to biotechnology research firms.

At the other pole, are the software consulting services, for which the patent game is as alien as a distant galaxy. Services are not patentable, even if they were, they would not be likely the basis of competition for software consultants, since programming solutions are so specialized they are one-of-a-kind transactions.

Since our surveyed companies tended to be service firms, not biotechnology companies, we expected that there would not be many patents. But the information plainly showed otherwise. The average number of patents currently held by each technology firm was 32.2—that indicates a high level of patent intensity. These patents were obtained by the companies themselves or through acquisition of patents developed by other companies.

There was also some revealing information about the recent rate of obtaining patents, indicating that the number of new patents awarded per year has significantly dropped in the past three years. In 1993, it averaged 3.4 per compa y; in 1994, the average was 2.8 and in 1995 it was 2.7.

This may suggest that the pursuit of patents has declined overall. It might also indicate a change in the composition of high technology company patent activity, with a shift to less patent acquisition. Although there has been an increase in innovation, there has also been a decrease in patents. We have heard CEOs mention the high cost of patent approval, and this may well be a factor.

Pricing

In our experience, there is a great deal of apprehension on the part of high technology companies when it comes to pricing. Typically, the pricing decision is the last product-management decision. A great deal may have been invested in a new product, and there is considerable pressure to recoup development costs.

If there is not much pricing flexibility, either because the industry has little room for price variation or because the customers have considerable market power, then a technology firm may have little in its pricing array. Barriers to price flexibility happen with low marginal profit opportunities. The technology firm may be able only to set a single price and hope for the best. If the price is too high, there will simply be no sale whatsoever. The arrow will fly past the target and be lost forever.

There can be little wonder why technology marketing managers in large companies and CEOs of small firms remark, "We got the bugs out. It went through the beta site great, now I just have to price it." Technical progress can be completely undone with a single bad number—the price.

Pricing in the software business has also exhibited some decidedly nonrational market expectations. Borland International bought a small company, Analytica, and sold a software package called Reflex for much less than Analytica did. Sales were much better when the price was tripled on a revised version. Executives were told that the low price hurt credibility.

We looked at how these companies do pricing, first in relation to the primary competitor and second in relation to the industry as a whole. Both of these are commonly held as being important pricing policy factors.

When asked how their price compared to primary competitors, the results were as shown in Table 14.7. A review of these results reveals an essentially conservative approach to pricing, with the mode being competitor parity pricing. It also indicates that the firms probably do know something about how their competitors price.

The next consideration was how pricing compares to the industry. The definition of industry was again left to the respondents, and the results are reported in Table 14.8.

The general pattern of responses was similar for competitor compar-

Table 14.7 Prices Compared to Primary Competitors

	% of firms
Price much higher than main competitor (>15%)	5.2
Price higher than the main competitor (5–14%)	32.2
In parity with the main competitor (±4%)	36.1
Lower than the main competitor (5–14%)	18.9
Much lower than the main competitor (<15%)	7.3

isons and industry comparisons. In both cases, the companies were slaved to statistical central tendency. Apparently, there is a dual pricing sensitivity among these technology companies, one to what competitors do and another to what the industry does. Comparing the two influences, we can said that the average response was very slightly higher than the industry but almost in complete parity with competitors.

Pack following may work for existing products, but for new products the heuristic will not work. For new products, the technology company CEOs indicated that pricing policy was varied. Some used cost plus pricing in an effort to recover investment cost, others picked the typical skimming pricing or penetration pricing.

The *skimming pricing* method sets a high price that cuts across a wide market. *Penetration* pricing implies a lower price in a more narrow market, a kind of vertical marketing. In all, pricing policies of high technology companies are surprisingly unremarkable and inherently conservative. The level of inventiveness that these companies exhibit in product innovation does not carry through to pricing.

Having reviewed where these technology companies stand concerning their potential competitors and the industry, we now discover what factors they considered in setting price. This should help show how they got to where they are in pricing policies.

The survey firms were asked to indicate which factors were considered the most in setting prices. The response choices were costs, competitor's prices, consumer demand, and other factors. The results are shown in Table 14.9. The "nonapplicable" and "no response" categories

Table 14.8 Prices Compared to Industry

	% of firms
Price much higher than the industry (>15%)	7.3
Higher than the industry (5–14%)	25.8
In parity with the industry (±4%)	37.3
Lower than the industry (5–14%)	14.6
Much lower than the industry (<15%)	4.3

Table 14.9   Factors Considered Most
in Pricing

	% Considered "Most"
Costs	33.0
Competitor's prices	15.9
Consumer demand	26.6
Other	15.9

accounted for 8.5 percent of the total. The "other" was a sizable collection of responses and captured comments such as combined factors.

What emerged from this is a substantial endorsement of the factor of costs in setting prices, followed by consumer demand and competitor prices. It is curious that, with cost being the most important factor, prices are still in parity with the industry and competitors. This may be an indication of pricing finesse as these firms work the margin between their costs and industry and competitor prices.

Price setting certainly does not seem to be a matter of solely parity or solely internal costs. There are other factors that impinge on price setting. One could be markups by middlemen, though there are few middlemen among these companies. The middleman markup process works by having each successive middleman add to certain percentage to the price with transfer to the next middlemen.

Another way of pricing could be discounting after the sale. A third could be setting prices compared to other products in the line. Fourth could be the extent the company has pricing autonomy over prices. Its complete independence from a corporate holding entity usually means considerable pricing flexibility.

Yet a fifth way of pricing might be experimentation, or variation of prices just to probe customer acceptance. In other words, pricing policy might stem from top-level executives rather than the more customary approaches. These are possible explanations that would benefit from additional exploration.

Our consideration of pricing now leads to that familiar and very important question of whether pricing practices have performance consequences. In this instance, we took pricing in comparison to the industry as our objective. Again, the performance variable was sales growth. In a variation of how this was done before, we ignored the middle range "in parity" responses and grouped the remaining "higher" and "lower price" responses.

Our chief finding was that pricing higher than the industry was related to higher sales growth. This finding was statistically significant. It was strongly supported in the industrial and commercial machinery and computer equipment industry. It was also strongly supported in the computer programming industry.

At a glance, it seems perfectly sensible that higher prices mean greater sales revenues, but assuming normal demand elasticity, sales could decline if prices climb too high. Taking this into account, there are consequences in the higher pricing–higher sales relationship.

Returning to the theme of factors considered in setting prices, we noted that costs are the most popular factor. When we considered firm performance and pricing factors, a much different story was told, however. Higher sales-growth firms used consumer demand and lower sales-growth firms used costs. The difference was statistically significant. This is an especially consequential finding that adds importance to the pricing decisions of the technology firm. That is, pricing seems to have performance consequences. Pricing factors for high and low sales-growth firms are presented in Figure 14.2.

As a summary of our observations and conclusions on pricing, we can say that pricing seems to be done with great sensitivity to external factors and internal costs. Put this way, pricing in technology businesses is a balancing art. It also makes a difference among these technology companies.

Promotion

In contrast to the reservations we encountered regarding pricing, there was enthusiasm about the subject of promotion. There was always an answer with promotion—more or less, this target market or another. That was different from our pricing questions, where there was only one answer—one price—and it could be wrong.

Our basis for describing the promotion activities of high technology companies was contact with marketing personnel, who spoke informally with us about promotion practices, rather than our survey research sample. In *promotion*, we include paid advertising, public relations, sales promotion, and personal selling. We do not elaborate on personal selling here, since aspects of it have been covered in the earlier review of distribution and sales. But we have several comments on sales.

Compared to other firms operating in other environments, especially consumer goods firms, technology companies are tangential and scattered in their use of promotion tools. They are tangential because they rarely use single promotion tool intensively, and they are scattered because they use several of the many available means of promotion.

There are exceptions—the multimedia rollout of Windows 95 by Microsoft—but the exceptions stand out because they are just that. Most technology product introductions occur far away from the general public's eye and ear.

This can be shown another way by asking what the impact might be if technology companies were to promote the way other companies do. If they did, we would hear and see much more of them than we do. But when was the last time an advertisement for Rockwell or Boeing appeared

High Sales Growth Firms

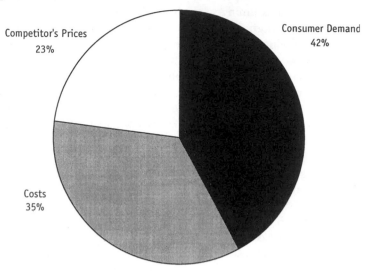

Competitor's Prices
23%

Consumer Demand
42%

Costs
35%

Low Sales Growth Firms

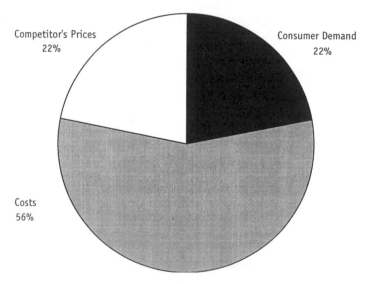

Competitor's Prices
22%

Consumer Demand
22%

Costs
56%

Figure 14.2 Factors considered in setting prices

in a national mass-market magazine? Yet these are the aerospace industry giants and they have extensive advertising budgets. They never appear on any listing of top advertisers, however.

Some companies, such as Intel (Intel Inside!), and more recently the bunny suit dancers have fought for identity in the consumer's mind by purchasing broadcast media, but most other component suppliers remain invisible. For the most part, this is by design, since technology audiences are effectively reached by other, less intrusive media channels.

It's worth observing just how these media-quiet companies do their promotion. The answer is very selectively. If the mass media are shunned, the specialized media are not. These companies advertise, but they do so in trade publications—those hundreds of titles such as *Aviation Week and Space Technology* for the aerospace business, *Electronics* for the electronics business, and *Computer World* for the computing business. Advertising expenditures are found here in page after page of display advertising, most of which is not directed to end users but to intermediate customers.

Promotion also occurs at trade shows, and each of the businesses we have explored in this book has its own national trade show. But again, these are invisible to the public. At these trade shows, technology companies display their products to prospective customers on a firsthand basis.

When we encounter high technology businesses using mass advertising, it is almost always what advertisers call image advertising—that is, it is not intended to show products but to create a positive public perception of the company. McDonnell Douglas used network television to convey the sense that it was building safe combat aircraft—to bring home "someone's father, someone's husband . . . daughter," runs the voice over a carrier landing shot.

For the most part though, 72 percent of high technology advertising goes to product advertising, while 23 percent goes to image advertising.[7]

Another part of the promotion mix is public relations. In this area, technology companies are either devoid of any systematic approach to public relations or are immersed in a particular form of it. An illustration of the latter is Apple Computer, which made use of publicist Regis McKenna. This relationship made a name for McKenna, who recognized that within the computer business there was a concentration of communications by a few publications and people. He worked the concentration well, identifying "opinion leaders" in the communications segment and suggesting stories about Apple that were later printed. His idea was to get the then unknown Apple mentioned by the great mentioners in the business. Standard public relations tools such as sending out press releases were insufficient for getting the attention those in a position to shed light on new upstart companies.

More recently, McKenna (1995) has made a case for using technology better for high technology promotion. He has concluded that consumers are beginning to view brand names with increasing indifference, as there is a proliferation of marketplace choices. Information technology can be a tool for rebuilding the power of brands. With emerging technologies such as high-speed communications, computer networks, and new software, companies can have real-time dialogues with customers and provide real-time interactive services. This enables the company to draw the customer into a conversation. Marketing managers will need to initiate and sustain such dialogues.

Within the field now, the large technology companies often contract with public relations agencies to perform services such as arranging editorial conferences, public events, and sponsorships of corporate citizenship events. Smaller technology companies do not have as many financial resources and consequently do little if anything in the area of public relations. More likely, they put their money into advertising in trade publications or into sales promotion materials. Our impression, based on doing high technology public relations as a profession, is that most small-firm public relations is done in an unsystematic and experimental way.

Larger technology firms usually hire a public relations agency and an advertising agency. Smaller firms most often hire an advertising agency that may or may not have a public relations unit.

Sales promotion materials—small-value giveaway items such as pens and clocks—are a means of promotion. Very often, we have seen small firms finance sales presentation materials such as interactive CDs, PC-run corporate profiles, videos, and slide presentations, used in conjunction with giveaways. Sales promotion materials are quite common in these businesses.

As a summary of the promotion activities of high technology companies, we can say that these firms are selective in the media they use. They do not employ intrusive media, such as television and radio, and consequently are not very visible to the public.

The way marketing is done by these organizations is more a function of their size than their nature. Very small, young, high technology businesses prepare their own marketing materials, even if they are biotechnology or electronics companies. As the organizations grow, their needs for marketing and coordination of media campaigns increase. Then they turn to advertising agencies. This doesn't usually happen until the company produces more than $1 million in revenues, though. The advertising agency lives off its 15 percent commission on the advertising it places for the company and any additional fees for services. High technology firms must hit about $100,000 in advertising expenditures before an agency will even be interested in them.

If the business grows even more, it faces a decision about whether to continue using an advertising agency or to develop its own in-house capability. The considerations are the costs of in-house vs. agency labor, the skills and capabilities of internal and external talents, the necessity of knowing the whole business well—something that often shifts the decision to the in-house side, and the quality of the creative services.

In addition to the expected location of agencies in and around Boston's Route 128 and California's Silicon Valley, there has been a growth in agencies around Washington, D.C. The June 14, 1996, *Washington Business Journal* reported that technology firms in the area found outside advertising agencies as cost-effective compared to in-house creative services.

High technology organizations that reach the pinnacle are in the happy state of being able to make the internal or external choice. The external part becomes sweeter, since they have advertising agencies pursue them rather than vice versa. The call that sounds the gong at the advertising agency is from a really huge technology company, asking the agency to "pitch" them with an "on spec" advertising campaign. The mere chance to spend money and energy on the possibility of securing an advertising account is very good news indeed. Even very large, well-established agencies will scramble to get a shot at a $10 million advertising account, which the larger technology companies represent.

Our personal experience with advertising agencies and high technology companies has shown an occasional conflicts between agencies and companies, mostly to do with people. Technology company staff express frustration with the lack of understanding of their products by agency account staff. For their part, agency people complain that company people overstress the technical features rather than the product benefits for the customer. This can be a wide chasm and it has given rise to many West Coast advertising agencies that hire engineers as account executives or otherwise differentiate themselves on the basis of some technical knowledge.

Despite these common-ground approaches, the bulk of software advertising emphasizes features rather than benefits. This may be because the advertising is in trade publications read by people in the business. It nonetheless strikes those advertisers in consumer product marketing as somewhat primitive and a little off the mark. As technology products move to a mass market—as they will with simplification—it will be more necessary for producers to emphasize benefits over features. This allows us to return to a concluding topic in this chapter on marketing: the relatively low level of sophistication in the use of marketing.

The Management of Marketing

The array of tools that high technology has for marketing—pricing, promotion, distribution, and product—are quite varied and powerful. They suggest the importance of managing marketing.

Marketing management is that process of planning, organizing, and controlling the elements of the marketing mix. With that definition, we reviewed this practice in technology firms. It is a practice having more the flavor of an art than a science, even in the milieu of science-based businesses. High technology marketing managers have little research on which to base their marketing decisions. Weiss and Heide (1993) showed that high technology buyer behavior research has considered only the limited outcomes of purchase or rejection. They go on to demonstrate how situational effects of buying behavior explain workstation purchases.

High technology marketing managers have adopted the product management techniques used in other business fields, mainly because product management for high technology developed from the examples used in consumer products.[8] As we described earlier in this and Chapter 14, the life cycle model is abbreviated for high technology products. This is taken as a truism in the field, and it serves to reinforce the usefulness of this management tool. In any event, we've seen many variations of these matrices used frequently in this field.

The management of marketing includes sales management. Marketing managers often manage the sales force. The sales staff of technology firms works in different ways, covering the whole spectrum of compensation methods, types of selling, and sales organization. But our first-hand experience with technology sales has led to a few observations that may be a departure from what is commonly believed. The first is that many small firms with thin capital bases do not have commissioned salespeople and use a salaried sales staff. The second is that the salary levels are very high, often on a par with the chief executive. The third is that small companies use salaried staff in conjunction with manufacturing representatives. Together, these observations indicate just how much is invested in sales efforts, a conclusion that echoes these companies' commitment to product development and pricing policy.

Also based also on our professional experience, we have noted the nature of technical sales in the software engineering business as having an emphasis on consultative sales for the successful salesperson. Time and time again, it was not the missionary salesperson or the relationship salesperson but the consultative salesperson who not only made the sale but was asked back by the client to help "solve" other problems—and in the process make another sale.

What do high technology companies in marketing do that should be improved? The views of Woods and Remondi (1996) are that many high technology companies still use outdated marketing techniques despite marketplace changes that require new techniques. These companies do not understand or use database marketing, and they don't measure marketing results. They are not aware that relationship marketing (where the marketer obtains a knowledge of the whole business) can help them. Companies can build relationship through creation of a flexible and customized sales and marketing database, use telemarketing, use lead management, and track leads to sales conversions. They should also integrate direct marketing and report results.

When high technology businesses do marketing well, they do not do it identically—Schlosberg (1993) describes two different but successful examples: Microsoft does forecasting through its sales force; challenger Apple tells customers about its product advantages over Microsoft. Other smaller firms favor the trial-and-error method of product introduction.

Kadanoff (1993) shows how Apple Computers has been successful in using a nontechnical approach to selling computers by using emotional appeals. It has been observed that limited budgets, intense competition, and short product life cycles cause high technology companies to replace intuitive marketing with structured, step-by-step plans (Flanagan, 1993).

Summary

Marketing for a technology company can be compared with the circulatory system of a living organism. Both are absolutely essential. Nevertheless, many smaller technology companies shy away from fully exploiting marketing techniques and some of the larger companies do not fully exploit marketing as a way to achieve competitive advantage. This review of marketing has been on a very broad scale, encompassing all aspects. In doing this, we found that there are things similar to conventional companies and things different.

High technology companies use fewer and simpler distribution channels. They are satisfied with their distribution systems. The average number of customers is 200. That's a considerable number for these predominantly small firms, but not all customers are equally important. There is a definite skew, with a much smaller number of customers accounting for most sales. Most customers are found more than 1,000 miles from the company.

It is not possible to generalize about the customers, since they are so varied. However, these companies tend to sell to other high technology companies, thus lending a vertical depth to this industry.

We found that these companies had wide geographic scope—they sell products on a nationwide basis. Although geographic scope ends at our shores, it is remarkable that these generally smaller companies consider their markets as transcontinental.[9]

When we explored the matter of whether the scope of operations made a difference in sales growth, we could find no significant differences between greater and lesser geographic market scope.

These firms take their customers seriously. Their sensitivity to external prices also shows an orientation to customers. Those that set prices base on consumer demand performed better than those that base prices on costs.

Product management is a vital element of these companies, as well. There is extensive product modification and innovation, as indicated in the number of patents held by separate companies. Over three-quarters of these firms use internal innovation to bring products to market. New product development is an important element of the product portfolio. The firms that offer customized products do better than those that offer standardized products.

Pricing policy produced the most revealing information about high technology companies. There was a strong element of pricing in parity with customers and the industry. There was also strong sensitivity to internal costs because these costs were the most popular way of determining price.

Promotion by high technology companies is very selective, with trade publication advertising, sales promotion, and presentation materials dominating. There is much less use of public relations and the broadcast media.

The sales forces of technology companies work in similar ways as the sales forces of traditional firms. Small technology companies often put the sales force on salary and they pay well. If there is a difference from traditional businesses, it is that high technology sales involve more direct knowledge of customer applications. Many sales in high technology have tended to move toward customer-defined problems and integrated solutions, sometimes involving nonproprietary hardware and software.

If there is a statement that best describes how these companies do marketing, it is that they have broad geographic scope, are attuned to their customers and competitive environments, and are innovative in their offerings.

In the next chapter, we will answer one of the most basic questions about high technology companies: What happens as they evolve? That is, do they become more alike or different because of size? Chapter 5 will enable us to pack together what we know about the venture and what we have learned about the mature company. This, in turn, will clear the way for overall analysis of the field and conclusions about the voyages of these companies.

Notes

1. This commentary is from Bradley Johnson, "Sculley the Brand Comes Undone" *Advertising Age*, February 28, 1994, p. 31.

2. Described in *Market Segmentation* by Art Weinstein (Chicago, IL: Probus Publishing, 1994).

3. One view of product customization and technology is offered by Faye Rice, "The New Rules of Superlative Services (The Tough New Customer)," *Fortune* 128, no. 3 (Autumn-Winter, Special Issue, 1993), pp. 50–53. Rice states that successful firms will produce more customized products and services. These products may be relatively high or low tech, the point being that customization may not mean customers want more technology in products. That is something technology company managers may want to remember.

4. As summarized by G. Dale Meyer, University of Colorado, October 1995, Washington University BYU Conference on Technology Entreprenueship.

5. "New Product Strategy in Small High Technology Firms: A Pilot Study," contained in MIT Sloan School of Management Working Paper No. 1428–1–84.

6. As reported by G. Dale Meyer in a summary of T. Pavia's "Product Growth Strategies in Young High Technology Firms," *Journal of Product Innovation Management* 7 (1990): 4, pp. 297–309.

7. As reported by Chuck Paustian, "Communicating is Essential: Branding, Market Strategy at Heart of Agency/Client Relationship," p. A–13. *Business Marketing*, September 1995.

8. One attempt to provide a marketing management tool for high technologists is found in G. David Hughes, "Managing High-Tech Product Cycles," pp. 44–55, *The Executive* 4, No. 2 (May 1990). Hughes combines the Boston Consulting Group growth-share matrix and the product life cycle concept. The four-quadrant grid plots appropriate management styles and strategies through the life cycle.

9. A particularly important and very recent development in international businesses operations has been strategic technology partnering. This partnering has marketing impacts. See John Hagedoorn, "Understanding the Rationale of Strategic Technology Partnering: Interorganizational Modes of Cooperation and Sectoral Differences," *Strategic Management Journal*, July 1993, pp. 371–385. Hagendoorn looks at the motivations for these partnerships and differences among the areas. In a later article, he and Jos Schakenraad gauge the effect of such partnering on the companies doing them. See J. Hagedoorn and J. Schakenroad, "The Effect of Strategic Technology Alliances on Company Performance," *Strategic Management Journal*, May, 1994.

References

Ali, A., R. Krapfel, and D. LaBahn, (1995). "Product Innovativeness and Entry Strategy: Impact on Cycle Time and Break Even Time." *Journal of Product Innovation Management* 12: 1, pp. 54–69.

Baumol, W. (1993). "Formal Entrepreneurship Theory in Economics: Existence and Bounds." *Journal of Business Venturing* 8: 3, pp. 197–210

Carmel, E. (1995). "Cycle Time in Packaged Software Firms." *Journal of Product Innovation Management,* March, pp. 110–123.

Flanagam, P. (1993). "Marketing High Tech: Lessons for Everyone." *Management Review*, March, pp. 27–29.

Kanandoff, M. (1993). "Nontechnical Approach to Marketing High Tech Has Benefits." *Marketing News,* April 26, pp. 10, 14.

Karakaya, F., and B. Kobu, (1994). "New Product Development Process: An Investigation of Success and Failure in High Technology and Non-High Technology Firms." *Journal of Business Venturing* 9: 1, pp. 49–66.

Lynn, F., and S. Heintz, (1992). "Where Does Your New Technology Fit into the Marketplace." *Journal of Product Innovation Management* 9: 1, pp. 19–25.

McKenna, R. (1995). "Real Time Marketing." *Harvard Business Review* July-August, pp. 87–95.

Perelman, M. (1991). *Information, Social Relations and the Economics of High Technology.* New York: St. Martin's Press.

Schlosberg, H. (1993). "High-tech Business marketers show a Lot of Creativity." *Marketing News*, May 10, pp. 2–3.

Schoonhoven, C., K. Eisenhardt, and K. Lyman, (1990). "Speeding Products to Market: Waiting Time to First Product Introduction in New Firms." *Administrative Sciences Quarterly* 35, pp. 177–207.

Weiss, A., and J. Heide, (1993). "The Nature of Organizational Search in High Technology Markets." *Journal of Marketing Research*, May, pp. 220–233.

Woods, T. and J. Remondi (1996). "Relationships Vital for High Tech Markets." *Marketing News*, May 20, pp. 8–9.

15

FIRM SIZE DIFFERENCES AMONG HIGH TECHNOLOGY COMPANIES

Now that we have concluded the review of high technology ventures and established companies, and we have considered the myriad component activities that make up these organizations, it is appropriate to draw conclusions about the nature of high technology companies by considering a simple remaining, but possibly quite important, factor about these businesses: size. This is way they can be unique or similar.

Because so many of our results were effects of company size we determined that a closer look at the size factor was warranted. The largest companies, such as Intel, Microsoft, Boeing, and DuPont, have different problems from the small one- or two-person consulting shops. That is apparent, but the extent of those different problems is also worth knowing. Is it orders of magnitude or a few points on the scale that separate the way distribution is done at the giant Cincinnati Milicron and the tiny Isthmus Engineering?

So far, we looked at company size indirectly. In Chapter 9, the subject was the growth of high technology firms. From ventures to established companies, our intent was to show that things change as a technology firm evolves. And they do change in important ways. In this chapter, our aim is to inspect the impact of size—not over time and for a single firm, but at one time for many firms. We want to know if size matters for technology firms, and if size has a linear or uneven effect on how business is done. For example, do the larger technology companies have more patents, the number of which generally follows the increase in corporate size class, or are there bumps and gaps in these figures, with fairly even growth according to size but then an abrupt change, either fewer or more patents as corporate size becomes larger? This is one of the many issues we could evaluate. However, our confinement to a couple dozen of these issues reflects practicality and the deliberate selection of aspects of technology operations that we have explored before.

Company Size Determination

Being complex, evolving, and very socialized institutions that consume varied resources and produce considerable value, high technology companies demand varied and complex measures to determine size. Sales revenue is certainly one way. Market share is another. Production numbers and productivity are still others. Our choice, however, was to use current employee numbers. Simply put, size of the company was gauged by number of employees. This is but one dimension of size, but it is a well-reported and meaningful measure. Indeed, it was one of the very few measures that all high technology companies used. It was also a measure we believed was accurate, since it could be verified more easily than most other size standards. Employee counts seem to mean something inside organizations, as well. We were told that at IBM, for example, the number of employees you had was one of the surest ways of establishing status. Other measures were regarded as less reliable within Big Blue.

With the measure of size determined, we applied it to our surveyed group of established high technology companies. The employee numbers were placed in ranges and the companies were divided into five classifications: less than 5 employees, 6 to 10, 11 to 24, 25 to 49, and 50 and greater. Various organizational factors, including the now familiar spectrum of leadership, marketing, and structural variables, were put against the size classifications to see if things changed with size. Not all factors were looked at because that would not be practical. And some for which it was obvious that size made a difference were excluded, such as sales, complexity of marketing operations, and product mix. It was not necessary to plow this ground.

Comparison Factors

Corporate Ownership. Our initial sweep of the subject concerned corporate ownership. Table 15.1 presents the number of firms (as the top term) and percent of firms (as the bottom term) for the classifications of either privately held or publicly held companies.

Private ownership was very highly and statistically significantly associated with size. Small companies were totally privately owned, while slightly over 69 percent of companies with more than 50 employees were privately owned. As the head count goes up, the march to public trading hastens. (Although we are not referring to individual firm evolution in

Table 15.1   Firm Ownership and Size

Employee Numbers	Privately Held	Publicly Held
0–5	53/100%	0/0%
6–10	32/94.1%	2/5.9%
11–24	44/91.7%	4/8.3%
25–49	22/88%	3/12%
50 and more	45/69.2%	20/30.8%

this comment, we make the point that we have immediately discovered a significant size difference.)

Control. The next determination was to connect size to whether the technology firm was independent or a unit of another company. The results are depicted in Table 15.2. Almost all companies with fewer than 50 employees are independent—exceeding 90 percent of our sample. However, there is a sudden dropoff of independent ownership for companies over 50 employees; only 60 percent of these are independent. This may indicate that spunoff divisions of parent companies are fairly sizable. It could also mean that independent technology firms become acquisition candidates after they have more than 50 employees. This result was statistically significant.

Scope. Do companies of different sizes tend to be regional or do they become international in scope? The surveyed technology companies fell into the four categories of geographic scope described in Chapter 14. The results are summarized in Table 15.3.

The most singularly pronounced discovery was that, no matter what the size of the technology firm, most of the larger firms believed the scope of their business was international. That is not to say that half their business sales were international. It is that they have a global view of their market.

The findings strongly supported the regional focus of smaller companies and the enlargement of that market with an increasing number of employees. While it is sensible that market horizons expand with size, there is no necessity that it occur. A growing technology company could

Table 15.2   Independent Firm or Corporate Unit

Employee Numbers	Unit of a Firm	Independent
1–5	1/2%	50/98%
6–10	3/8.8%	31/91.2%
11–24	3/6.3%	45/93.8%
25–49	1/4.2%	23/95.8%
50 and more	25/39.7%	38/60.3%

Table 15.3   Geographic Scope and Size

Employee Numbers	Local	Regional in U.S.	All of the U.S.	More than 1 country
1–5	7/13%	3/5.9%	16/31.4%	25/49%
6–10	2/5.9%	2/5.9%	7/20.6%	23/67.6%
11–24	1/2.1%	2/4.2%	11/22.9%	34/70.8%
25–49	0/0%	2/8%	4/16%	19/76%
50 and more	0/0%	6/9.1%	9/13.6%	51/77.3%

keep its regional focus, especially if it is located near customers, but that is not the case here.

Hierarchical vs. Flat Organization. Turning to another possible differentiation based on sized, we looked at two structural dimensions: how flat or how hierarchical the organizations are. The responding firms placed themselves in either of these categories or in the category of "other." The results are shown in Table 15.4.

Larger companies were much more hierarchical than flat and this was expected. The result was highly significant. This is certainly no surprise, but it is an especially strong finding. Only 6 percent of the very small firms were hierarchical while 25 percent of the largest firms were.

While it may seem absurd that very small companies can be hierarchical, some have actually reported that they are. What's more, we knew an eight-person software consulting firm that had a minimum of three layers (or more properly people) from the president to the software engineer. The engineer would have to report up the ranks to get virtually any kind of work approvals. This was probably the highest density bureaucracy we have ever seen, if density is a matter of bureaucracy per employee.

Overall Structure. Since technology companies are found in different sizes, do they show differences in the way they structure themselves? Part of this issue was addressed in Chapter 13, and we did find different structures for technology companies. Here, we found a shift from customer or product organization for small firms to functional or divisional forms

Table 15.4   Organization Hierarchy and Size

Employee Numbers	Hierarchical	Flat	Other
1–5	3/6%	44/88%	3/6%
6–10	2/6.5%	28/90.3%	1/3.2%
11–24	1/2.2%	43/93.5%	2/4.3%
25–49	4/16.7%	18/75%	2/8.3%
50 and more	16/25.4%	41/65.1%	6/9.5%

Table 15.5 Organization Structure and Size

Employee Number	Customer/ Product	Geographic	Function/ Division
1–5	28/60.9%	1/2.2%	17/37%
6–10	21/70%	0/0%	9/30%
11–24	21/46.7%	0/0%	24/53.3%
25–49	13/54.2%	1/4.2%	10/41.7%
50 and more	27/43.5%	3/4.8%	32/51.6%

with growth into the over 50 employee size firm. The results are shown in Table 15.5.

The shift clearly does not happen until the company crosses the 50-employee threshold. The shift toward less independence also happens at the 50-employee threshold because companies tend to be units of other organizations rather than freestanding firms. The 50-employee mark is a convenient demarcation between the smaller and larger high technology firms. The size of the technology firm does make a difference in the kind of organizational form employed.

Chief Executive's Characteristics. Can something be concluded about the size of the high technology company and the characteristics of the chief executive? On our question as to whether the CEO was a founder of the company, the replies were categorized by size of firm. A summary of the results is contained in Table 15.6.

When we explored the responding companies, we found that 92.5 percent of the smallest firms had CEOs who were founders, but this percentage dropped off linearly to 39.7 percent for firms with more than 50 employees. This was highly statistically significant. The size of the company gives a good indication of whether the founder is still CEO.

CEO Education. Another possible size consequence is the educational level of the CEO. The educational level of the CEO was matched to size of the company. The results are depicted in Table 15.7.

In looking further into the CEO's background and the size of the company, we found that the smallest companies had the highest level of

Table 15.6 CEOs as Founders

Employee Numbers	CEO a Founder	CEO not a Founder
1–5	49/92.5%	4/7.5%
6–10	26/83.9%	5/16.1%
11–24	36/78.3%	10/21.7%
25–49	15/62.5%	9/37.5%
50 and more	25/39.7%	38/60.3%

Table 15.7   CEO Education Level and Size

Employee Numbers	High School	Some College	College Degree	Graduate Work	Post-graduate
1–5	1/1.9%	6/11.3%	7/13.2%	13/24.5%	26/49.1%
6–10	3/9.7%	1/3.2%	4/12.9%	12/38.7%	11/35.5%
11–24	0/0%	6/12.8%	12/25.5%	16/34%	13/27.7%
24–49	0/0%	3/12.5%	5/20.8%	10/41.7%	6/25%
50+	0/0%	1/1.6%	21/33.9%	20/32.3%	20/32.3%

education among their CEOs, with 49.1 percent having postgraduate work. Only 25 percent of medium-size company CEOs (25–49 employees) had postgraduate educations. This, too, was statistically significant.

To find the most educated technology leaders within a business grouping well populated with well-educated leadership, look at the smallest high technology companies. In this group, we find the newly minted genetics PhD or the assistant professor of electrical engineering launching a new technology company. The companies are built around this core of newly acquired knowledge.

CEO Experience. Past work experience of the CEO was another factor we considered, and the result was that the smallest companies had CEOs with the greatest industry-related experience. This is summarized in Table 15.8.

Industry-related experience drops as size increases, to a minimum of 62.5 percent for companies in the 25–49 employee range, then it rebounds to 72.1 percent in the largest companies. The differences among employee categories were again found to be statistically significant.

It is an interesting note that the smallest companies had the CEOs with the most industry-related experience. The expectation was that the larger companies would have CEOs with more industry-related experience. Yet the small companies reported having CEOs with considerable education and experience, a doubly rich base to tap into. The expectation that there was more industry-related experience in the larger firms may not have been met because the larger companies brought in CEOs from other business fields. This is speculation, but it is possible.

The industry-related experience factor is another reason to state that there are consequential differences between and among the different size technology companies. The smallest companies are apparent treasuries of experience and education.

Industry Life Cycle. The idea that these technology businesses occupy different positions within industries that rise and fall is a theme throughout the book. To view this in the context of corporation size, we sorted the companies by the number and percentage occupying each size classes. When the phase of the industry life cycle was considered, the small com-

Table 15.8   CEO Experience and Size

Employee Numbers	Experience Not Industry Related	Experience Industry Related
1–5	5/9.6%	47/90.4%
6–10	5/16.1%	26/83.9%
11–24	8/17%	39/83%
25–49	9/37.5%	15/62.5%
50 and more	17/27.9%	44/72.1%

panies spread across the four stages while the largest companies were mainly found in the growth stage. Again, the differences were statistically significant. The results of industry life cycle and size considerations are in Table 15.9.

Product Type. Our review was now concerned with several technology marketing factors where there had been no previous resolution of marketing factors and company size. One dimension for which there were no real differences was standardized vs. customized products. This dimension captures the results of making either very similar products or one-of-a-kind products for customers, as discussed in Chapter 14. The results of the product standardization or customization can be seen in Table 15.10.

Distribution. The number of distribution channels might be associated with the size of the technology company. It would be reasonable to speculate that smaller firms would have fewer channels. Smaller firms not only are likely to need fewer distribution channels but also are less likely to have them available, since they are not in a position to attract distributors owing to their small scale of operations.

Ranges for the number of channels were arranged against the size of the company. The results are shown in Table 15.11.

No matter what the size of the technology firm, no less than 70 percent use three or fewer distribution channels. The majority of small companies had a single distribution channel as expected, but in the largest companies, the mode (40 percent) of the companies had two to three

Table 15.9   Industry Life Cycle and Size

Employee Numbers	Introduction Stage	Growth Stage	Maturity Stage	Decline Stage
1–5	11/21.2%	12/23.1%	19/36.5%	10/19.2%
6–10	3/9.7%	14/45.2%	9/29%	5/16.1%
11–24	3/6.5%	29/63%	6/13%	8/17.4%
25–49	1/4.2%	13/54.2%	6/25%	4/16.7%
50 or more	2/3.6%	33/58.9%	14/25%	7/12.5%

Table 15.10   Product Standardization/Customization and Size

Employee Numbers	Standard Products	Custom Products
1–5	22/45.8%	26/54.2%
6–10	14/42.4%	19/57.6%
11–24	24/52.2%	22/47.8%
25–49	14/56%	11/44%
50 or more	27/42.2%	37/57.8%

channels. The most interesting finding was that distribution channels do not appear to grow with size. Three channels is the upper limit for the great majority of firms. Although not statistically significant under the standards we employed, the results were close to being significant ($P = .113$).

Middlemen. Related to the number of distributors is the matter of middlemen within each distribution channel. As we described in Chapter 14, the middlemen are those who take and resell product. Computer wholesalers are middlemen when they buy from the manufacturer and sell to the retail store. We expected the smaller firms to have fewer middlemen because they tend to have simpler distribution in general. The results showed different outcomes, however, as displayed in Table 15.12.

One point of commonality among these different size companies is the number of middlemen they have enroute to the final customer. At least two-thirds of the companies in any of the five size categories had single middlemen. Fewer companies had two to three middlemen. Differences were not statistically significant. In all, then, the number of middlemen does not provide a basis for distinguishing among different size technology companies.

Growth Rate. Rate of growth is one aspect of that has thus far been a strong and reliable marker of high technology businesses. As such, it was certainly worth determining if there are ramifications as far as size is concerned. Our interest was in how these companies believe they are growing compared to how the industries in which they do business are

Table 15.11   Distribution Channels and Size

Employee Numbers	No Channels	One Channel	2–3 Channels	4–5 Channels	6 or More
1–5	2/4.8%	23/54.8%	11/26.2%	4/9.5%	2/4.8%
6–10	1/3.4%	9/31%	14/48.3%	2/6.9%	3/10.3%
11–24	1/2.4%	14/33.3%	13/31%	6/14.3%	8/19%
25–49	1/4.2%	4/16.7%	12/50%	0/0%	7/29.2%
50+	1/1.8%	21/38.2%	22/40%	4/7.3%	7/12.7%

Table 15.12   Distribution Middlemen and Size

Employee Numbers	No Middlemen	One Middleman	2–3 Middlemen	4 or More Middlemen
1–5	2/7.4%	22/81.5%	3/11.1%	0/0%
6–10	3/11.5%	18/69.2%	4/15.4%	1/3.8%
11–24	3/9.1%	22/66.7%	7/21.2%	1/3%
25–49	1/4.8%	18/85.7%	2/9.5%	0/0%
50+	1/2.3%	37/86%	4/9.3%	1/2.3%

growing. We used the same qualitative measures of growth (faster, the same rate, or slower) as in Chapter 14 and compared the measures to the size of the firm. What did these companies say about their rate of growth compared to their industries? The results are summarized in Table 15.13.

In essence, the smallest companies believe they are growing slower than their industries, the medium companies at about the same rate, and the largest companies believe they are growing faster. In this case the differences were statistically significant.

Product Life Cycle. High technology products experience a life cycle of their own. Concerning the product life cycle of their products, size of company seems to matter. In Table 15.14 the typical stages of the product life cycle are compared to size of firm.

In the smaller firms, products were believed to be spread across five stages (introduction, rapid growth, growth, maturity, decline) while the majority of larger firms reported their products to be in the growth or maturity stage. These differences, though notable, were not statistically significant. The smaller companies may have fewer high technology products, but there is more scattering of these fewer products across the product life cycle.

Pricing vs. Industry. The issue of pricing was found to be an especially useful way of characterizing these businesses. Using the same pricing classification as in Chapter 14—much higher than industry, higher than the industry, in parity with the industry, lower than the industry, and much

Table 15.13   Growth Rate and Size

Employee Numbers	Growing Faster	At the Same Rate	Growing Slower
1–5	8/16.3%	12/28.6%	27/55.1%
6–10	8/23.5%	18/52.9%	8/23.5%
11–24	19/41.3%	23/50%	4/8.7%
25–49	14/58.3%	5/20.8%	5/20.8%
More than 50	40/63.5%	16/25.4%	7/11.1%

Table 15.14   Product Life Cycles and Size

Employee Numbers	Intro- duction	Early Growth	Growth	Maturity	Decline
1–5	9/20.5%	11/25%	7/15.9%	13/29.5%	4/9.1%
6–10	7/25%	4/14.3%	8/28.6%	8/28.6%	1/3.6%
11–24	10/22.7%	6/13.6%	16/36.4%	9/20.5%	3/6.8%
25–49	5/20%	3/12%	13/52%	4/16%	0/0%
50+	10/16.7%	3/5%	22/36.7%	18/30%	7/11.7%

lower than the industry—we compared the size-subdivided firms to the pricing classifications. Table 15.15 shows the results.

There were significant differences among the firm sizes. The larger companies set prices higher than the industry while the smaller companies tended to set prices in parity or lower than the industry. The most popular response for all sizes of company was to set prices in parity with the industry. Because the differences were significant, our conclusion is that price setting practices further differentiate these technology firms and that difference is related to size.

Pricing vs. Competitors. What about pricing practices in relation to competitors? The results of this question were similar to the industry pricing question and are shown in Table 15.16.

The pattern is toward larger firms setting higher prices than competitors and smaller companies setting prices in parity or lower than competitors. These results were statistically significant. In these two pricing considerations, we have significant agreement and reason to believe that the size of the company makes a difference in pricing.

Price-setting Criteria. The bases for setting prices was also investigated. As in Chapter 14, we set the possible criteria as cost, competitors, consumer demand, and other. We again discovered differences based on size, though they were not statistically significant in this instance. The results are summarized in Table 15.17.

For the smallest companies, costs were the most frequent answer, but that shifts to consumer demand for a medium-size company and then

Table 15.15   Pricing Compared to Industry

Employee Numbers	Much Higher	Higher	In Parity	Lower	Much Lower
1–5	4/8.7%	5/10.9%	17/37%	16/34.8%	4/8.7%
6–10	3/10.7%	7/25%	13/46.4%	4/14.3%	1/3.6%
11–24	3/7.0%	11/25.6%	22/51.2%	4/9.3%	3/7%
25–49	1/4.2%	12/50%	6/25%	5/20.8%	0/0%
50+	4/6.6%	24/39.3%	28/45.9%	4/6.6%	1/1.6%

Table 15.16   Pricing Compared to Competitors

Employee Numbers	Much Higher	Higher	In Parity	Lower	Much Lower
1–5	2/4.8%	3/7.1%	17/40.5%	13/31%	7/16.7%
6–10	2/7.1%	7/25%	9/32.1%	7/25%	3/10.7%
11–24	2/4.4%	9/20%	20/44.4%	13/28.9%	1/2.2%
25–49	1/4.3%	6/26.1%	8/34.8%	6/26.1%	2/8.7%
50+	5/8.1%	22/35.5%	29/46.8%	4/6.5%	2/3.2%

back to costs for the two largest classifications of technology company size.

The anomaly in the 11- to 24-employee firm is a weak departure from the overall pattern of cost-driven pricing. There was a difference of only one company between these two categories. In all, the comparison of differences among the groups was not determined to be statistically significant.

The largest companies had the greatest dispersion on pricing-setting criteria. This could reflect a greater array of products upon which different pricing strategies are used. The smaller firms have fewer options if they have fewer products.

The high technology companies we have worked with or know about view pricing as an art rather than a science. These may be science-based companies, but they do not use formulas for pricing. No software pricing models to speak of are used as the main determinants of price. Some modeling is done in the larger firms, but it is mainly confined to forecasting demand. Instead, pricing is done on the basis of judgment, with some use of decision-support methodology, but that was only in companies with more than 50 employees.

Price-setting Responsibility. As to who makes the pricing decisions, we found that in the very small companies we talked to it is the province of the company head. In firms with 50 or more employees, there is usually a sales or marketing manager who consults with the CEO to make prod-

Table 15.17   Pricing Factors

Employee Numbers	Costs	Competitor Prices	Consumer Demand	Other
1–5	18/38.3%	9/19.1%	11/23.4%	9/19.1%
6–10	12/41.4%	2/6.9%	11/37.9%	4/13.8%
11–24	16/35.6%	5/11.1%	17/37.8%	7/15.6%
25–49	9/36%	7/28%	5/20%	4/16%
50+	19/31.1%	13/21.3%	17/27.9%	12/19.7%

uct pricing decisions. Usually there is input from sales staff, especially about what competitors are charging.

Product Changes. The frequency of change in products might be way in which these technology companies exhibit differences or similarities. We subdivided the firms by size and compared them in product change intervals of annually, one to five years, and more than five years. The results are shown in Table 15.18.

The results showed that most of the companies did product changes between one to five years. The exception was the 11- to 24-employee group, where nearly half of the firms made product changes annually. The results did not pass statistical significance, however.

Competitor Response. The time it takes for the technology firm to respond to a move by a competitor could take many forms: competing product, price reduction, and counterattack by targeting certain competitor's customers. When considering the competitor's reaction time to new products, we found consistent agreement that it was between one and three years, with most responding firms putting competitor reaction in that category, as depicted in Table 15.19. The finding did not pass statistical significance, however, so we did not make conclusions about the competitor reaction and size matter.

Response to Competition. A much different observation can be made concerning the reaction time of the different-size companies to their competitor's new products. When technology companies looked inward, they called themselves more responsive and the time dimension of responsiveness is connected to size. The pattern of response is shown in Table 15.20.

The smaller companies react much faster than the larger companies. This establishes that there are differences in firms based on their size, although there is not a progressively slower reaction with larger firms. The 11- to 24-employee size companies reacted fastest. The differences were also found to be statistically significant.

Supplier Location. The location of suppliers—specifically how far away they are in relation to the technology company—was another aspect that could be a consequential discriminator among different sizes of technology firms. The distance classifications were done exactly in Chapter 14

Table 15.18 Frequency of Product Changes

Employee Numbers	Annual Product Changes	1–5 Year Product Changes	More than 5 Year Product Changes
1–5	13/29.5%	21/47.7%	10/22.7%
6–10	7/25.9%	14/51.9%	6/22.2%
11–24	19/46.3%	15/36.6%	7/17.1%
25–49	8/36.4%	8/36.4%	6/27.3%
50+	10/18.5%	24/44.4%	20/37.0%

Table 15.19   Competitor Reaction Time

Employee Numbers	Less than 1 Year	1–3 Years	More than 3 Years
1–5	12/32.4%	18/48.6%	7/18.9%
6–10	6/27.3%	14/63.6%	2/9.1%
11–24	10/27.8%	20/55.6%	6/16.7%
25–49	5/22.7%	11/50%	6/27.3%
50+	12/20.3%	38/64.4%	9/15.3%

less than 10 miles, 11 to 20 miles, 21 to 30 miles, 31 to 40 miles, and more than 40 miles.

The expectation is that the smaller companies would have closer suppliers than the larger companies because they may not have the resources to go far afield. The larger firms might have this opportunity, however, as suppliers start courting them instead of vice versa. The results of this review are depicted in Table 15.21.

Although the smallest companies showed the greatest dispersion and the highest percent of nearby suppliers, the most distinct pattern showed a majority of suppliers for all the size classifications were located more than 40 miles away. Although the differences were not statistically significant, they came close ($P = .126$). The location of a majority of suppliers does not provide a basis for saying that there are differences based on firm size. Our expectation of a difference did not hold up.

Technology transfer. While technology transfers were not found to be a critical matter for the technology companies, it may still be a basis for differentiating companies. Respondents assessed the significance of technology transfers from outside their company if they accepted such transfers. They categorized their opinions within five response choices, and the results are shown in Table 15.22.

The smallest firms placed most importance on technology transfers. Companies with more than six employees believed technology transfers

Table 15.20   Own Company Reaction Time to Competitors

Employee Numbers	Less than 1 Year	1–3 Years	More than 3 Years
1–5	25/67.6%	12/32.4%	0/0%
6–10	11/55.0%	9/45.0%	0/0%
11–24	25/73.5%	5/14.7%	4/11.8%
25–49	6/28.6%	11/52.4%	4/19%
50+	22/40.0%	29/52.7%	4/7.3%

Table 15.21   Supplier Distance

Employee Numbers	0–10 Miles	11–20 Miles	21–30 Miles	31–40 Miles	41+ Miles
1–5	7/17.5%	4/10%	2/5%	2/5%	24/60%
6–10	3/10%	0/0%	6/20%	2/6.7%	19/63.3%
11–24	1/2.2%	3/6.5%	2/4.3%	1/2.2%	39/84.4%
25–49	2/8.3%	3/12.5%	0/0%	1/4.2%	18/75%
50+	7/12.5%	2/3.6%	4/7.1%	4/7.1%	39/69.6%

were mainly somewhat important or neither important nor unimportant. There was no statistical significance in this finding.

That does leave us noting that technology transfer is a mild basis for separating high technology firms in this group, although the proof standard was not met. We can't dismiss the overall pattern of responses, however, as a general indication of difference.

In one area of technology transference, there was resounding agreement among the companies, no matter what their size. That was on the question of whether the company had transferred technology to a non-affiliated organization. Remembering that these companies not only use the scientific breakthroughs of others but also create them themselves, it's an area worth exploring. An example of external technology transfer is the Ampex Corporation, which developed and passed on the video recording technologies used in every home camcorder today.

The results of this consideration are summarized in Table 15.23. No less than 68% of the firms of any size reported that such transfers were not done. Outside technology transfers are not a pronounced behavior, though the smallest and largest companies do them more frequently than other size firms. These results were not statistically significant.

Patents. The number of patents offered is way that differences among the companies might be found. As expected, the number of patents currently held is generally related to size, as shown in Table 15.24.

In addition to patents held, the patent award rate was reviewed. The

Table 15.22   Significance of Technology Transfers

Employee Numbers	Very Significant	Somewhat Significant	Neither	Somewhat Insignificant	Very Insignificant
1–5	9/36%	7/28%	5/20%	1/4%	3/12%
6–10	1/7.1%	7/50%	4/28.6%	1/7.1%	1/7.1%
11–24	4/22.2%	9/50%	4/22.2%	0/0%	1/5.6%
25–49	3/23.1%	6/46.2%	1/7.7%	0/0%	3/23.1%
50+	8/25%	13/40.6%	4/12.5%	2/6.3%	5/15.6%

Table 15.23   Technology Transfers Outside the Firm

Employee Numbers	Did Transfer Technologies	Did Not Do Transfer Technologies
1–5	16/32%	34/68%
6–10	6/18.2%	27/81.8%
11–24	9/19.1%	38/80.9%
24–49	5/20%	20/80%
50 or more	15/24.2%	47/75.8%

companies reported fewer than one patent per year were companies under 25 employees. But that figure dropped to one patent every two years for companies with between 25 and 49 employees, and then takes a major leap to nine per year for the largest companies. This was not found to be statistically significant. In the 50+ employee category, there was one firm that had hundreds of patents. This pulled the average higher.

Thus, size of the technology company is no sure indicator of patent acquisition activity. Put this way, patent acquisition is way of discovering differences based on organizational size rather than similarities.

Business Strategy. When the business-level strategy of the technology company was considered, there were some performance consequences for both ventures and established companies, as we described in Chapter 7 for the ventures and Chapter 12 for the established companies. We turned now to look at strategy from the perspective of company size.

In our review of the different basic strategies of market share, growth, profitability, market concentration, turnaround, and liquidation or other, there are differences based on the size of the technology company. In the two categories of smallest companies, the predominant strategy choice was profitability. But that shifted to a market-share increasing strategy for all three remaining categories of business size.

The extent of these differences is presented in Table 15.25. Although the differences were not found to be statistically significant in the typical range of acceptability, there is slight statistical support ($p = .3$) for recognizing differences. The fact remains a uniform strategy is not evident

Table 15.24   Patents and Size

Employee Numbers	Average Number of Patents Held
1–5	1.9
6–10	0.75
11–24	4.1
25–49	12.66
50 or more	107.0

in these companies. Strategy does appear to change with size, though it is not entirely provable for this group. What's more, the kinds of strategies are consistent with what we have been told have been used by technology heads. Initial strategies are aimed at recouping initial investments. Later strategies are aimed at keeping market share or at building market share. These strategies were those actually reported. They also fit the particular circumstances of the environment and competitive position of the technology company.

Education. The subject of education is one that might show differences based on the size of the technology company. The findings were especially interesting in this regard because the smallest companies had the best-educated employees. Slightly over 51% of the smallest firms had employees with a graduate degree or higher. That dropped to 24% for the companies in the 6- to 10-employee range, 19% for the 11- to 24-employee range, and then increased to 32% for the 25- to 49-employee range. The lowest mean was 16% in the largest companies. The differences were found to be significant.

The smallest firms were likely to have considerable technical expertise among a few people—that expertise is assumed to be related to their educational level. That educational level is lower in larger organizations reflect their more diverse needs, some of which do not come from the university.

Unionization. With the issue of unionization, there was a very clear break by size of firm, with no unionization in companies under 50 employees. Above that, only 4% of the technology firms had union workers. This was a statistically significant finding.

If there is one description of the high technology worker that best applies to small companies it is that he or she is not a member of a union. It also appears that it is not until a company is fairly sizable (for high technology firms) that unionization is a factor, and then it is a small one.

Number of Customers. What can be said about the number of customers and the size of the technology firm? The customers are unevenly distributed. The smallest companies had about 350 customers, but that number dropped to 131 for medium-size companies (between 25 and 49 employees) and jumped to 2,300 customers for companies with more than 50 employees. A somewhat similar pattern applied to foreign customers. While we cannot demonstrate these differences as statistically significant, they suggest the reasonable conclusion that customer numbers vary. They also give an indication of just how dramatic that variation is.

Summary of Results

The objective of this chapter was to determine if size makes high technology companies more different from one another or more the same. In this section we highlights and summarized the findings. In order to draw these assorted pieces of information to some overall conclusion, we have summarized the results in Table 15.26, which simply indicates

Table 15.25   Strategy and Size

Employee Number	Market Share Increasing	Growth	Profita- bility	Market Con- centra- tion	Turna- round	Liqui- dation	Other
1–5	10	9	14	2	7	2	3
	21.3%	19.1%	29.8%	4.3%	14.9%	4.3%	6.4%
6–10	7	6	11	3	4	0	2
	21.2%	18.2%	33.3%	9.1%	12.1%	0%	6.1%
11–24	20	11	6	3	4	0	2
	43.5%	23.9%	13%	6.5%	8.7%	0%	4.3%
25–49	10	6	4	1	2	0	2
	40%	24%	16%	4%	8%	0%	8%
50+	28	16	9	1	6	0	3
	44.4%	25.4%	14.3%	1.6%	9.5%	0%	4.8%

points of similarity and differences among the firms. The organizational factors are opposite the conclusions regarding organizational size.

Do things change or remain the same for different size companies? Our responses fell very much on the side of things being different for different size technology companies. Our summary results show that of 22 organizational factors, all but five showed differences based on size. While the size of the firm explains much about how companies differ in any industry, it is especially pronounced in high technology. We found substantial differences even in the small employee number ranges. And there were additional dimensions, such as patent activity, which are not common across all size companies in which size differences were found.

Not only are there differences in technology firms because of the nature of their business, there are also differences among companies in the same line of business because of their size. Things change considerably as technology companies fall into size classes—so much so that size must be a factor in explaining the essential nature of these companies.

The revelation that technology company size makes a difference gets to the crux of the problem in defining *high technology*. Conventional definitions are based on workforce percentages and percentages of research and development intensity. With these parameters, size is inconsequential. A very large technology company and a very small one could have the same workforce and research percentages and so be placed in the same high technology bin. Yet we have seen how size really separates these companies more than it units them. The percentage definitions are inadequate.

This then leads to questioning the value of proportion-based methods in defining high technology business. From what we have found out, this method needs considerable supplementation or abandonment be-

Table 15.26 Summary Table on Firm Size

Organizational Factor	Summary
Private/public ownership	All small companies privately held, 31% of larger companies publicly held
Independent or corporate unit	Almost all small companies independent, 40% of larger companies are corporate units
Geographic market scope	More regional scope for smaller companies, national scope for larger firms. All firms see international scope
Hierarchical or flat structure	Small firms have flat structures, larger firms more hierarchical
Company basic organization	Small companies have customer/product organizations
	Larger firms have functional/divisional form
Educational level of employees	highest for small firms, lowest for largest firms
Extent of unionization	No differences associated with size since unionization largely absent in all firms
CEO characteristics	More smaller firms have founder CEOs; smallest firms most educated CEOs; smaller firm CEOs have more industry-related experience
Industry life cycle stage	Small companies found across different stages; larger firms in growth stage
Product standardization/ customization	No differences based on size
Number of middlemen	No difference based on size
Supplier location	Smaller companies have nearby suppliers but most suppliers more than 40 miles away for all size ranges—no major differences based on size
Number of distributors	Small firms have single distributors, larger firms with 2–3 distributors
Growth rate of company	Smaller firms with slower growth than larger firms
Product life cycle	Small companies have product in range of stages; larger firm products more in growth stage
Customer numbers	Smaller companies with fewer customers; medium-size firms with fewer customers than larger firms
Pricing practices	Smaller companies pricing in parity with industry and in parity or lower than competitors; larger firms pricing higher than industry and competitors; small firm price basis is cost, same for larger firms.
Product change frequency	No differences found in size

Table 15.26   (continued)

Organizational Factor	Summary
Value of technology transfers	Smaller companies believe transfers more important
External technology transfers	No differences in size, external transfers uncommon
Patent activity	Substantial patent activity only in largest firms
Basic business level strategy	Smaller companies use profitability strategies and larger companies use market share increasing strategies

cause it does not presently capture the character of this business grouping.

We will now turn to answering the questions of where high technology companies are found. That is a topic of constant interest and the answer almost always changes. Nevertheless, it is consequential information for the technology company and for the federal, state, and local governments. It is therefore consequential for the technology stakeholder and taxpayer alike, as well as many others. It is our next subject.

IV

LOCATION, SUPPORT, AND CONCLUSIONS

16

HIGH TECHNOLOGY
BUSINESS
CONCENTRATIONS

The great epicenters of high technology development, Boston's Route 128 and California's Silicon Valley, are the extreme continental limits of high technology business locations. The trading limits of these poles are even greater. The companies are essentially global in scope and thus not constrained by geography. But our look here is at the locations of high technology businesses—where they locate their headquarters or major divisions and where the high technology they do originates. Although these businesses span the continent, they do not do so with uniform density. There are pockets of high technology prominence, some of which happen in expected areas, such as near research universities, but they may also be found in unexpected places, especially in emerging high technology development areas. That is what we look at in this chapter, along with a characterization of these sites.

Depending on how high technology locations are determined, there are at least 21 and as many as 30 centers for high technology business. Not all high technology firms are located in these areas, but it is in these centers where we can see how vibrant this enterprise is. The geographic centers contain the firms themselves. They also hold specialized suppliers and, in some cases, most of the customers.

How are the centers defined? Various formulations have different criteria, but they commonly (and the way the term is implemented here) use the number of high technology firms in relation to all local firms and existence of a high technology employment base as essential definers.

Where are the centers? There are broad regional multistate development areas all along the Atlantic Coast, in the Southwest, and along the Pacific Coast. A few other states pepper the county with high technology sites, but for the most part this broad generalization is about the most that can be said regarding geographic aspects of high technology. What follows are brief depictions of high technology concentrations based mainly on our contacts with local business development offices.[1] Our focus here is on communities rather than regions or states. This helps better pinpoint the concentration and show, at the lowest possible level, differences between communities.

The principal locations are enumerated, followed by some distinctive aspects of each location. Our launching point is at the very origin of high technology, the Boston area. From there we weave across and up and down the country to describe other focal points.

Boston and Route 128

The first wave, and now the most diversified area for advanced technology, is the greater Boston area, inhabited by computer and semiconductor firms, telecommunications, aerospace, and medical electronics companies. These are the core high technology firms, though other forms of business dominate the city to the point that Boston does not have a distinctively high technology flavor. What it does have is high technology endurance. Having endured economic boom and bust periods in over forty years of high technology business lends a respectable robustness to the proper Bostonian high technology environment.

With such a large number of very different companies, can we fairly characterize this as a concentration? The answer is yes because this area has the greatest intrusiveness of high technology into other, more traditional forms of business. About one-third of manufacturing employment is high technology employment, where computer-integrated manufacturing abounds. The publishing companies here make considerable use of digital typesetting, scanning, and layout systems and the many hospitals are sites for state-of-the-art medical diagnostic equipment.

The focus is mainly in Boston, but high technology is located all around the area, most famously along its surrounding highways. While much of the concentration remains along Highway 128, current development is on I-495. This represents a second generation of the computer and software companies (led by DEC) that made Highway 128 a technology landmark.

The educational strengths of Boston have been among the major reasons for creation of technology firms here. The academic connection is MIT, Harvard University, Boston University, the University of Massachusetts at Boston, and other area colleges. Not only do faculty from these institutions start new businesses but they advise ongoing technology firms. There is no area of the country where university based expertise, investors, and entrepreneurs have more successfully interacted to form new companies. We gave a fuller description of this sacred technology ground in Chapter 2, in our review of the history of high technology.

Boston is the first settlement for American high technology, the point from which Silicon Valley emerged and from where Atlantic Coast emigrants left to inhabit the Research Triangle and other parts of the country. The homeland has not always prospered, however. More recently, the technology companies of Boston have hit on hard times. The heyday of defense spending is long gone and a combination of high taxes, high wages, expensive leases, and high housing costs has caused technology companies to relocate.

Even though the amount of defense-supported research has diminished, the 31 hospitals and research institutions in the area draw more than $500 million in federal research grants, ranking Boston as the leading research grant recipient city. If there is a cast to the high technology done in Boston, it is basic and applied research supported by federal funding rather than in-house corporate technology development.

There has been a mid-1990s revival of technology of sorts. Resuscitation of the city's technology base has come through the Boston Redevelopment Authority and the creation of the Technopolis, a center for technology development that brings together academics, research institutions, and private individuals in an effort to initiate and support technology business. Another injection of energy came when Gentyme, a pharmaceutical firm, built a $110 million plant in the city.

Because Boston's longstanding history of high technology, its educational and transportation infrastructure, and the availability of highly educated (and surprisingly low-cost) technical labor continue to attract high technology businesses, the city is still a magnet. A further and very important attraction is its strong technology investment heritage. Investment bankers look for technology companies in Boston because they have found them to be good investments.

Overall, the biggest companies are in insurance, banking, and publications, not the obvious high technology firms. It is thus not so apparent where the high technology is happening, though this certainly has been the case for nearly half a century. In Boston, the high technology is in service-supplier firms, the software shops, the consultants, and the electronic forms processors that subcontract with the largest companies.

The Greater New York Metropolitan Area

This concentration is one of both population and industry—so much so that the high technology sector may be overlooked. All kinds of major businesses are located in this area, and the huge population base provides the skilled labor for the high technology sector.

The area includes northern New Jersey and western Connecticut. Long Island is a high technology manufacturing area, led largely by the

Grumman Aircraft facility in Bethpage, with telecommunications operations also present. Northern New Jersey is the scene for pharmaceuticals and telecommunications. Many IBM facilities stretch north through the Hudson Valley. New York City itself is not a concentration point, but it is the funnel for high technology finance and commerce. High technology transactions happen in the city: venture capital operations and mergers and acquisitions.

The largest sectors of business in New York are telecommunications, aerospace, medical electronics, and phamaceuticals. Major technology companies include Singer-Kearfott, Hoffman-LaRoche, and Western Electric. High technology in the area is slower than it has been in most other areas in terms of number of firms started. The firms that begin here tend to be smaller (less than 100 employees) than in other areas.

New York City has, in the words of the state's economic development commissioner Charles Gargano, "one of the world's greatest concentrations of research and development institutions and firms." Particularly prominent in the city are investors in high technology ventures, a natural outcome of being the world's financial center.

While the city per se might not be thought of as a high technology site, technology creation and exploitation occurs in the headquarters operations of the many, large city-based businesses. Corporate decisions about technology development are made at headquarters offices; though it may be obscured, the high technology done in corporate research and development most often originates in the city.

Building a stronger technology base is part of the state's economic development strategic plan. More support of technology development and transfer, greater investment in key existing industries, insuring that assistance to universities provides benefits for New York firms, and facilitation of ways to generate seed capital for startups are the elements of the plan.

Baltimore and Central Maryland

Telecommunications, pharmaceuticals, and computers are the predominant high technology businesses in the Baltimore and central Maryland area. Westinghouse and Square D (electrical equipment) are two leading firms. A mark of high technology in this area is its diverse character. Large manufacturing companies and much smaller support service businesses represent the range of businesses.

The region has been able to link technology companies with financial resources because of the presence of many financial services companies. In this way, the area is similar to New York City. Much of the research and development is corporate based, but high technology is not a standout activity, mostly owing to the size and number of traditional companies.

In Baltimore, Baltimore Development Corporation President M. J.

Brodie has indicated that high technology companies locate in Baltimore area because it has the highest national concentration of scientists and engineers. These technical workers are in the Baltimore-Washington corridor. Financial services and ongoing technology operations in the area are other attractions.

About 500 venture capital–backed technology firms are located in the area, a clear indication of the availability of venture capital for technology entrepreneurs. The Port of Baltimore, major airports, and overnight trucking to two-thirds of the region's population are transportation factors. Additionally, federal laboratories such as the National Institutes of Health, the Federal Drug Administration Center for Biologics Evaluation and Research, and the Walter Reed Army Institute of Research are indigenous technical resources. The National Security Agency and the Social Security Administration are nearby. Johns Hopkins University and the University of Maryland are biological research centers.

The tendency for having bursts of high technology startups has reoccured in this region. Writer Janet Novack reported in the August 20, 1996, issue of *Forbes* that proximity to the federal government is especially attractive to telecommunications software startups. Some of the startup companies are Transaction Network Services, CyberCash, PSI Net, UUNet, and America Online. There was 25 percent growth in the number of technology companies between early 1995 and mid-1996, bringing the total of technology firms in this area to 1,600.

Research Triangle, North Carolina

The Research Triangle is located at the midpoint of three major research universities in North Carolina. That makes it unique among the high technology areas. In a sense, it is the only deliberate high technology concentration. The Research Triangle is sited between Duke University, North Carolina State University, and the University of North Carolina. Each institution has its own academic slant, but all contribute to the vitality of the region with technology transfers, technical expertise, and new engineers and scientists.

In this area, the concentration is on the computer business. That concentration and the heavy university involvement are the major definers high technology here. The Research Triangle's top technology companies are IBM with 12,000 employees, Northern Telecom, the SAS Institute (business analysis and statistical software), GTE, and the Research Triangle Institute (provides contract research).

The state commerce department reported that motor vehicles and equipment was the industry category with the highest number of workers (28,920), but that the computer programming business had the largest number of firms, constituting 44 percent of the total.

What might most likely draw new technology ventures or established firms to this area are the modernity of the infrastructure and the intimate educational resources, both essential for high technology growth. The Research Triangle is designed for high technology, and thus it attracts and retains these businesses.

This is the third most prominent high technology region, just behind Silicon Valley and greater Boston, and, as mentioned, probably the only designed center of the three.

Orlando

Tourism is the major business of Orlando, but the space business is also present nearby. Occupying the belt between the Kennedy Space Center and the Gulf Coast, the high technology firms of this area have a definite aerospace bent. Lockheed Martin has had a longstanding presence in the area, but there are more—a new generation of nonaerospace technology business has grown up. What's unusual about this area is a largely absent academic connection. There are no major research universities in this area, but there are university–high technology connections. The University of Central Florida's Institute for Simulation and Training is where military simulation, dynamic terrain research, and emergency response training are developed. The University of Central Florida has created a laser research center. Computer-based instructional programs and digitized voice and print technologies are researched and developed as well. There are also many photonics companies that are developing medical, industrial, and military uses for laser technology.

There are also telecommunications giants in the area. AT&T has a Complementary Metal Oxide Semiconductor Wafer Fabrication plant—a true high technology production facility—and Siemens Stromberg-Carlson, which produces electronic switching systems, has an operation in Orlando. Lures for high technology companies are the university connections, a fiber optic network, over 50 industrial parks, and no state or local personal income taxes.

Tampa-St. Petersburg

The Tampa Bay area is contiguous with the Orlando high technology belt. It, too, has a predominant form of business, but it is telecommunications, not aerospace. GTE Florida, the largest technology employer, has a workforce of 4,500. The manufacturing of defense telecommunications systems was the impetus for high technology growth.

The campus links are to the University of South Florida and the University of Tampa. More sources of young engineers than technology transfers, these institutions are nonetheless important high technology development factors.

The particular attractions of this area are the educated labor base, though much of this comes from immigration rather than the indigenous labor supply. The costs of doing business are relatively low. Additionally, low taxes and no personal income tax are economic advantages of this region. Overall, there are varied technology enterprises in the region.

Atlanta

This major metropolitan center has a large number of high technology companies. There are fewer "feeder" consulting firms and more larger and somewhat more conventional technology firms. The leading-edge technology giants are not found in Atlanta, but there are business units of large firms. As with other centers, support for high technology comes from major universities—Georgia State University and the nearby University of Georgia in Athens. The very strong engineering program at Georgia State is a source of engineers and academic consultants.

The largest technology-related businesses are AT&T and Delta Airlines. Beltronics and Hitachi Home Electronics are leading high technology businesses. There is a division of Hewlett-Packard in Atlanta, and the Institute of Nuclear Power is also based there. Nearby is Martin Marietta.

There is no dominant type of technology firm, largely because Atlanta is a major metropolitan center with many technology and general business employers. But even the conventional giants such as Coca-Cola do research and development work on automated process control operations and other technology applications. The Center for Disease Control also is a technology user; it uses testing methodology that was in the area's experimental laboratories only months earlier.

An attraction for prospective firms is the pro-business attitude of city leaders. The Beltronics electronics company is a large transplant to Atlanta. Transportation access and the lower than average cost of living (for the United States) are also attractions.

The special character of high technology in Atlanta is its broad base and modern outlook, owing to the aggressive efforts of city leaders to attract technology firms. In fact, just twenty years ago there were only a few major technology companies. Now there are hundreds.

Ann Arbor, Michigan

Find a major research university and you will immediately locate a nest of high technology spinoffs. Such is the case with the University of Michigan and Ann Arbor. The dominance of the University of Michigan is felt throughout this high technology area. Businesses have spun off from faculty and graduates of Ann Arbor.

The top high technology employer is Warner-Lambert/Parke Davis, with 885 employees. This pharmaceutical firm is followed by University Microfilms International, which employs 851. It produces microfilmed documents and computer database products. The third largest employer is ADP Network Services, providing client data center services through 800 employees. Comshare Inc. is the largest software development and services firm, with 686 employees. Other major firms are Chrysler Proving Grounds, Gelman Sciences (biomedical devices), ERIM (R&D imaging technology) and the Medstat Group (health care information).

The real story of Ann Arbor high technology may be in the smaller firms, however. There are over 300 software firms in the area, most of which sprung up because of the Ann Arbor campus connection.

As far as high technology attractions, the Washtenaw Development Council cited Detroit's Metropolitan Airport, a well-educated and skilled workforce, the public schools system, a good quality of life, and high technology resource support. The resource support includes technology transfers from the University of Michigan. Not-for-profit agencies such as the Industrial Technology Institute, the National Center for Manufacturing Sciences, and MERRA (an economic development agency) all provide research and other assistance for high technology companies.

Madison, Wisconsin

High technology in Madison is a substantial but not a major economic part of the city. The major economic mainstays are state government, the University of Wisconsin, manufacturing, health care, and insurance. The largest private employers are Oscar Mayer, American Family Insurance, and CUNA Mutual Group (insurance), but these are outdistanced by the state and University of Wisconsin. The most prominent technology companies are Nicolet Instruments (analytical instruments) and TDS (telecommunications and data systems). Promega and Agracetus are biotechnology firms located near Madison. Astronautics is located here as well. There are dozens of other minor firms and small consulting and training companies. There are also a few engineering companies.

Many business startups have come from University of Wisconsin faculty. In addition to this resource, the city received certification of its claim as a high quality of life community, when *Money* magazine in 1996 named it the number one city in the United States. An educated workforce and natural beauty also make Madison attractive. There are two high technology parks, one operated by the Gialamas Company and another by the University of Wisconsin.

Minneapolis-St. Paul

This Midwestern region's largest metropolitan base has the largest infrastructure as well. Biomedical and computer companies are the major examples of its technology inhabitants. The major companies with technology bents are 3M, Honeywell, Valspar, and Cray Research. There are also many smaller software development and computer graphics companies. These support the main firms on an as-needed basis.

The Minneapolis Community Development Agency pointed to the University of Minnesota as a factor that attracts high technology firms. The mid-continent location, the good reputation of the workforce, its educational system, and a good quality of life are also mentioned as possible reasons for high technology business development.

Minneapolis has a large number of high technology firms, but they tend to be focused in computers and software. It is the largest Midwestern city with a technology character and also one of the older technology centers.

Dallas–Fort Worth

The main high technology concentration is in Dallas, with fewer firms in Fort Worth. The diversity of firms is pronounced: telecommunications, computers, semiconductors, pharmaceuticals, and aerospace companies.

Of all the high technology centers described here, the Dallas–Fort Worth center has the most diversified high technology base, including factory automation, biotechnology, chemicals, computer hardware, defense, energy and advanced materials, phamaceuticals, photonics, computer software, test and measurement, and telecommunications. The largest technology firms are Lennox in factory automation, Occidental Chemicals, the giant EDS with 72,000 employees in computers and data processing, Fina in energy, Nielsen North American Merchandising Solutions in computer software, and the Tandy Corporation. Texas Instruments and Vaught Aircraft are major defense contractors, General Dynamics was also a major aerospace employer.

The advantages of this area are the high quality of the workforce, an outstanding educational infrastructure, and excellent accessibility to customers and suppliers, according to the Dallas Chamber Economic Development Group. Technology firms of all sizes are spread around and within the cities. Like most metropolitan areas, there are many small support business, such as energy consulting and software consulting companies. High technology companies are a major business in this area and are found at every level, from small support services to the huge EDS and Texas Instruments.

Austin, Texas

The University of Texas plays a major role in high technology business formations because of the research done there, especially federally sponsored research. Some of the immediate beneficiaries are regional technology businesses. Industries here are telecommunications, computers, and semiconductors.

Austin is one of the smallest technology centers with the largest number of large high technology firms. IBM employs 7,500 and divisions of Motorola (which has a wafer fabricator located there) and Lockheed are also present. Texas Instruments, Radian, Unisys, and Intera Technologies are located in Austin.

The Sematech business consortium, which located here, was a recent award for the community. National competition among major technology sites resulted in Austin being chosen as a development location for the next generation of semiconductors.

Of the 700 manufacturing firms in Austin, 200 are high technology companies. Computers, computer peripherals, software, electronic instruments, biotechnology, semiconductors, and pharmaceuticals are predominant types of businesses. High technology is conspicuous and a core business in Austin.

Denver-Boulder

Telecommunications, computers, and aerospace are the major high technology firms in and around Denver and Boulder. Federal government grants and the University of Colorado at Boulder are two main reasons for the development of firms in this area. The largest employers are U.S. West, Columbia Health-ONE, AT&T, Lockheed Martin, United Airlines, Coors Brewing, and the Rocky Flats Environmental Technology Site. Half of these companies could be considered technologically-intensive firms; this shows just how important technology-based businesses are for the greater Denver economy.

In answer to the question of why a high technology business would originate or relocate to Denver, the chamber of commerce's economic development office stated it was because of excellent access to the University of Colorado and the Colorado School of Mines. The good quality of life helps in recruiting professionals, according to the chamber.

Colorado Springs

Farther south of Denver is the high technology area of Colorado Springs. Historically, the major high technology concentrations were in the telecommunications and semiconductor fields. There was considerable defense spending, which encouraged these kinds of firms. Digital Equipment, TRW, and Hewlett-Packard were some of the high technology names in the boom days. There is considerable service support of high technology here, as defense contractors provided maintenance and contract software engineering for the Defense Department. Fort Carson, the NORAD Center, and the Air Force Academy were the mainstay federal benefactors until the end of the cold war.

The Colorado Springs Economic Development Corporation publishes a directory entitled "High Technology Companies." The summer 1996 edition lists 239 high technology companies, most of which are software companies, followed by electronic manufacturers and semiconductor supply and support. The slant is definitely toward the computer side of high technology business.

Reasons cited for the area being attractive for high technology are its highly skilled but competitively priced labor, a culture supporting technology transfer through such programs as the Colorado Institute for Technology Transfer and Implementation, ten local colleges, the variety of opportunities for software development, a telecommunications infrastructure, a centralized location, and the beauty of the Pikes Peak region.

Some familiar names remain in the area—Apple, DEC, MCI, and Rockwell—while newcomer companies like Sun Microsystems, Oracle, and Geodynamics are also found here. The character of this region can be summed up as a mature development region with software and semiconductor businesses leading the way.

Salt Lake City-Provo

High technology here is telecommunications, computer equipment, pharmaceuticals, and semiconductors. Sperry Univac and National Semiconductor are perhaps the most familiar companies. High technology firms are located south of Salt Lake City along the I-15 corridor; concentrations are in Orem and Provo, with the struggling Novell-WordPerfect Company in Orem and other firms near and around Brigham Young University in Provo.

Brigham Young University adopted the Stanford model for stimulating adjacent high technology development—that is why startup companies are found nearby. The University of Utah in Salt Lake City has a research park for high technology startups.

Provo has the second largest concentration of software technology companies and the third largest concentration of high technology companies. There are 150 companies that support Novell-WordPerfect alone. Major technology firms are Gazelle Systems, Erying Research, 386 Systems, Folio Corporation, and Systems Connection. There are also corporate divisions of IBM, Sanyo, R. R. Donnelly, Intel, Smiths Megadiamond, and Ameritech in the Provo-Orem area.

A very strong work ethic and the nearby mountains are given as reasons for technology companies to locate in the area, according to the local economic development agency. To this can be added the cooperative academic connection; this relationship has fostered most of the new businesses.

In Salt Lake City, there is a slant toward aerospace, with the major technology firms being Morton (8,000 employees), Thiokol (8,000 employees), and Hercules Aerospace (6,000 employees). The largest employer is Hill Air Force Base, with 20,000 armed services and civilian employees.

Technology transfers come from the University of Utah. The first users of technology are the many new software, testing, and electronic technology firms that organize themselves around the transported innovations and speedily commercialize them.

Phoenix

Telecommunications and electronics are the most notable high technology firms in Phoenix. There is also considerable semiconductor and computer manufacturing. Where there's sand there's silicon, and that may be a way of remembering the dominant high technology businesses of the Southwest.

Motorola is the largest high technology employer with 20,000 employees. In Phoenix, Motorola is involved in producing semiconductor products, systems technology, and computer products. The second largest company is Allied Signal with 8,900 employees. It does work in the areas of aerospace, automotive, and engineered materials. Intel is the third largest employer with 7,280 workers in the fields of microcomputer components and related products. Closely following is Honeywell with 7,000 employees. The Phoenix branch of Honeywell produces automation and control systems, products, and services.

The locations for these firms are along the interstates, near Arizona State University and Sky Harbor Airport. The transportation system, both roads and air, is excellent. This is an often cited community attraction. The stated advantages of Phoenix are a competitive cost environment and one of the most modern infrastructures in the nation, according to the Phoenix business development administrator. The stable weather; scenic environment, existence of many other high technology companies with their labor, supply, and services bases; Arizona State University and colleges; and good quality of life were all mentioned as attractions. Like Tampa, Phoenix is a net importer of qualified labor as experienced engineers relocate to Phoenix.

Albuquerque

This city has come to be known as Silicon Mesa, largely though the efforts of its development advocates. There is a large electronics manufacturing segment in Albuquerque, and telecommunications and semiconductors are leading high technology businesses. This is the particular character of high technology, Albuquerque style.

Government projects led the development of high technology companies as they have in other locations. Two national laboratories dating back to World War II give the city a relatively long tradition in high technology, at least as long as Boston and Los Angeles.

Engineers for the local firms and laboratories are trained by the University of New Mexico at Albuquerque. The Albuquerque Technical Vocational Institute has a manufacturing technology program concentration in semiconductor manufacturing and hands-on high technology manufacturing training for students.

Albuquerque Economic Development has identified the largest high technology employer as Intel, which not only manufacturers the Pentium microprocessor here but also is launching a major new plant expansion, making it the largest semiconductor factory in the world. Motorola Ceramic Products, Phillips Semiconductors, General Electric Aircraft Engines, and Ethicon Endo-Surgery are also major employers.

The development office identified skilled workers, a university system, and national laboratories as high technology draws. Sandia and Los Alamos National Laboratories and the Air Force's Phillips Laboratory are the chief examples.

The area companies have started technology transfer ventures in laser technologies, so photonics is an emerging field in the region. The ubiquitous software consulting companies start frequently (and end just as frequently). They are also found by the dozens in Albuquerque.

Los Angeles

The Los Angeles area has virtually every major type of high technology with the possible exception of major biotechnology firm concentrations. Telecommunications, pharmaceuticals, computers, semiconductors, and aerospace are all here in numbers and force. Perhaps we don't think of this as a high technology concentration because there is simply so much of it within the largest population concentration.

If there is any distinction, it is the heavy imprint of aerospace companies. Rockwell is an example. The Los Angeles area was the nation's aviation manufacturing center (by number of companies), and it has been this way since the 1930s. In a way, it is the oldest high technology center in the United States. The space business followed, and aerospace still holds its dominance in the area.

Most high technology growth is in Orange County; newer growth areas are in Riverside and San Bernadino. Like the New York area and the other major metropolitan centers, there are many small high technology companies that are diverse in scope. Hughes Aircraft, McDonnell-Douglas, Rockwell, and Burroughs are some of the larger firms. The stamp of aerospace is clearly indicated with these companies. The business of high technology has been fostered by the universities as well. UCLA and USC are major labor and innovation resources that have contributed to the technology eminence of the area.

Special attractions for high technology are the huge labor base and pleasant climate of southern California, though the costs of living and doing business are high.

Silicon Valley

The best known geographical identifier for high technology is Silicon Valley. In many ways, it is our most contemporary geographic anchor because our modern understanding of what high technology has come to be emerged here. A broad stretch of land from San Francisco to San Jose is what is commonly defined as Silicon Valley.

Over half of the manufacturing in this area is high technology manufacturing. That's an extraordinary percent considering that San Jose area is the home of traditional manufacturing as well. It has a large concentration of scientists and engineers.

The history of this region is described earlier in Chapter 2. Concentrations of high technology are in telecommunications, computers, semiconductors, and aerospace. These developments came in waves, led by computers and aerospace and followed by telecommunications and semiconductor manufacturing.

Representative firms are Hewlett-Packard, Apple, Intel, Advanced Micro Devices, Oracle, and Lockheed. As large as these firms are, they are in a sense overshadowed by the small companies—those hundreds of computer and software companies that are found everywhere in the valley. These firms are where new software is developed and proven, a vast killing field for the unsuccessful and a catapult for the truly commercial. The particular fury and fight for survival in Silicon Valley is the dynamic of small and mid-sized firms whose particular efficiency is putting computer and software advances into business. Seemingly constrained because they focus on a single product, they are truly malleable because they shed structure as needed and leap into superior technologies. It may be the same people, but it is never the same technology. Their forte is not to change but to dissolve. Some of these workers have told us that they are in their sixth company and expect to be in their tenth by the time they turn 30 years old.

Silicon Valley has a huge and well-trained labor infrastructure. That is an advantage, along with the climate and scenery. As in Los Angeles, living and business operations costs are high.

In spite of high labor costs, suburbanization, and smog, Silicon Valley is still a mecca in the view Richard Brandt, writing in the 1993 Annual of *Business Week*. He notes the valley has the highest concentration of technical, financial, managerial talent, and knowledgeable people in the world. This is combined with an active spirit of entrepreneurship; 22 percent of U.S. venture capital firms are in the Bay area. A third of 1992's 152 nationwide venture startups happened here. Another interesting trend is the transference of non-U.S. firms to the area. The May 9, 1992, issue of *The Economist* noted the relocation of three British-owned high technology companies that wanted to establish a marketing base in northern California: Madge Networks, IXI Ltd., and Tadpole Technology.

Portland

Portland is a large urban city that, unlike New York, has not hidden its high technology companies in a huge metropolitan complex. The large high technology companies in Portland are prominent. The largest, Intel, employs over 8,000 people, and the second largest, Tektronix (electronic measurement, color printing, video systems, and network displays products) employs 4,600. Other major firms are U.S. West, Hewlett-Packard, Epson Portland Inc. (printing terminals and related products), SEH America, and Wacker Silitronic. The last two are silicon wafer manufacturers. Integrated circuit manufacturers, disk drive makers, and manufacturers of power supplies, as well as fiber optic transmission equipment manufacturers, are also among the larger high technology firms here.

The Portland Development Commission stated that the city and state have "strategically focused our business environment, our workforce and our educational system, and our planning and development processes to support the semiconductor and electronics industry." In all, there are over 1,100 high technology firms employing over 48,000 people. Semiconductor manufacturing is the largest component of the high technology base, though computer peripherals are also major businesses.

The commission has cited several reasons why high technology companies find advantages in Portland. One is that the area is has a rapidly growing population. State population growth is twice the national average. Other reasons are the emergence of transpacific trade, a stable and skilled workforce, and the nearby scenery and outdoor resources.

Seattle

Seattle is the high technology center with the single largest technology firm: the Boeing Corporation with 104,000 employees. There are hundreds of support firms for aircraft subcontracting and software support. The second largest employer is the University of Washington, a training ground for Boeing engineers. As prominent as nearby Microsoft is on the national technology scene, it is not as dominant as Boeing is locally. Boeing is everywhere in and around Seattle.

Attractions of the city are the educated workforce and the good quality of life. Seattle is an international port and an international airport. Like Portland, Seattle will likely benefit from Asian trade and is probably the largest city with greatest potential for growing more technology companies.

Conclusions on Locations

There are both commonalities and great distinctions among these locations. A commonality is their proximity and occasional adjacency to research universities. They are sizable but for the most part not immense population areas. They are crossed by interstates and have airports that can handle commercial airliners.

What is unusual is how many major cities do not have these centers. Chicago, Cleveland, Detroit, Jacksonville, Philadelphia, Pittsburgh, St. Louis, and Washington D.C. do not have high technology business concentrations in the immediate metropolitan area. Locating in a major metropolitan area is not one of the distinctive characteristics of this kind of business. On the other hand, the giant technology companies such as IBM, Intel, and Motorola have branches in the major cities.

By and large, we have found concentrations of high technology along the coasts and scattered in a few midwestern, southern, and western states. This pattern has been observed by others as well. Giese and Testa (1988) compared the national average of high technology employment to census geographic divisions and found the greatest concentration of high technology industries in the Pacific Coast states of California and Washington and the eastern States of New Hampshire, Massachusetts, Connecticut, and New Jersey. Employment concentrations were reported higher in the eastern states, but the western states have a greater percentage of the largest high technology businesses. Between the coasts, there is a wide scattering of high technology businesses, with only Utah, Arizona, Colorado, Kansas, Missouri, Minnesota, and Indiana showing concentrations on the three-range concentration devised by the authors. All other states have below-average concentrations. (Differences between these authors and our listing of concentrations are due to different selection criteria. We did not use the state as a primary basis for establishing concentrations, preferring sub-state concentrations.)

Emerging Areas

High technology development areas have been emerging for over forty years. To think of these business activities as having a home for nearly half a century is to introduce an incongruity into what has been reported as a contemporary business. Yet Hewlett-Packard has resided in Silicon Valley for that long, and Digital Equipment has been on the Route 128 corridor for almost as long.

With an ever lengthening history, high technology has its second and third generations of firms that have staked their claims in new places. Some of these locations are, as expected, near or in the well-known geographic locations, but others are where they would not be expected.

Bozeman, Montana, is one of these areas. There are about 45 high technology firms in Bozeman. Montana State University is much of the reason, since academic research has been spun off to commercial ventures. Bozeman BioTech, for example, is developing ways of using bacteria to protect seeds from fungus rot. Other biological science ventures have sprung up in the area as well.

Some of the emerging areas have features in common with Bozeman,

Figure 16.1

while others do not. South Dakota has become a computer manufacturing center for Gateway Computers, which has striven to put the touch of customization on PC production. A host of cities in an ever-expanding ring outside Denver are home to Geographic Information System (GIS) firms and software consulting companies. Newer technologies seem to find new locations.

Even though there are emerging areas for high technologies, the mature areas do not become extinct. All the older technology centers continue to thrive as creators of technology firms, in spite of economic ups and downs.

Summary

High technology has a different complexion in each location, most often taking the form of some leading business, such as software in Silicon Valley or aviation in Seattle. As a way of our showing the differences, Figure 16.1 places the technology centers within the United States.

There was also commonality of location features articulated by the development directors of these areas. Part of that was the linkage of high technology to educational institutions, for both innovation sources and trained workers. Another common theme was that the regions' particular attractions may have less to do with having other high technology firms or physical networks (such as fiber optics) and more to do with general inducements that attract any kind of business—quality of life, transportation, taxes, and so on. The regional development directors also tended to be liberal in their use of the term *high technology*, with the tendency to include large companies that are only marginally technology firms.

What can be said about factors that draw high technology firms as opposed to those influencing traditional firms? Apparently, the factors are the same but with a different priorities for high technology manufacturing plants. Stafford (1982) divided 104 plants into high technology and traditional plants based on SIC codes. For high technology plants, the priority of factors was labor, transportation availability, quality of life, market access, and utilities. For, traditional plants, it was labor, market access, transportation availability, materials access, and utilities.

Transportation availability shows up on both lists. A separate study of transportation needs for technology concentrations in Great Britain, quaintly called Silicon Glens, placed the transportation issue low on the importance continuum. Button (1988) argued that evidence in the United States and Silicon Glen showed that the provision of appropriate transport services was not a sufficient factor for attracting high technology enterprises. These services were a prerequisite for their consideration of their locating in an area. But many other factors found important to high technology firms, such as labor availability, are themselves influenced by transportation considerations.

One conclusion about technology siting is that states will be largely reactive, but that the national and international levels of governmental support are more important. States encourage and provide networks but are not the prime movers.[2] The competition among municipalities and states for high technology businesses has been well recognized—the competition for Sematech is an example of the competition among high technology–hungry municipalities. Galbraith and DeNoble (1992) state that past local development strategies may now be inappropriate for attracting technology firms. Because of progress in manufacturing processes, the infrastructure requirements of high technology firms vary. Factors such as strategies, community infrastructure needs, and location behavior were considered, and community infrastructure requirements, location intentions, and actual spatial tendencies were found to be significantly related to successful community high technology development strategies.

In summary, we have profiled the features of the leading high technology centers. The surrounding resource base was found to be important in explaining aspects of business operations. In our next chapter, we'll see what states have to offer as resource bases for high technology business development.

Notes

1. An initial effort at capturing the location patterns of high technology companies was made by the Maryland Department of Economic and Community Development. Entitled "A Comparison of High Technology Centers in the United State," it was published by the state of Maryland in July 1983. This doc-

ument formed part of the basis for this chapter. However, we contacted 21 local chambers of commerce and development agencies in late 1996 to obtain more up-to-date and inclusive material on local high technology business activities. These results are also reported in this chapter. In addition, some information about cities, such as Boston, Tampa, Atlanta, Austin, Salt Lake City-Provo, and Seattle, is taken from *Cities of the United States* by L. Schmittroth, and M. K. Rasteck (Detroit: Gale Research, 1994).

2. This is a concluding point of Jurgen Schmandt and Robert Wilson, eds., *Growth Policy in the Age of High Technology* (Boston: Unwin Hyman, 1990).

References

Button, K. (1988). "High Technology Companies: An Examination of Their Transport Needs." *Progress in Planning* 29, Part 2 (Oxford: Pergamon Press), pp. 77–146.

Galbraith, C., and A. De Noble (1992). "Competitive Strategy and Flexible Manufacturing: New Dimensions in High Technology Venture Based Economic Development." *Journal of Business Venturing* 7: 5, pp. 387–404.

Giese, A., and W. Testa (1988). "Targeting High Technology." *American Demographics*, May 1988, pp. 38–41.

Stafford, H. A. (1982). "Location of High Technology Firms and Regional Economic Development." *Joint Economic Committee: United States Congress* 6: 1.

17

STATE SUPPORT FOR
HIGH TECHNOLOGY
BUSINESS

This chapter describes state programs that support high technology business development. Virtually every state makes some effort to draw or internally nurture high technology businesses. The reason states compete for these businesses is that they are attractive and they grow fast. They are attractive because they are seen as "clean" businesses that employ high-caliber workers. They can grow exponentially. They are industrial willows compared to the slow-growing oaks of traditional businesses. Plant them in your state and they are likely to contribute to the tax base quickly. The wages of high technology workers are greater overall, so states with lower per capita resident incomes see high technology as a way of catching up.

While much attention has been on state efforts to draw in new technology businesses, the plain fact is that most success comes from helping home-grown technology businesses. Providing there is an appropriate infrastructure, business development directors have said that watching for and assisting promising new technology ventures is the better route to economic development than a nationwide technology chase.

Under our federal system, states are free to independently pursue economic development, and they do so. All you have to do is visit the trade shows of the technology business. There are the states, hawking their advantages. The result is that we have 50 ways to win high technology businesses. Development directors apparently do not feel they can quit the competition, even if they may prefer in-state development.

In this chapter, we review those ways of supporting high technology businesses, state by state, using a survey. Our survey first asked what programs and laws existed to stimulate the creation or relocation of high technology businesses. Second, we asked how each state coordinated its support for high technology businesses with any federal programs, as we discuss in Chapter 18. The survey was sent to the heads of all 50 state economic development offices. The written replies and follow-up phone calls were combined as a basis for this chapter. Some information was obtained from *The States and Small Business*, published in 1993 by the U.S. Small Business Administration. This is the source for information about Michigan, Pennsylvania, Rhode Island and West Virginia. In some cases, this information was combined with informal contacts with business contacts in these states to help provide a more complete picture of state support. Going state-by-state shows how different and how similar state high technology programs are. A summary extracts some major points.

The following are the main support programs offered by the states.

Overview on Terms

In this and the following chapter, there are a number of terms used to describe state and federal programs. This glossary defines some of the more common programs and agencies that will be encountered:

ATP   The Advanced Technology Program is a federally sponsored partnership between government and private industry to accelerate the development of high risk technologies and not yet possible technologies that promise major commercial developments and widespread economic benefits. The program encourages a change in how industry approaches R&D by providing a way for industries to extend technological reach.

DARPA   The Defense Advanced Research Projects Agency is a federal agency that helps maintain U.S. technological superiority and guard against technological advances by adversaries. Its mission is to develop innovation and often high-risk research ideas.

EPSCoR   The Experimental Program to Stimulate Competitive Research is a program of the National Science Foundation designed to improve the competitiveness of state universities' research capabilities and assure geographic distribution of research awards. It has been used by states that have traditionally lacked major research university linkages to compete for federal research aid.

MEP   Manufacturing Extension Partnership is a network of local centers in all states that offers technical and business assistance to smaller manufacturers. The centers use field agents to work with firms. Typical help is locating new equipment, cost reduction help, and market expansion assistance.

NIST   The National Institute of Standards and Technology is a federal agency that aims at helping industries develop technology needed to improve product quality and reliability, modernize facilities, foster rapid commercialization of science-based products, and other activities. It administers the ATP and MEP.

SBIR   Small Business Innovation Research is a federal program which provides awards to innovative small firms for developing marketable technologies. Awards are made for phases of projects and are often supplemented with other grants.

SBA   The Small Business Administration is the major federal agency dedicated to support small and medium-sized businesses. It provides loans and information for technology entrepreneurs and other business managers.

In the state survey, "programs" refers to those programs that high technology firms may be able to use as sources of support and "coordination" means how the states interact with federal and other agencies to deliver assistance to technology firms.

Alabama

Programs

The Alabama Department of Economic and Community Affairs lists numerous programs and tax incentives that may be used by high technology firms.

The Alabama Commission on Aerospace Science and Industry receives state support for business expansion of the state aerospace industry, and the Alabama Research Institute provides grants to research universities to develop the state's economy. The Economic Diversification Program provides assistance to local economies affected by defense reductions. Also important is EPSCoR, an effort to bring research and development to the state. The Technology Assistance Program is a National Institute of Standards and Technology–funded program to coordinate technology resources in the state. The NASA Technology Transfer Initiative is a multi–federal agency program to transfer NASA technology to the state's business community.

A capital investment tax credit lets businesses claim tax credits against their income taxes for certain businesses and qualified investments of 5 percent of the investment per year for twenty years. The corporate income tax is 5 percent. There are inventory tax exemptions, reduced sales and use taxes on new manufacturing machinery, property tax exemptions for industrial projects, and other programs.

The state has enterprise zones where special tax incentives are offered for new or expanding businesses hiring targeted employment groups and a tax incentive for businesses that sponsor basic skills education programs.

This information was provided by Russell Moore of the Department of Economic and Community Affairs.

Coordination

There is coordination with the National Institute of Standards and Technology as well as other informal program coordination with federal agencies.

Alaska

Programs

Alaska has a technology development agency, the Alaska Science and Technology Foundation. It is a grant agency that works with the state's industrial community. It funds particular projects for industries that may or may not include high technology businesses. There are no special tax credits or deductions for high technology. The foundation's forte is cash grants, an endowment for people who are looking to start Alaska businesses. Other grants are Knowledge Grants given to universities and Technology Grants. The grants are paid back only if the venture is successful.

Coordination

The state coordinates its support with National NIST. There are NIST manufacturing centers in the state. There is also a cooperative arrangement between the state and NASA in the form of a research database creation effort.

This information was provided by Robert Chaney, Technology Administrator for the Alaska Service and Technology Foundation.

Arizona

Programs

The state Department of Commerce has reported that there are no major programs offered only to high technology businesses. High technology business development efforts are included in the package of programs available for new and emerging businesses.

The Small Business Financing Program of the department provides staff analysis of loan requests and the assembly of application packages for small businesses. There are Small Business Administration (SBA) loans available for economically distressed counties. A state Business Assistance Center is available for consulting assistance. There is a statewide SBA 504 loan program that offers loans to small businesses with net worths of less than $6 million.

Arkansas

Programs

The Arkansas Industrial Development Commission used our survey to cite acts passed by the 1995 legislative session. As a result of these acts, the state boosted the scope of activities done by the Arkansas Science and Technology Authority, its main technology arm. The authority engages in activities that exert leadership in and give direction to programs and services designed to help Arkansas gain the benefits and opportunities that come from advanced science and technology. It is a quasi-public agency with both private and public members. The authority was recently authorized to undertake programs involving a wide range of industries that include establishment and encouragement of science and technological research. For the first time, the role of technological advocacy was given to the authority by the legislature. The authority can make investments for technological purposes. It can also finance research for business purposes, assist small businesses in fund applications, and develop technologies.

Arkansas has targeted certain kinds of technology firms. Legislation was also enacted that directed all appropriate state agencies to provide research assistance to help encourage qualified medical companies to locate in the state. Another law revised and made more specific the qualifications for loans and other funding from the aviation and aerospace development fund, a potential high technology business funding source. Still another law clarifies "new permanent employees" and other terms in the Arkansas Enterprise Zone Act. For a state not usually known for high technology, Arkansas has an organization and purpose in its high technology efforts.

This information was provided by Del Boyette of the Arkansas Industrial Development Commission.

California

Programs

The State Office of Strategic Technology is the main institution for state high technology business development activity. Its mission is to accelerate the development of new products and services in the high technology sector. It also works with the private sector to do this.

California has a matching grant program as part of the California Technology Investment Partnership. The state invests grants of up to $250,000 per project, which can be leveraged with federal grants. These are matching grants. The private sector may also put in a match. The state forms partnerships as needed with the federal government and private sector so that all partners can contribute to a qualified project.

Tax credits are awarded in the Enterprise Zones program, which are scattered throughout the state. Defense conversion funds are available. A variety of other training and loan programs are also available for high technology companies, as well as other firms.

Coordination

California coordinates its support with many federal high technology programs. For example, the state works with the NIST in the Manufacturing Technology Partnership. It also works with the Economic Development Administration, as well as the National Science Foundation, through which six grants have been awarded to universities. There is also an approach to technology business development called the I Quad Technology Development Philosophy. One I means investments under matching grants, another I means infrastructure for the Gold Strike Partnership with three regional technology alliances. The third I is for initiatives (actual industry cluster development), and the last I is for integration.

The Gold Strike Partnership is a nonprofit organization that administers state programs. It acts as a platform for the private sector and provides services that the state cannot. It is somewhat unique since it is not a state agency but is affiliated with the state, yet works in the private sector. Coordination is done through the matching grant program. If a university contacts the Partnership for NSF support, the partnership facilitates the funding process.

This information was provided by Richard Keeler of the Office of Strategic Technology.

Colorado

Programs

The Colorado Advanced Technology Institute is the main state agency that deals with high technology businesses. The state has several important business incubators for high technology, namely the Boulder Technology Incubator, the Colorado Biomedical Venture Center, and the Colorado Innovation Foundation. Legislation has been proposed to create tax credits for biotechnology companies by exempting them from the state's personal property tax on purchases of equipment.

Coordination

The state coordinates its support chiefly in the SBIR program. Colorado has attempted to develop networks to support SBIR activities with a National Science Foundation grant. Informal coordination with phone calls and World Wide Web contacts also happens.

This information was provided by Dr. Phil Bradford of the Colorado Advanced Technology Institute.

Connecticut

Programs

Connecticut Innovations Inc. (CII) is the state's lead agency in high technology development. Connecticut joins the fast developing list of states that have high technology point agencies. CCI develops new technologies through funding, access to university, manufacturing resources, and other professionals. It is a 15-member quasi-public agency that is especially interested in developing these technologies: advanced marine applications, advanced materials, aerospace, applied optics and microelectronics, biotechnology, computer applications, energy and environmental systems, medical products and instrumentation, and telecommunications.

Entrepreneurial firms with college and university affiliations may be able to take indirect advantage of Yankee Ingenuity Initiative Grants and Elias Howe and Charles Goodyear Grants, which go to public colleges and finance basic research that business partners may commercialize.

There are also Advanced Technology Centers supported by CII. They include Yale's Neuroengineering and Neuroscience Center and the University of Connecticut's Precision Manufacturing Center, which aims at discovery and business collaborations and applications. Critical Technologies Grants are also directed to the University of Connecticut.

The CII helps small businesses prepare SBIR and Small Business Technology Transfer proposals. Under the Connecticut Small Business Innovation Research Assistance Program, SBIR marketing efforts can be funded up to $50,000 per project. This is an unusual type of assistance among state programs. Normally, high technology marketing efforts are not funded.

The Connecticut Federal Technology Partnership Program is another source of federal research dollars that the CII is involved in. Companies that have been in business for at least a year can tap into a whole new set of resources. CII has Product Development Investments and Product Marketing Investments available. Funding between $100,000 and $1 million is available per project for product development. This can cover labor, overhead, patents, raw materials, and other costs. Product marketing can cover advertising, inventory costs, trade shows, and related costs.

In Connecticut, public support for technology venturing is considerable and, very important, it follows along as businesses grow.

Coordination

Coordination of initial inquiries is done through the Technology Assistance Center of the CCI.

This information was provided by Laura Rose of Connecticut Innovations Inc.

Delaware

Programs

Tax advantages are what Delaware has long used as a corporate lure. This is, after all, a state that places advertisements in flying magazines urging pilots to register their planes in Delaware. There are no specific high technology tax breaks, but the general tax advantages include a three-fifths majority legislative vote for new or increased taxes, a constitutional limit on state appropriations, no state or local general sales taxes, and tax credits on corporate income and reduction of gross receipts tax for new and expanding businesses.

There are additional tax credits on corporate income and a reduction of gross receipts taxes for new and expanding businesses locating in 30 targeted areas. There is property tax relief for new construction and improvements of existing property. There are no personal property, fixtures, or inventory taxes.

Targeted Industry Tax Incentives are available. Corporate Income Tax Credits can be obtained by laboratories, computer processors, engineering firms, and manufacturers. The program provides credits of $250 for each new employee and each $100,000 investment. Other types of credit are available.

Regarding technology related programs, there are two research grants. The Delaware Research Partnership provides matching grants for joint university-industry research projects. Total funds are $1 million. Projects are generally limited to $100,000 for the total project cost and require a 50 percent industry match.

SBIR Bridge Grants are provided to SBIR Phase 1 winners who have not accessed Phase 2 awards. Up to $50,000 is available to grantees.

Private financing is offered through the Delaware Innovation Fund. Delaware Venture Partners is a venture capital pool that is a particularly inventive way of making capital available. The Delaware Access Program is a state tool for making risky business loans available; these may be tapped by qualified technology companies.

Technical assistance is available from the Delaware Manufacturing Alliance and the Small Business Development Centers. The Delaware Technology Park near the University of Delaware is a private, nonprofit corporation dedicated to high technology industry growth in Delaware. There are seven technology research centers; all are affiliated with the University of Delaware. They are the Agricultural Biotechnology Center, the Center for Applied Coastal Research, the Center for Catalytic Science and Technology, the Center for Composite Materials, the Center for Molecular and Engineering Thermodynamics, the Institute of Energy Conversion, and the Orthopedic and Biomechanical Engineering Center. The names give an indication of the research area.

This information was provided by Susan Rhoades of the Delaware Development Office.

Florida

Programs

Florida has been a low tax state for quite a while. That's a well-known fact among its individual immigrants, but it applies to corporations as well. Its published business tax advantages include a corporate income tax rate for C corporations of 5.5 percent with an initial $5,000 exemption. There are no subchapter S corporate income taxes, nor are there limited partnership corporate income taxes. There is no state personal income tax. That can be considered an advantage for high technology employees as long as they do not expect a high level of social services and education.

Property taxes are assessed at the county level and exemptions of up to ten years may be granted to new or expanding businesses. That feature is especially meaningful for new technology ventures. There is also no sales tax on raw materials incorporated in resale of final products.

The State of Florida Department of Commerce Division of Economic Development replied to the questions by noting that Florida has become very active in attracting high technology businesses. Economic Development Policy Coordinator Wynnelle Wilson commented, "Several years ago, the state created the Innovation Partnership under Enterprise Florida Inc. This partnership was formed based on recommendations developed by the Florida Department of Commerce and the Florida Chamber of Commerce, and focuses on better utilizing the technological strengths of the state to make existing firms more competitive and to create new firms." The Innovation Partnership is used in conjunction with the typical set of loans, credits, and training programs employed by other states to build their high technology bases.

The Innovation Partnership operates the Florida Manufacturing Technology Center, which provides manufacturing extension and related engineering and business services to small and medium-size manufacturers. Also, Innovation and Commercialization Corporations are found in six different locations. These are joint ventures with community resources designed to provide entrepreneurial assistance and services to move technologies from universities and federal and industry laboratories to the marketplace.

Another program of note is the Technology Research Investment Fund, which is a revolving investment fund for company technologies with emphasis on applied research and development leading to marketable products and processes. Yet another is the Defense Conversion Program, which allocates defense conversion matching funds to federal defense conversion programs including the Technology Reinvestment

Project. Job training programs, venture capital funds, and upgrading of small and mid-size manufacturers (by way of the Florida Advanced Manufacturing Consortium) are also available to technology firms.

The Florida Department of Commerce operates a business hotline. The department coordinates a business contribution program where business contributions to community development programs may be eligible for a 50 percent credit on corporate income tax or insurance premium taxes.

One innovative dimension of high technology support is lab lending. The State of Florida Cooperative Research and Development Agreement lets businesses get access to Martin Marietta laboratory resources. Defense Based Realignment and Closure provides matching grant funds to communities affected by 1995 federal closures. The Florida Energy Loan Program helps small businesses in conducting projects resulting in a reduction of energy costs. There is also an Enterprise Zone Program encouraging businesses to locate in zones. Also there are incentives available for creating new jobs.

One interesting development has been the Gulf Coast Alliance for Technology Transfer, in which 11 federal laboratories have made their facilities available for commercial use. Spaceport Florida Authority works to make launches available for commercial interests. NASA sponsors technology transfers.

Like some other states, Florida offers minority business programs and support for international commercialization of Florida business goods.

The Center for Health Technologies Inc. is an incubator for moving developments from the laboratory into the marketplace. The facility coordinates technology development and transfer, assists in grant applications, and provides shared services and health technology incubator facilities. In all, Florida has the most diversified high technology support programs and substantial tax inducements as well.

This information was provided by Brent Gregory, Vice President, Enterprise Florida and the Florida Department of Commerce.

Georgia

A distinctive element of Georgia's business development efforts is the cooperation between business and government that is manifested in the many and, in some ways, creative programs. Again, high technology businesses are not directly targeted by broad scale, integrated efforts, but can benefit from general programs and tax incentives.

The corporate tax rate is 6 percent. There are property tax exemptions for inventory and software (a unique and possibly significant exemption for high technology businesses). There are sales tax exemptions for aircraft engine remanufacturing, pollution control equipment, raw materials, resale purchases, and manufacturing machinery. Certain other exemptions are available for material handling equipment, electricity, and government contracts.

There are job-creation tax credits that are tiered in value by county and job numbers. Additional credits are available for multicounty job creation and when it occurs in less developed areas. Tax credits are offered for established companies making capital investments.

The Quick Start no-cost employee training program is delivered through 32 area technical institutes. A one-stop environmental permitting program is also available.

Technology support is found in the Advanced Technology Development Center at Georgia Tech, where incubator space is offered. The center helps corporate research, and development teams interface with Georgia Tech researchers and facilities.

At the University of Georgia, the Intellectual Capital Partnership Program provides an entry point to the educational programs, faculty, and research and development programs of the university. One of the main aims is to make highly educated employees available through university resource access, multiacademic discipline education, and the development of new educational programs.

This information was provided by Randolph Cardoza, the Commissioner of the Georgia Department of Industry, Trade and Tourism.

Hawaii

Programs

The state reports it has many high technology development support programs. It helps support incubator facilities for science and technology. They are currently working on a federally granted program to improve the telecommunications infrastructure. To support technology growth, the state provides capital incentives and loans to stimulate entrepreneurial growth in science and technology. They are also attempting to expand the venture capital base in Hawaii. The state has developed enterprise zones to give preferential treatment to entrepreneurial firms along with tax breaks (about 6 percent above the prime rate). There are also foreign trade zones that can spur the development of industries with no fees being assessed on importation while the firm is developing the foreign trade zone. On the business attraction side, the state has a high technology development corporation that is assigned the responsibility of attracting technology companies to Hawaii.

Coordination

Hawaii is a member of the United States Innovation Partnership, which is a link to develop ways to synergize federal and state support. The state is also working on a collaborative program encouraging the federal government to do more in support of high technology programs. The SBIR process is very pronounced in Hawaii. There are state agencies that coordinate support with the federal government. There is a high technology development corporation that works closely with the federal government, as well as the Energy Resources and Technology Division for coordination activities.

This information was provided by Jim Crisafulli of the Energy Resources and Technology Division and Dennis Ling.

Idaho

Programs

Idaho has no special incentives for high technology businesses. There has been an increase in the number of high technology businesses in any case, according to the Idaho Department of Commerce, which attributes this to market-driven reasons. The incentives that are available are those available to most any business that can qualify.

The governor has pushed for legislation to create a large workforce development training fund that would assist businesses in upgrading the skills of their workers. This is consequential for high technology businesses, which need trained workers.

Idaho has a corporate income tax, sales and use tax, and property tax. The aggregate net tax for individuals is lower than the U.S. average, however. There are credits and exemptions that some high technology firms may use. For example, business inventories are exempt from taxation, as are goods in transit. Required pollution control equipment is exempt from sales tax. Registered motor vehicles, vessels, and aircraft are exempt from sales and use tax, as are utilities.

Sources of capital for business development in Idaho are fairly conventional. One is the SBA 504 long-term fixed-asset loan program. Another is the SBA Guaranteed Loan program for up to $750,000. A revolving loan program is available if the business has job creation potential. FmHA business and industrial loans are issued for more than $750,000 but less than $10 million.

Other loans available are energy conservation loans for businesses undertaking energy efficiency programs with paybacks shorter than ten years. Industrial Revenue Bonds, Economic Development Administration (EDA) Loans, and Grants Tax Increment Financing complete the business finance offerings. We can consider Idaho an emerging high technology state that is just starting to organize its public technology support programs.

This information was provided by Richard Twight of the Idaho Department of Commerce.

Illinois

Programs

In its response to the survey, Illinois chose to emphasize its financial industry and financial resources. It has the third highest rate of commercial and industrial loans and SBIC financing in the nation. This represents a strong financial infrastructure to help with high technology business development.

The Small Business Development Loan Program provides direct financing to small businesses with negotiable interest rates in cooperation with private and public sector lenders. The program can fund up to 25 percent of the total project cost or up to $750,000.

The Minority, Women and Disability Business Loan Program is aimed at helping special-needs entrepreneurs by providing direct financing to businesses at a negotiable interest rate in cooperation with private and public sector lenders. The program can fund up to 50 percent of the total project costs or up to $50,000.

The Participation Loan Program is also available. It works through banks and other conventional lenders to provide financial assistance to small businesses that will employ Illinois workers. The state will participate in loans up to 25 percent of the total amount of a project but not less than $10,000 nor more than $750,000.

The Development Corporation Participation Loan Program provides financial assistance to small businesses that provide jobs to workers in the region served by the Development Corporation. The state will participate in loans up to 25 percent of the total amount of a project but not less than $10,000 nor more than $750,000.

The Illinois Export Finance Partnership Program is a collaborative effort between the Department of Commerce and Community Affairs and the Illinois Development Finance Authority to help small and medium-size exporters gain access to working capital loans. The Surety Bond Guaranty Program helps minority and women contractors obtain information on how to bid for government contracts.

Illinois participates in the Enterprise Zone Program. In 91 certified zones, businesses are given tax incentives to make investments. The Industrial Training Program assists Illinois companies in training new workers or retraining workers. The Illinois Development Finance Authority Program provides numerous financial programs targeted to growth of Illinois businesses.

This information was provided by Dennis Whetstone, Director, Illinois Department of Commerce and Community Affairs.

Indiana

Programs

The business incentives in the state of Indiana are not segregated for high technology. Technology companies are eligible for general business incentives including job creation incentives.

For relocation, Indiana has a staff that fields leads from potential corporate relocations. "Attraction Reps" help find sites for incoming businesses. The reps also provide information about tax rates, utility costs, local development packages, and other information. Other staff members help with businesses that plan on in-state expansion or out-of-state expansion. These activities are done by the Business Development Office. The staff likes to work as an agent for the expanding or incoming business.

Concerning incentives in general, the Department of Commerce reports they do not use a top-down approach where the department tells the local units what to do to attract businesses. Instead, the local units drive economic development using Tax Increment Financing (TIF) and other types of local funding. The state has a supporting role, bringing together training, infrastructure, and tax incentives.

There are state training funds available, sometimes linked to Community Development Block Grants. When a strong company does not quite meet criteria for state training dollars, the state will sometimes use the additional federal aid to help. The state also partners with workforce development efforts.

With respect to infrastructure, the state will help a community prepare for a company moving in. If the company has a specific need, such as sewers, the state sometimes helps fund the project with local units on a fifty-fifty basis. This is driven by the local units and spurred by the likelihood of additional jobs and improved tax base.

The third area of state support is the Economic Development for a Growing Economy program, which is a reimbursement of state income tax. This is used as a gap filling inducement.

Coordination

The state legislature formed a private nonprofit agency called the Indiana Business Modernization and Technology Corporation. Recently, they have spun off and they became self-funded. They operate like a high technology manufacturing extension office. They have many offices in the state and they provide business assistance. The agency is a support agency for organizations involved in technology transfers and high technology company partnerships. The organization is affiliated with the Manufacturing Extension Partnership. It also operates the Wright Technology Network funded through Wright Patterson Air Force Base. This is a Defense Conversion program. There are also Economic Development Administration programs coordinated through the office.

Program coordination is done by staff people who call on small and medium-size manufacturers. They attempt to discover their technology transfer and operational needs, and then go to the federal laboratories and Air Force to find solutions.

The state funds a group at the Indiana University that does all SBIR outreach (seminars, proposal writing, etc.). It also has a Small Business Bridge Fund that makes loans for later phase innovations. The state does not deal directly with SBIR, however, but instead funds an Indiana University–sponsored grant writing program, which is called the Industrial Research Liaison Program.

This information was provided by the Department of Commerce and associates particularly Doug Moses, Bill Glennon, and Del Schuh.

Iowa

Programs

Iowa has a number of tax-based inducements for small businesses that would apply to high technology businesses. One of the more recent has been the elimination of property tax on manufacturing machinery and equipment assessed on or before January 1, 1995. The state has no corporate income tax on profits from sales outside Iowa. There is no sales and use tax on purchases of manufacturing machinery and equipment.

Concerning jobs, the state has state-funded, flexible new employee training programs designed by the company. A corporate income tax credit of $852 is offered for each new job. Iowa is also a right-to-work state.

There is no inventory or personal property tax and relatively low worker's compensation and unemployment insurance costs.

Coordination

To some extent, the Wallace Technology Transfer Foundation, created by the legislature, is a coordination mechanism. It helps technology firms plan for the future and obtain information about advancing technologies.

This information was provided by Bob Henningsen, Iowa Department of Economic Development.

Kansas

Programs

Kansas is a real surprise in the sophistication and breadth of its high technology business support. Kansas has a development corporation, the

Kansas Technology Enterprise Corporation, that coordinates economic development activities in the high technology area. Its mission is to "foster technological innovation and the creation, growth and expansion of Kansas enterprises." The set of programs it offers span the development cycle of young businesses.

One key program is the Advanced Manufacturing Institute at Kansas State University. This program focuses on research in automated design and manufacturing systems. The Center for Excellence in Computer Aided Systems Engineering at the University of Kansas aims at designing and tailoring software to meet a company's needs. The Center for Design, Development and Production at Pittsburgh State University helps businesses in solving technology-related problems, prototyping, and expanding technical capabilities. The Higuchi Biosciences Center at the University of Kansas is a pharmaceutical research center. Bioanalytical analysis, drug delivery research, and neurobiology and immunology are constituent centers. The National Institute of Aviation Research at Wichita State University conducts research on aircraft.

These are the major research centers that could be used as resources by incipient technology firms. There are also technology programs that are really funding sources for college researchers. This includes the National Science Foundation EPSCoR program, which aims at improving the competitiveness of state university research programs.

There are a large number of other resources available. The Applied Research Matching Awards aims at academic-business partnerships and finances research that leads to new or improved products. Training and Equipment Grants provide matching money to two-year academic institutions for new student training equipment. General business consulting services are offered as well as both SBIR and SBIR Bridge funding.

An Industrial Liaison Program provides expert consultation for Kansas businesses. Experienced business managers are available to help emerging businesses with a variety of problems.

There are at least three venture capital sources available for seed capital and follow-up funds. The providers can be contacted through the Kansas Technology Enterprise Corporation.

Several other resources must be mentioned: The Mid America Manufacturing Technology Center, which assists small and mid-size manufacturers in obtaining more operational efficiency; the Manufacturers Enterprise Corporation, which helps manufacturers put new products in the market; and the Kansas Value Added Center, which promotes the development of agricultural value-added products and services. Lastly, commercialization corporations include the Kansas Innovation Corporation, Mid-America Commercialization Corporation, and The Wichita Technology Corporation. Technology referral services are also offered.

Coordination

Coordination is done through the Kansas Technology Enterprise Corporation whose President, Richard Bendis provided this information.

Kentucky

Programs

The Kentucky Cabinet for Economic Development assists new and existing businesses through the Kentucky Industrial Development Act. In addition, the Business and Technology Branch has SBIR Bridge Grants. These facilitate the continuation of research between Phase 1 and Phase 2. The cabinet is creating the Innovation Fund, which is a loan program to support innovation in technology.

The key device for high technology businesses is the Kentucky Industrial Development Act. Corporations that build new manufacturing plants or expand existing facilities are eligible for tax credits. Projects that are eligible are land, building, and fixtures. Eligible costs are land, site development, utility extensions, and various services. The company receives a 100 percent credit against its income tax liability. The credit can be used for ten years or the term of the financing.

Coordination

The state established of the Kentucky Technology Service (KTS). The KTS receives funding from National Institute of Science and Technology Manufacturing Extension Partnership. The cabinet secretary has a seat on the KTS board. The Business and Technology Branch in the cabinet coordinates an annual SBIR conference, which includes speakers from the various SBIR programs. The SBIR conference is coordinated through both the University of Kentucky and the University of Louisville.

The Business and Technology Branch Manager is also on the EPSCoR committee. The Branch also maintains a Web site with links to technology information on the Internet.

This information was provided by the Kentucky Cabinet for Economic Development.

Louisiana

Programs

The Board of Regents for Higher Education administers the Louisiana Educational Quality Support Fund, which has an applied research component that requires industrial-business participation with the university. The Louisiana Economic Development Corporation in the Department of Economic Development provides guaranteed loans, an interest subsidy program, investment in and coinvestment with venture capital companies, and the City-State program of EX-IM Bank. The department's Technology Transfer Office at NASA's Stennis Space Center assists in the transfer of technology from the federal government to business and provides SBIR/STTR project assistance.

There are several Centers of Excellence in the three university systems that have a definite applied mission for high technology: The Pennington Biomedical Research Center, the Center for Advanced Microstructure Devices, the Micro-Manufacturing Institute, the Louisiana Productivity Center, and the Biomedical Research Foundation of Northwest Louisiana. There are also other Centers of Excellence that can provide assistance.

Within Louisiana, there are three high technology incubators and several university-related research parks. Tax abatement programs available to any other industry or business are also available to high technology businesses. These include a ten-year property tax exemption and the Enterprise Zone Program, which rebates sales tax and forgives corporate income and franchise tax under certain conditions. There is a similar program for businesses locating in a university related research park. The state also provides funding for the Small Business Development Centers that provide management and technical assistance.

Coordination

The Director of Policy, Planning and Technology Programs in the Department of Economic Development is the designated contact for the Advanced Research Projects Agency, the Advanced Technology Project, and the National Institute of Standards and Technology. The state offers assistance to business applicants for federal grants and loans and will act as the liaison for the businesses with the federal agency.

This information was provided by Mrs. Nadia Goodman, Director, Policy, Planning and Technology Programs.

Maine

Programs

Like some other states, Maine does not have programs exclusively designed for high technology businesses. It does have a set of programs and tax credits that high technology companies can use.

The Business Equipment Property Tax Reimbursement Program provides reimbursement to taxpayers for local property taxes paid on business machinery and equipment first placed in service or constituting construction in progress after April 1, 1995. Property taxes on that property may be reimbursed for 12 years. When combined with Tax Increment Financing, it has the potential of reducing new investment property taxes to zero over the 12-year life of the program.

The Maine Jobs and Investment Tax Credit provides an income tax credit for investments in most types of personal property that generate at least 100 new jobs within two years when the investment is placed in service. The credit is tied to the federal investment tax credit and is limited to $500,000 per year with carry-forward provisions. Credit will not exceed $3.5 million unless there are multiple qualifying investments in successive years.

The Municipal Tax Increment Financing is a locally determined program that allows municipalities to finance certain costs associated with a business development project by "capturing" and dedicating the new property taxes that will arise from those investments. This may include the municipality returning all or a portion of those incremental taxes to the business for application against the business's project costs. In most cases, the maximum term is 30 years.

State Tax Increment Financing is also available for high technology companies. Used in conjunction with Municipal Tax Increment Financing, this program is a mechanism under which up to 25 percent of the incremental increase in individual income tax withholding resulting from new employment activity within a municipal tax increment financing district is returned to the business within the district to offset the costs of improvements generating those increases.

High technology businesses may utilize the state sales tax exemption for machinery and equipment that is used directly in production. Many essential fixtures are included in the exemption. Also, 95 percent of the fuel and electricity used in a manufacturing facility is exempt from state sales tax.

The Maine Investment Tax Credit is also available. It is an income tax credit for machinery and equipment used directly in production worth 1 percent of the cost of such machinery and equipment per year over five years.

The Central Maine Power Corporation's Maine-Made Incentive provides a 1.5 cent per kilowatt rebate on incremental energy usage available for a three-year contract period. (These industrial rebates are offered by other power companies in other states as well.)

The legislature recently passed a tax credit of particular interest to high technology businesses. The Research and Development Tax Credit provides a two-tiered state income tax credit for state-based new research and development activities. It allows a credit of 5 percent of qualified research expenses over a three year average base amount and a credit of 7.5 percent of basic research payments.

Other assistance Maine provides is confidential site selection, cabinet-level major project economic development consultation, and a supplier network. Cabinet-level consultation must be considered a unique way of helping technology businesses.

Coordination

The state coordinates its technology initiatives with federal technology support programs. Coordination is mainly the responsibility of the Maine Science and Technology Foundation.

This information was provided by Susanne Pilgrim, Research Analyst, Maine Department of Economic and Community Development.

Maryland

Programs

The state has an incentive program that aims at business siting in the state and business expansion through low-interest loans and grants. The state assists both startup and established companies.

There are two incentives specifically for high technology companies, the Challenge Investment Fund and the Enterprise Investment for startups. The state also participates in various public-private partnerships and offers grants in some of the partnerships. One of them is the Suburban Maryland High Technology Council. There is also the University of Maryland, which has several programs, and there are a number of business incubators in the state.

Coordination

The state works with the Technology Transfer Program and the Community Development Block Grant program. There is an inter-agency agreement in which the federal government places a staff person in a state office to help the state work with businesses. The state also has a liaison for agriculture and aerospace. The coordination is done through each individual department by contract agreement.

This information was provided by Michelle Wright of the Department of Business and Economic Development.

Massachusetts

Programs

The Massachusetts Office of Business Development is a key agency that works with all kinds of businesses, including technology-based businesses, to help them move to Massachusetts or to help them expand in the state. The agency assists in finding sites, accessing training, and locating funding. They do not provide business funding, but do help companies get access to funding sources. The state has the highest research and development tax credit in the country at 12 percent. There are tax breaks for companies that have van pools. There is also a program called the Economic Development Incentive Program. The state has targeted certain communities that are interested in bringing in businesses to their community. In exchange, they can take advantage of up to a 20 percent tax deduction for the first five years of business. Local tax deductions may be available. There is also an abandoned building deduction and an investment tax credit of about 5 percent. Other loan assistance programs are available.

This information was provided by Steve Johnson of the Massachusetts Office of Business Development.

Michigan

Programs

The Michigan Department of Commerce provides a variety of services to small businesses, including high technology businesses. An ombudsman office negotiates and advocates regulatory issues for small businesses and also researches and lobbies business issues. The Development Services Bureau provides siting services, along with the Development Services Division.

The state Small Business Development Center provides consultation and educational programs. Of particular interest to high technology businesses are two programs. The Technology Services Office supports technical programs for small businesses, such as R&D funding and marketing assistance. It also partners with grant programs. The Technology Transfer Network lets Michigan firms get access to academic research and development resources. The Michigan Energy and Resource Research Association Small Business Development Center aims at helping economic development through increased R&D activities. The University of Michigan operates the Innovation Center to help state businesses commercialize new products. Finally, the Metropolitan Center for High Technology runs a small business incubator and develops technology-related products.

Minnesota

Programs

Minnesota has many programs for high technology businesses, offered by the Minnesota Department of Trade and Economic Development.

The Small Business Development Loan Program is an industrial revenue bonding program that may be available to finance portions of land, building, machinery, and equipment. Projects must enhance the local tax base and create or retain jobs.

The Economic Recovery Grant Program awards grants to local governments that in turn make loans for business development projects or to finance public improvements that support business development projects. The program awards up to $500,000.

The Capital Access Program provides limited guarantees to private lenders for loans made to small to medium-size businesses for startup or expansion costs.

The Rural Challenge Grant Program provides grants to the six Minnesota Initiative Funds that provide loans to new or expanding businesses to stimulate job creation, private investment, and economic growth in the 80 counties outside the Twin Cities metropolitan area.

The Urban Challenge Grant Program provides matching grants to nonprofit organizations that in turn make low-interest loans to businesses located in distressed areas of the Minneapolis-St. Paul area.

The Minnesota Job Skills Partnership Board awards grants of up to $200,000 to training projects designed by Minnesota educational institutions for Minnesota businesses. The program is aimed at businesses that are experiencing skill shortages or are anticipating expansion that will require newly trained employees. The employer makes a matching contribution in the form of funding, loaned equipment, or faculty.

The Targeted Jobs Tax Credit program can provide companies with federal tax credits as an incentive for hiring employees who are of minority status, economically disadvantaged, or disabled. The tax credit is 40 percent of up to $6,000 of wages for a maximum of $2,400 per employee in the first year of employment.

Several options from various sources may be able to provide a training reimbursement to companies for a portion of an employee's wage during limited on-the-job training or industry-based training. Training contracts allow trainees to earn regular wages while limiting the company's hiring and training costs.

Minnesota technical schools provide customized training for individual companies.

The Minnesota tax system features such provisions as: no capital value tax regardless of company profitability. The state's corporate income tax employs an income apportionment formula (sales, 70 percent; property, 15 percent; and payroll, 15 percent) that is advantageous for businesses with a significant share of sales outside the state. There is an R&D corporate tax credit available for an increase in qualified expenditures for research and development performed solely in Minnesota. Expenditures include wages, supply costs, computer costs, and 65 percent of contract costs paid to others performing research. Capital equipment for new or expanding manufacturing facilities is exempt and replacement capital equipment for manufacturers is eligible for a reduced sales tax rate.

Materials used to produce products for sale and many special tools, dies, and jigs are exempt from the sales tax. Most business personal property is exempt from a property tax, which is very advantageous for businesses with a high proportion of personal property.

Coordination

Minnesota Project Innovation held a conference in 1995 for small businesses. The conference focused on how small Minnesota businesses might be able to conduct business with the federal government, specifically on how procurement has been reformed. The state is also involved in the Small Business Innovation Research Program. The Minnesota Technical Assistance Program also provides coordination help.

This information was provided by Diane Knutson a specialist at the Minnesota Department of Trade and Economic Development.

Mississippi

Programs

The state offers many of the same kinds of incentives that other states have for manufacturing innovation. There is also the Rural Economic Development Act, which provides job tax credits, job training tax credits, and ad valorem tax exemptions. The minimum investment is $1 million capital investment or ten jobs or a minimum capital investment of $100,000 for an enterprise headquarters.

Coordination

Mississippi coordinates its programs "informally" with SBIR. If the state receives a request for assistance, the request is sent to the Mississippi Enterprise for Technology at the Stennis Space Center, which has technology transfer responsibilities for the state. They provide technical assistance for SBIR proposals. The state provides general technical assistance through the enterprise and state universities. The state's Science and Technology Commission Board is made up of the four major research institutions, along with private-sector members and economic development professionals. It seeks to coordinate the research activities that are going on in the state.

Information was provided by Denton Gibbes and Clay Lewis of the Department of Economic and Community Development and the Mississippi Science and Technology Commission.

Missouri

Programs

High technology companies may be qualified for a variety of programs available in Missouri. There is a Research Development Tax Credit, a five-year state credit for average amount of research expenses, and capital tax credits that provide capital to small innovative businesses. It entitles contributors to the fund tax credits of 30 percent of the contributions, as well as shares, distribution earnings, and dividends from the company that is receiving the funding. This is to encourage investments into technology businesses. There is also a new program called Certified Capital Companies, or CAPCOs. Its purpose is to fund early-stage high-growth businesses that are frequently high technology businesses. It provides a 100 percent credit per year against the state insurance premium tax for cash investments in certified capital companies that are venture capital companies. Missouri has done this in order to correct the state's inadequate venture capital base.

Coordination

Within the last few years, Missouri created the Missouri Technology Corporation, whose purpose is to work with the Department of Economic Development on technology issues and help coordinate state activity in business and technology programs. Under this umbrella, the state operates Innovation Centers, Centers for Advanced Technology, field offices for the NIST, and Manufacturing Technology Centers that aim at improving Missouri manufacturing through technology improvements, facilitation of technology transfers, and developing funded applied research projects. Through state field offices, Missouri provides consulting to businesses concerning the application of workplace technology. It also works with ventures searching for capital.

This information was provided by Jim Gardner of the Missouri Department of Economic Development.

Montana

Programs

The main high technology activity agency in Montana is the Montana Science and Technology Alliance, which offers equity investments instead of loans for high technology investments. The state is willing to invest as an equity partner to get businesses up and running. The assumption of an equity interest is unusual among the many different state programs to support high technology businesses

There are a number of statutory tax programs, some of which are applicable for technology companies. Many of these are related to manufacturing processes. There are also property tax assessment reductions for research and development firms. In order to attract relocations, the state offers loans and tax reductions based on creation of jobs (50 or more jobs for an income tax reduction to pay for loans).

Coordination

Montana is involved in the SBIR program. There are several programs in the state that help businesses make SBIR applications. The state also works with NASA and NIST. The NASA program is operated through Montana State University.

This information was provided by Andy Poole of the Department of Commerce.

Nebraska

Programs

The Employment and Investment Growth Act is one example of Nebraska's many support programs. Under this, a business investing at least $3 million and creating 30 jobs gets a sales and use tax refund for property and building materials, a 5 percent tax credit on new employee compensation, and a 10 percent investment tax credit. Businesses investing $10 million and creating 100 jobs get personal property tax exemptions for fifteen years. Greater investments result in more refunds.

The Employment Expansion and Investment Act helps small business startups. If they create new jobs, an investment tax credit can be applied to corporate tax liability. The credit is $1,500 per employee and $1,000 for each $75,000 of investment.

The Enterprise Zone Act provides tax credits for qualified businesses investing in certain areas. The credit is available if the investment is at least $75,000 and employment increases by two people.

Wage benefit credits also go to companies that add 500 or more new jobs and $50 million in new investment. Utilities can also offer lowered rates to businesses adding new investments. Blighted areas near cities have been designated as such so that property taxes can be used to reimburse companies for land purchases. There is also a version of TIF available and the state provides training programs.

As far as tax policy is concerned, there are no property taxes on business inventories, no tax on intangibles, and no sales tax or use tax on ingredients used in manufacturing. There is a sales tax refund for pollution control equipment and no sales tax on water usage.

Nebraska has also created a Manufacturing Extension Partnership to help manufacturers gain access to high technology and modern manufacturing methods. This is provided through the Nebraska Industrial Competitiveness Service. It is affiliated with the national Manufacturing Extension Program of the Department of Commerce National Institute of Standards and Technology.

This information about the many areas of support for high technology and other businesses was provided by Tommy Thorpe, Director of the Nebraska Industrial Competitiveness Service.

Nevada

Programs

The Nevada Commission on Economic Development reported that the state offers a tax structure that is less burdensome to businesses and employees with no corporate or personal income tax. There are no franchise, inventory, inheritance, estate, or gift taxes. While there are no specific high technology business programs, there are incentives and financing programs available to new and expanding businesses. Sales and use tax abatement are given to businesses that meet the state economic diversification and development plan. Property tax abatement is available for recycling businesses when at least 50 percent of the material or product has been recycled. A personal property tax exemption is available if the property is used as a facility for the production of electrical energy from waste material. Certain property tax exemptions are offered, and there is a sales tax deferral for industries that buy certain types of capital equipment in excess of $100,000.

A job training program is offered that pays up to 75 percent of training costs. It can be used before and 90 days after a plant opening.

The Nevada Development Capital Corporation is a private, for-profit development fund for Nevada companies that are not new ventures. The Southern Nevada Certified Development Company is a nonprofit company that packages SBA loans for southern Nevada. The Nevada Self-Employment Trust is a nonprofit program aimed at improving economic self-sufficiency of low and moderate income men and women through entrepreneurial training and credit access. The Nevada Revolving Loan Fund Program helps small businesses in rural Nevada with business expansion loans. The Nevada State Development Corporation offers SBA long-term, fixed-rate, low down payment, second-mortgage financing for real estate and fixed assets for owner-operated businesses. The Rural Development Corporation offers loans to rural small businesses for startups and expansion.

Overall, the attractions of Nevada are the low tax rate and the numerous state-sponsored support programs for growing businesses.

New Hampshire

There are a limited number of programs available, but high technology firms are eligible for them. The New Hampshire Industrial Development Authority assists businesses in some types of financial needs, including loans and getting credit. Industrial Revenue Bond Financing can be used for financing manufacturing facilities and purchasing equipment. The Guarantee Plan for Machinery and Equipment can guarantee up to 35 percent of a loan for certain machinery and equipment. There is also a state Office of Business and Industrial Development that facilitates tasks for incoming businesses and business expansions in typical development office fashion.

New Jersey

Programs

New Jersey provides the same type of program support for high technology that other urban states do. But what is especially true of this state is the extensive infrastructure available for emerging or entering high technology firms. A transportation system that interconnects major population centers and the availability of scientists and engineers are parts of that strong infrastructure. The state has more scientists and engineers per capita than any other state. Since there are 57 state colleges and universities, there is a rich stream of close-by educated talent. New Jersey ranks fourth nationally in research and development expenditures.

There is an existing corporate technology base led by the pharmaceuticals—Merck, Hoffman-LaRoche, Bristol-Meyers Squibb, Schering-Plough, and in health care, American Home Products, but there are many other firms as well. The state has the third highest concentration of corporate headquarters.

Overall, New Jersey is not noted for being a tax haven for businesses, but exemptions and credits can be found that may be used by leading firms. New Jersey has not created an organized, broadly based organization to attract high technology businesses.

Employment in two high technology areas, electronic and computers, as well as pharmaceuticals, has dipped slightly between 1993 and 1995, but in computer services it grew in the same period.

It is interesting that the volatility in the high technology employment base and the strong infrastructure has not provoked a more concerted effort to attract public attention within New Jersey.

This information was provided by Cardella Ansley, Principal Research Analyst for the New Jersey Department of Commerce and Economic Development.

New Mexico

Programs

The director of science and technology in the New Mexico Economic Development Department, Durand Smaith, pointed out that New Mexico is basically a high technology state with its number of PhDs the highest in the nation; it is the second highest (with $3 billion) in research and development expenditures per capita.

The state has a program called the New Mexico Technology Commercialization Alliance, which offers support for startup and high technology businesses. Assistance is available for business planning, funding, and technology problems. The alliance is composed of members of New Mexico's three national laboratories. The three research universities, University of New Mexico, New Mexico State University, and New Mexico Tech, also offer support.

Recently the state doubled the size of a venture capital fund to make loans to new technology businesses. There is even a tax break connected with the reusable launch vehicles at the missile range, the historic White Sands Range.

Business relocations may be eligible for industrial revenue bonds that exempt property taxes during loan repayments. The loans are up to three years at slightly above the prime rate, depending on the credit rating. Intel used this program.

The Implant Training Program is available. The state helps train employees of companies coming into New Mexico. The program pays up to 50 percent of the training costs and wages for a six-month period.

Coordination

The state ranks first in SBIR program awards, a sure indication of its SBIR intensity. There are also many federal grants awarded to the state. Coordination is done through the Department of Economic Development and the Commercialization Technology Alliance. The state has a group that meets regularly to work with small existing businesses to solve problems with startups, funding, and technical assistance.

New York

Programs

There are many programs available, as would be expected in this state. New York's Science and Technology Foundation plays a large part in the support of high technology business. It has a venture capital fund for financing businesses. There is also a Federal Technology Program involving 13 universities in technology development. The Technology Development Organization Program is a means for the state, through the foundation, to contract with ten nonprofit organizations to do business assistance specifically for high technology businesses. For business relocation efforts, the state uses the Economic Development Zones Program, where by tax incentives are made available for business relocations. The governor's office is involved in recruiting businesses for New York.

Coordination

The state has a contract with NIST and it brings in about $8 million annually. That includes support for growing high technology companies.

New York also works with NIST and the Advanced Technology Program, where competition for grants has begun. Requests for grants from the program involve projects such as integrated manufacturing. The state helps companies prepare for the proposal solicitations.

New York also obtains approximately $25 to $30 million in SBIR awards per year. The state has business development specialists in state offices who help firms with applications, coordinate with the federal offices, and screen solicitations for matches between firms and grants.

This information was provided by Christopher Burke of the New York State Science and Technologies Foundation.

North Carolina

Programs

North Carolina had an early start in developing high technology programs. It created an Industrial Extension Service in 1952 and started putting together the Research Triangle Park shortly thereafter. Although North Carolina has the typical array of tax credits, job training, and loans as other states, its particular forte is in the creation and coordination of research centers.

The MCNC is one of the centers. It is a private nonprofit corporation aimed at the proliferation of new electronic technologies. The MCNC brings together universities, industry, and government and provides for sharing of the latest developments in electronics, supercomputing, and telecommunications. The organization pulls leadership from the state chancellors, college representatives, and governor's appointees.

Another major center is the North Carolina Supercomputing Center. The special mission of this organization is to speed commercialization of supercomputing within the business community. A unique provision is that half the center's supercomputing resources be available for state businesses.

The North Carolina Biotechnology Center is designed to support biological research. Improving the research capabilities of state universities, further cooperation among government, industry, and academia and business development assistance is among its goals. There are no laboratory facilities because it is a coordinating center.

Establishing technology centers are not the only thing different about North Carolina. The state also shows a brand of foresightedness. A broad-based state panel, the North Carolina Alliance for Competitive Technologies, produced a booklet entitled "Strategies for a Competitive Future" in 1995. The panel assessed strengths and weaknesses of the state's technology base and concluded with recommendations to support modernization, improve the entrepreneurial culture, increase technological deployment, maintain and strengthen research and development, invest in the infrastructure, educate workers, and aim technology at consumer needs.

Prospective technology companies should be able to learn more about North Carolina's technology infrastructure through the centers.

Coordination

The main coordination is done through the North Carolina Alliance for Competitive Technologies, according to the executive director, Walter Plosila, who supplied this information.

Some information reported here was provided by Christopher Coburn, *Partnerships: A Compendium of State and Federal Cooperative Technology Programs* (Columbus, Ohio: Battelle Publishing, 1995).

North Dakota

Programs

North Dakota does not offer tax credits exclusively for high technology companies, although a variety of tax incentives are available to startup and expanding companies in North Dakota. In that, it is very similar to other states. The only program specifically designed to stimulate technology development is vested in Technology Transfer Inc. (TTI), a nonprofit development company. The agency provides funding and leadership for the commercialization of new technology in North Dakota. It also promotes the use of modern manufacturing processes among North Dakota's 400 production manufacturers. Most of TTI's investments are royalty based and most of its sources for investments are tax dollars appropriated to TTI by the North Dakota legislature on a biennial basis at about $1 million per year.

Coordination

TTI attempts to leverage its investments in research and development with the use of SBIR, STTR, and other federal research sources. TTI coordinates this effort by sharing in the cost of preparation of applications to these federal sources and by co-investing in the subject research project. In addition, TTI uses resources found within the North Dakota University System to apply to private industry. On a more limited basis, TTI provides federal laboratory technologies to companies with specific technology needs.

Information was provided by Warren Enyart, CEO, Technology Transfer Inc.

Ohio

Programs

Ohio has a large number of programs and tax incentives for high technology businesses, though they are not targeted only for these businesses. The state highlights its job-creation tax credit, sales tax exemption for research and development equipment, and tax-increment financing as technology support programs. There are direct loans for part of land or building costs and the Ohio Enterprise Board Fund lends for construction and business acquisitions. Community Development Block Grants are also issued for businesses.

The Community Development Corporation Program offers competitive grants of up to $50,000 for projects that may help meet needs of a defined low-and moderate-income neighborhood or target population. The state development corporation and Small Business Administration offer 504 type loans for fixed assets for small businesses.

The 166 Regional Loan Program is available for land and building acquisition, expansion or renovation, and equipment purchase. Up to 50 percent or $200,000 loans are available. Revolving Loan Funds are available for job creation projects emphasizing employees who are low-and moderate-income households. Minority direct loans and a minority contractor bonding program are also available.

Tax incentives can be found in the Enterprise Zones program, where up to 100 percent abatement of real estate or tangible personal property taxes for up to ten years can occur for businesses creating or retaining employment in a facility in an enterprise zone. The Community Reinvestment Areas are similar except the abatement is good for fifteen years and the local legislative body establishes the terms.

Ohio has a comprehensive employment and training program including industrial training—"Enterprise Ohio" links business and industry to the training resources of two-year community and technical colleges. A state job placement agency; a targeted jobs tax credit program; veteran's employment incentives; and the Ohio Training Exchange, which trains companies that supply Ohio's major manufacturing firms in technologies, can also be accessed. The exchange will be of interest to high technology training organizations.

The unique and directly related resource that Ohio has for technology firms is its eight Edison Technology Centers. These offer business access to state-of-the-art basic and applied research performed in-house or obtained through links with universities, federal laboratories, and other institutions. They create opportunities for business spinoffs by sharing technology and research findings. The centers are jointly funded by business, government, and universities. Each center has individual focus on polymers, industrial systems, advanced manufacturing, welding, biotechnology, or advanced materials.

Coordination

Much of the coordination is done through the Thomas Edison Centers, which receive federal funding. The SBIR program was also cited as a coordinating mechanism.

This information was provided by the Ohio Department of Development.

Oklahoma

Programs

Oklahoma has a varied set of economic incentives for high technology and other firms. One particularly aimed at high technology businesses are sales tax refunds. They can be used by high technology businesses for sales taxes paid and are a refund for sales and use taxes. Qualified companies must apply to the tax commission, which sets up an account. Sales taxes are refundable from the account. There is also an investment and new jobs tax credit.

A major program is the Oklahoma Quality Jobs Program, which provides quarterly cash payments of up to 5 percent of payroll for qualified companies for ten years. The company must fall into certain SIC codes (computers, research and development laboratories, engineering, and business services are examples) and must meet other requirements.

There is a technology transfer income tax exemption and new product development income tax exemption, which could be used by high technology businesses. A host of sales tax exemptions, enterprise zone programs, and training programs fill out Oklahoma's broadly based and inventive array of state programs.

Concerning high technology, there is a Basic and Applied Research and Technology Program administered by the Oklahoma Center for Advancement in Science and Technology. It is a grants program for high technology companies. There is also an Incubator for New Technology Businesses and the sponsor of incubator facilities is exempt from rental income taxes if the business is new. Although Oklahoma is a largely rural state, there are a considerable number of programs that could assist high technology business enterprises.

Coordination

The Oklahoma Council for Science and Technology is the main agency that works to coordinate federal support programs.

This information was provided by Karla Graham and Victoria Armstrong of the Oklahoma Department of Commerce.

Oregon

Programs

Oregon has business incentives that are not exclusively for high technology businesses, but they are applicable to them. High technology is a key industry in Oregon.

The state has an active business recruitment and expansion program. Unlike other states, it is not aimed at getting businesses to relocate in Oregon. Instead, it is based on internal business expansion. This makes Oregon somewhat different from other states that are in the technology business chase.

There are active industry associations that help high technology businesses. These associations focus on different industries.

Oregon has the Enterprise Zone Program and the Strategic Investment program, the latter which exempts large investments from property taxes in exchange for a community service fee of 25 percent of abated taxes. Corporate tax credits are available for pollution control, energy efficiency, reclaimed plastics, research and computers, and scientific equipment donations. These last two are related to high technology. There are financing options such as industrial development revenue bonds and business loans. Workforce development assistance is also available.

Coordination

The state coordinates its support for research and development contracts with SBIR, NASA, NIST and the Defense Department. It also tries to link Oregon businesses with those resources. Oregon makes information available to Oregon companies on developments in these areas.

Pennsylvania

Programs

This large state has a wide variety of typical business development programs and a few for high technology businesses.

The main small business advocacy agency is the Office of Enterprise Development. The Economic Development Partnership puts together major private and public interests to address state economic concerns. The Business Resource Network is a small business advocacy office and information center. There is an Industrial Development Authority that offers long-term low-interest business loans.

Of special interest to technology companies is the Ben Franklin Partnership, which promotes advanced technology to make conventional industries more competitive internationally and to spin off new technology intensive businesses. The partnership has four Advanced Technology Centers, which are private-public consortia. Each center has joint research and development activities in certain areas, education and training programs, and entrepreneurial assistance. There is an incubator program and SBIR assistance.

Rhode Island

Programs

Led by the Department of Economic Development, there are a number of support services and tax incentives for high technology and other businesses operating in Rhode Island. A federal procurement program helps businesses that want to sell to the federal government. The Small Business Development Centers (SBDCs) offer counseling and education and other assistance.

With respect to technology businesses, the Rhode Island Partnership for Science and Technology offers grants to businesses for applied research done with universities, colleges, and hospitals. Up to 60 percent of research budgets can be funded. The SBIR program is also available.

South Carolina

Programs

South Carolina does not have any laws or programs that are specifically targeted to the high technology industry. The state has the Seed Capital Fund, a privately capitalized fund that makes loans. There are also technology incubators. These are limited to technologies connected to the research universities and they are not commercial incubators. They are aimed at maturing technologies that are linked to the core competencies of the universities.

Coordination

The state has two programs. One is the EPSCoR program principally from the National Science Foundation. Another is an EPSCoR program supported by NASA, the EPA, and a couple of other federal agencies. There is also the South Carolina University Research Foundation program funded by the Department of Energy, as well as SBIR and NIST programs. Coordination with EPSCoR is informal. The state provides copies of its strategic plan and they distribute copies of the plan. There will soon be coordination of SBIR efforts as well.

This information was provided by Doug McKay, the director of Technology and Special Assignments for the South Carolina Department of Commerce.

South Dakota

Programs

There are no separate tax programs for high technology businesses. For all businesses, there are no corporate income taxes, no personal income taxes, no personal property taxes, no business inventory tax. Because of the lack of taxes for all industries, South Dakota does not offer tax breaks or credits. For new and expanding industry, the state offers low interest (3 percent) loans and taxable and/or tax-exempt bond financing.

Coordination

South Dakota has an active SBIR program based on the campus of the Dakota State University in Madison. South Dakota is involved in EPSCoR and those efforts are based in University of South Dakota at Vermillion. The Governor's Office of Economic Development works closely with the state's university system to enhance startups and recruitment efforts in high technology firms.

Information was provided by Mary Cerney, Governor's Office of Economic Development.

Tennessee

Programs

The Tennessee Department of Economic and Community Development is the lead agency for high technology. It has the mission of promoting economic development and growth on behalf of Tennessee citizens. The department works with local and regional organizations. It markets the state's high technology hot spots to the outside. It also carries out a marketing program to recruit new industries to the state. Within the department, the role of the state in science and technology is coordinated.

Coordination

The Science and Technology office is the coordinating body. The advisor in the office works closely with a statewide network of technology leaders in university research laboratories and businesses. Primary functions of this office are coordination and facilitation of programs and policies in science and technology as they contribute to technology-based economic development. This includes SBIR, STTR, ATP, and other federal grant programs. This is principally done by serving as a source of information on such programs.

In 1993, legislation created a Science and Technology Advisory

Council. This council advises the governor and legislature on the state's scientific and technological goals, policies, and programs to spur job creation and growth and technology transfer.

This information was provided by the Tennessee Department of Economic and Community Development.

Texas

Programs

The state of Texas has several programs available to stimulate the creation of high technology businesses or business relocations. The Texas Department of Commerce's Office of Growth and Retention does this by promoting economic development in Texas, assisting in the recruitment of targeted industries such as telecommunications and electronics from out of state, and the expansion of existing Texas businesses. The office helps with site selection assistance, trade shows, and marketing trips. The same department's Business Services Division provides financial assistance through various loan programs, including the Texas Enterprise Zone Program, Historically Underutilized and Small Business Linked Deposit Program, and the Texas Capital Fund Programs—TCF Infrastructure Grant Program, TCF Real Estate Development Program, TCF Main Street Improvements Program, Texas Leverage Fund, and Statewide SBA 504 Loan Fund.

Another support program is the Smart Jobs Fund Program, which was established in 1993. The fund provides matching grants to pay for customized, industry-specific training to new or existing Texas companies. The Texas Department of Commerce Business Information and Referral Program staffs a toll-free information line for businesses and individuals seeking information on starting a business or expanding current business in Texas.

Coordination

The Texas Manufacturing Assistance Center, an economic development service of the Texas Department of Commerce affiliated with the Manufacturing Extension Partnership of the U.S. Department of Commerce National Institute of Standards and Technology, assists manufacturers by tapping into the technology of federal labs and expertise of the federal government. The center is one of the largest manufacturing extension projects in the country and provides assistance to small Texas manufacturers seeking to upgrade their operations to become more competitive.

This information provided by Kevin G. Walker, Director, Texas Office of Community Assistance/Small Business.

Utah

Programs

The most significant program aimed at stimulating the creation of high technology businesses and the relocation of businesses in Utah is the Centers for Excellence Program. Through this program, grants are awarded to college and university researchers in the state to help them complete late-stage research into innovative technologies and to commercialize them through licensing agreements with existing companies or the creation of new businesses.

Within Agriculture and Natural Resources as a category, there are five individual centers for meat processing, dairy foods, value-added seed technology, environmental technology, and solid waste recycling. Within Computer/Information Technologies, there are nine separate centers for Engineering Design, Advanced Archaeological and Paleontologic Imaging and Modeling, Computer Graphics and Scientific Visualization, 3D Computer Graphics, Inverse Problems, Imaging and Tomography, Software Science, Self-Organizing Intelligent Systems, and Multimedia Education and Technology at both the University of Utah and Utah Valley State College.

Another group of Biomedical/Biotechnology is composed of the National Center for Design of Molecular Function the Centers for Biopolymers at Interfaces, Cancer Genetic Epidemiology, Controlled Chemical Delivery, Development and Molecular Biology, and Center for Chemical Technology. The last group is Aerospace/Advanced Materials. It contains the Centers for Aerospace Technology, Advanced Composites Manufacturing and Engineering, and the National Center for Advanced Combustion Engineering Research.

Another program is the Industrial Assistance Fund in which loans may be offered to relocating or expanding businesses to help with relocation and setup cost. Other incentive programs are available depending on the needs and circumstances of the new business.

Private-level entities involved in this matter are the Economic Development Corporation of Utah, the Utah Technology Finance Corporation, and several venture groups offering various forms of assistance for high technology businesses.

Cooperation

The state works with and coordinates its support for high technology businesses with several high technology programs, including the Advanced Technology Program and the Department of Commerce's National Institute's of Standards and Technology. This coordination is often done through applications for grants from these and other federal programs. If the grants are awarded, they help Utah aid high technology businesses in expansion and technology transfer. One such grant helped establish the Utah Manufacturing Extension Partnership. This set up a statewide network of engineers to assist companies, including high technology companies, in enhancing their productivity and technological performance.

This information was provided by Darrell Kirby of the Utah Office of Technology Development.

Vermont

Programs

Vermont offers research and development tax credits and low-cost financing for high technology companies. The state offers training grants to assure a productive workforce. A new Vermont manufacturing extension center will provide business with technical assistance in installing the latest manufacturing advances. The Vermont Technology Council and the state have developed four Centers of Excellence: Applied Biotechnology, Food Science, Environmental Science, and Advanced Materials. Although this is a small state, it does have an active technology business development effort.

Coordination

Vermont has an EPSCoR program that provides Small Business Innovation Research training and grants to develop technology applications.

Information was provided by the Vermont Department of Economic Development.

Virginia

Programs

Virginia's Center for Innovative Technology (CIT) is a state chartered nonprofit entity that exists to enhance Virginia's competitiveness in providing businesses with access to the state's technology resources and assisting in the creation, retention, and attraction of high technology jobs and businesses. The center's programs include technology development assistance, business assistance (e.g., incubators), and financial assistance—primarily co-funding of research and development projects.

Co-funding of technology projects has been done by the CIT with a total of 836 projects funded thus far. Thirteen technology development centers have been established.

The Virginia group is more recently focused on job creation, competitiveness improvement, supporting high technology business efforts to commercialize their products, and several other directions, all of which have the aim of improving the state's competitiveness.

Coordination

The Center for Innovative Technology coordinates a program to inform Virginia businesses about SBIR, ATP, TRP, and other major programs through a bimonthly newsletter, occasional conferences, and seminars. The center offers some matching dollars for these grants.

Information was provided by Virginia's Center for Innovative Technology.

Washington

Programs

Washington has two main tax incentives applicable for technology businesses. One is a credit on the business and occupation tax for research and development done by high technology businesses, which may be up to 2 percent of their research and development as long as it does not exceed $2 million per year. The second incentive is a sales tax exemption from state and local taxes of 8.2 percent for all new construction and equipment. There is also the Washington Technology Center, which does prototype research development between the universities and industries. The center acts as a broker between companies that want to do research and development but can't afford it and academic contacts who may fill the gap.

Coordination

Washington does not coordinate its support with any federal high technology support programs.

This information was provided by Tom Campbell of the State Department of Community Trade and Development.

West Virginia

Programs

There are no high technology programs reported. The Small Business Development Center helps small and minority businesses in filing state and federal forms. Startup counseling, loan packaging, and training programs are also offered.

The West Virginia Economic Development Authority provides low-interest loans for land and construction. The loans are aimed at manufacturing firms which are likely to create new jobs.

Wisconsin

Programs

Wisconsin reported that there were no laws that were specific to technology-based firms, with the exception of a law creating the Technology Development Fund. The program provides two phases of loans. Phase 1 provides up to 40 percent of the cost of the research and development of a new product or process. The loan is repaid with a 3 percent royalty on net sales of the product if the research is successful. If it is not successful, the award becomes a grant. Phase 2 provides a loan for up to 40 percent of the cost for commercialization of the new product. The funds can be used for fixed assets and working capital. It is a regular loan rather than a royalty repayment.

The state also provides assistance related to the federal Small Business Innovation Research program. It provides free SBIR technical reviews for applicants. Reviewers critique applications and provide comments. Also provided are SBIR Bridge Financing loans to businesses that have completed a Phase 1 SBIR and have submitted a Phase 2 application. The business can receive up to $40,000 for a one-year period to help maintain the research activity. The loan ends when the applicant receives notification of the results of the Phase 2 application. This loan is repaid by a 3 percent royalty upon commercialization.

In addition, the state provides matching funds for a National Institute of Standards and Technology (NIST) grant to the state. The grant was awarded for the establishment of a Manufacturing Extension Partnership. Technology-based companies are also eligible for other general loan and grant programs but these are not targeted for high technology companies.

Coordination

Support is not coordinated with the federal government per se. Matching funds were provided for the NIST grant. The state also allows businesses to use Technology Development Funds as a match for a SBIR award.

This information was provided by Louie Rech, Technology Development Coordinator, Wisconsin Department of Commerce.

Wyoming

Programs

The state has created a Science Technology and Energy Authority to provide financial assistance to high technology companies in the form of five-to ten-year loans, at or near the prime rate. This is the only program specially targeted to high technology businesses. Wyoming does not have an income tax so derived tax breaks are not offered.

Coordination

The state works with the SBIR program. There are contracted consultants who assist firms in earning SBIR grants. The coordination is between the authority and the company. If the company obtains an SBIR grant, it may get additional funding from the authority, though that is not guaranteed.

This information was provided by Paul Howard of the Division of Economic and Community Development.

Summary

This review of state programs related to business support for high technology businesses has demonstrated that, for the most part, states have not devised whole sets of initiatives for this business activity. Most often, the support rests within other business support activities, including creation of TIF programs, enterprise zones, training tax breaks, investment tax credits, and certain kinds of sales tax exemptions. It is also important to note that many states have also attempted to address the matter of high technology business development by establishing special councils or authorities, which are aimed at better coordinating grant activities or consulting services for technology-driven businesses. Some states have also developed incubator areas for brand new technology firms. For the states that have made an additional effort in the technology business development area, there is considerable variety in what is done, ranging from research and development tax credits to whole financing packages.

The findings from this chapter may be put together with the information on technology business concentrations in Chapter 16 for a depiction of what local and state developments have occurred. In the next chapter, we will look at the federal role in this issue and thus complete our review of the governmental infrastructure for high technology businesses.

18

FEDERAL GOVERNMENT
SUPPORT FOR HIGH
TECHNOLOGY BUSINESS

This chapter offers a perspective on federal government actions affecting high technology business. It also covers coordinating activities of the federal government with high technology and other private-sector businesses. We also consider the wide variety of interrelationships between high technology businesses and the federal government. This, then, is the last chapter in our three-chapter review of high technology concentration and infrastructure.

Information for this chapter was based on an Internet search of federal agency databases. It included Clinton Administration policy papers and public addresses.

The perspective here is exclusively domestic: we consider U.S. government policies. There are many actions taken by other governments in support of their own domestic high technology businesses. The United States is not unique in its development of high technology businesses; nor is it the only country that programs and policies for this type of business. Many countries have confidence in their technology progeny and are not shy about the accomplishments of their technology venturers. To catch the flavor of how some other countries have dealt with their science-based businesses, we include in Appendix C a description of business developments in a few nations.

What this book excludes is local government's role in high technology development. Even though local concentrations of high technology operations were presented in Chapter 16, the specific role of local government was only indirectly mentioned. This role is important but varies from one jurisdiction to another. Local units of government have zoning powers, property taxation powers, and the ability to create special development areas. Because of such highly localized variations, it is difficult to summarize and draw conclusions about local governmental laws and policies affecting high technology businesses.

These local powers are real, nonetheless. An emergent high technology venture that has left the owner's basement and is looking for its first real location will have an encounter with local government; local zoning laws will establish where it can locate and what it can do there. Sewer connections to the business will also be a local affair, as will refuse collection, police protection, and roads. For high technology businesses, the most immediate interface is with local units of government. Often, if these relationships work well, the venture will grow in the same location. If not, it may amble on to another place before it has heavily invested in a community. That makes this earliest interface with local government critical for state governments as well. Businesses satisfied with their local governments are more likely to stay in the state.

The Federal Presence

When there is high technology support by the national government, we find it nested within other initiatives, programs, and departments. The federal government has not aggregated its budget allocations into a category of high technology expenditures. And that is completely sensible, considering the illusory nature of this activity. Detection is a matter of locating particular programs with technology support elements and sorting through these programs to find high technology support within the broader array of federal programs. This is aided, however, by the efforts of the Clinton administration to highlight programs the president believes demonstrate the value of high technology. Thus far, support for high technology has not been a partisan issue, although the potential for partisanship abounds especially as it concerns investment policies and tax treatment of capital gains. Instead, House Speaker Newt Gingrich's general endorsement of high technology extends to his social policy of arming the impoverished with personal computers. To one degree or another, the parties in office have made positive statements about the value of high technology businesses. Verification of this comes from the plant-site speeches of major officeholders.

The beginnings of any concerted high technology policy in the United States have been recent, mostly in the post-Vietnam war era and stemming from other Defense Department stimuli as discussed in Chapter 2. Even so, there has been no explicit high technology agenda in our time. However, since the collapse of the Soviet Union, there has been a "swords to plowshare" transformation as defense technologies have been converted to commercial technologies. A classic but still contemporary example of that conversion can be found at the Oak Ridge National

Laboratory. There, the formerly highly classified Y-12 plant made parts of the nuclear warheads and optical devices for the Strategic Defense Initiative "Star Wars" program. Researchers outside Oak Ridge can now use the plant's concurrent engineering center, prototyping facility, and ultraprecision manufacturing equipment. The plant itself is being converted to a center for defense and manufacturing technology. At the Department of Energy's Lawrence Livermore Laboratory, industry parties can now gain access to scientists, as well as some of their research.

With this example of federal sharing of facilities in mind, it is appropriate to review just how such assistance can take form. This is not to say that all these devices are used, only that there is a considerable range of theoretically available tools, some of which may be in actual use. Federal government support for high technology can come in many forms. We briefly describe them to present at least a partial overview of their nature, which may lead to a more complete understanding of their potential.

One form is direct *financial aid* to businesses. This includes grants of the type we will describe later that fund high technology programs. There is also a potential role for the federal government to underwrite loans to private high technology businesses, either partially or totally. These are the primary investment devices.

The federal government also has an abundance of *tax powers* that affect high technology businesses. In very basic form, there are deductions from corporation taxes and tax credits. A three-year extension of the R&D tax credit has been signed into law, aimed at creating incentives for private investment in research and development and in new technology firm formation. Tax deductions for investments and expenses derived from new technologies in the workplace are the other main form of federal tax assistance. A suggested investment policy device is to have no capital gains tax on capital that remains in a business startup.[1]

The federal government is a *buyer of high technology services.* Its contracting powers are significant and must certainly be considered as an element of potential federal support for high technology firms. The federal government is the immense and sole customer for defense firms. There are also the standout examples of government purchasing of the vast array of technologies that make the space program possible. But there are also less dramatic occasions when a federal purchasing agent might buy a batch of less proven but technically superior microprocessors. This action serves to support technical achievement by directing tax dollars to innovative companies, even if some risk might be taken. Although seemingly conservative and incremental in its buying practices, the federal government's scale of operations is so large that the lucky high technology suppliers often enjoy a long and prosperous relationship with government customers.

At first sight, the interests of U.S. purchasing agents and high technology firms may seem incompatible. The federal purchaser is driven by reliability, standardization, and volume discounts. He or she wants long mean-time-between-failure parts that are invariable and have the lowest unit costs. The high technology firm may offer untested, customized products that are hemorrhaging with high investment costs that the producer wants back. Yet we see in the chapter how some federal programs form conduits between government and high technology that end up mitigating some of these differences.

The traditional mode of high technology support was in the area of *federal research projects* through federal agencies. Unlike aid for private companies, the tried-and-true method to develop technology was to pay for it in the budgets of federal agencies doing research. Most of the major science-related federal agencies have active research programs directly funded by the U.S. budget. These agencies also subcontract for private-sector research.

But there are nonfinancial forms of assistance as well. One of the most notable ways is to put the *spotlight on new technical companies*, a practice observed more frequently in election years and as important trade agreements loom. Media stories appear about these "breakthrough" and "cutting-edge" businesses that cast them as Jacob-like redemptions of American competitiveness. The effects of such attention can also linger. The companies can also be buoyed by promises of federal contracts that seem to understandably coincide with political campaigns.

An additional way the federal government fosters the business of high technology is when administrations and office holders *provide information bridges* to the media. A president may have assistants pass on the word about good high technology examples from one U.S. region or another. The same is true for Senate and House members, who pass on news about successful high technology businesses in their districts.

Yet another form of support is to provide a "good office" for negotiations between parties on high technology matters. For example, a member of the House of Representatives might offer to negotiate a dispute between two local governments over where a new software engineering firm may be located. This may also take the form of a U.S. senator's offering his or her staff assistance as a high technology ombudsman. Senate and House staff members may offer to help guide the technology entrepreneur through what is often a maze of permits and licenses.

Finally, there is the technology knowledge base available among congressional staffs, especially those of members on science and technology committees. There is often sufficient expertise here for technology founders to obtain advice on the feasibility of their ventures, along with some indication of their commercial prospects.

Recent Federal High Technology Support

The preceding has shown how significant the federal government can be in fostering high technology. Realistic thinking suggests that not all these possibilities can become actualities, so we take a much more narrower range of support by the federal government. The programs we describe reflect the political policies and preferences of different administrations, as well as the particular macroeconomic conditions of the times—which are the constraint side of the help equation. For example, high interest rates impede capital investment for technology firms; that is a major macroeconomic factor.

Each administration puts its stamp on technology programs. For the Reagan and Bush administrations, it was indirect defense spinoff civilian technology development, while the Clinton administration has more actively redirected defense spending toward civilian technology programs. We turn now to the programs most recently offered by the Clinton administration. The objective is to show what is being emphasized and, by default, what isn't.

Policy pronouncements by the Clinton administration have supported the notion that improving international competitiveness rests at least in part on fostering both high technology ventures and the use of advanced technology in established businesses, especially manufacturing businesses. Federal support thus consists of a certain level of moral support for the idea and benefit of high technology, along with varied but modestly funded direct grants, educational programs, technology transfer activities, and other "good office" gestures. The result is not so much an inferno as a kindling.

The administration cites high technology products as crucial for international competitiveness. These products account for 35 percent of world manufacturing. The problem is that the United States has increased its high technology exports only about 4 percent over the last decade. Imports of high technology products have grown and the result is a one-tenth decline in the U.S. share of this market—from 40 to 36 percent. A comparative competitive assessment done by the administration showed that the United States was behind in 13 critical electronic technologies. In areas such as optical information storage, multichip packaging systems, and display technologies, the United States was judged as lagging.

The administration's technology program, as part of its National Economic Strategy, is intended to accelerate technology development and commercialization. It also aims at dissemination of technology into the general business sector, shifting expenditures from defense to civilian uses, institution of a "dual-use" strategy of applying civilian technology to military usages; fostering partnerships between the private and public agencies; creating an economic environment favoring capitalization of technologies; improving technology education; and targeting some assistance to small and medium-size businesses.

Defense had long dominated the federal government's research and development budget, and we have long been bound to the technology streams connected with defense. One of the side effects has been a rigid acquisition process that lengthened production cycles and increased costs at exactly the same time that the private sector was being driven to reduce its costs and product commercialization schedules. A parting of the ways happened as there was less mutually beneficial technology development. According to administration policy pronouncements, the Defense Department began to experience a lag in innovation capacity, and that prompted more federal action.

What are some of the specific ways the federal government supports technology development? The answer varies and changes. As mentioned earlier in the chapter, federal government support for high technology reflects the changing priorities and administrations between 1940 and today. In Chapter 2, we demonstrated that federal support early on was for the specific purpose of winning a war. This restricted many different streams of technology (an example being basic research on genetics), but it pushed forward certain defense technology initiatives. Thus federal support was often in the form of substantial contracts for technologically superior weapons; aircraft and nuclear weapons serve as sterling examples. The defense business continued to fund private technology research well into the 1960s, when universities and some private business, as well as hundreds of thousands of dissenters, questioned the purpose of this funding stream.

Today the tenor of federal support has a much more pronounced emphasis on civilian uses of technology. The Clinton administration, and to an extent the Bush administration, put the transformation of the defense industry in motion with a series of programs that are outlined next. The nature and purposes of these programs should help provide an understanding of how far one of the basic drivers of high technology has changed direction.

Major Federal Programs

Conversion of defense industry machinery to the service of nonmilitary economic objectives has been a theme, to one degree or another, of all presidents since Dwight Eisenhower. But the theme did not have a real mandate until very recently. The key program for doing this in the Clinton Administration is the *Technology Reinvestment Program* (TRP).

The program has three key areas for funding: technology development, technology deployment, and manufacturing education and training. It addresses both technology development and the creation of a mechanism to make defense conversion longlasting.

There were 2,800 proposals submitted for the offer of $472 million in matching federal grants, a response the Clinton administration considered enthusiastic. Ideas were submitted by all kinds of firms from all across the country.

The fund provides grants for what are considered to be dual-use technologies—those that meet both consumer and national defense needs. It also provides support for defense firms seeking to make a successful transition to commercial markets.

A primary source of support for existing businesses seeking to develop new technologies with a potential for market success is the *Advanced Technology Program* (ATP), administered by the National Institute of Standards and Technology (NIST). The program acts as an accelerant for technologies that result in new products, processes, or services. It does not fund new product development per se but it does alleviate some of the development risk of new products by helping push very innovative products and services to market. The purposes of the program are to commercialize scientific discoveries and refine manufacturing methods, so products and services break into the market quicker, with lower cost or higher quality. The broad scope of the program means many different types of innovations can be considered.

This program is not solely a federal subsidization of innovation, no matter how well demonstrated the benefits might be. Characteristically, more than half the private research and development costs are paid by the recipient organization.

Selection of projects is based on the technical and business merits of competing proposals. The technical review by scientists and engineers is followed by an economic review by business experts without a vested interest in the outcome. Proposals are rank-ordered and awarded grants on the basis of widely publicized criteria.

In deciding where to focus its funding resources, the program solicits and selects research areas from industry, professional, and academic sources. Thus, funding decisions incorporate views from outside the funding agency itself. Ideas for possible funding are measured against four criteria: potential for impact on the U.S. economy, cutting-edge important technical ideas, strong industry commitment to share costs, and an opportunity for the ATP to make a difference in funding work that leads to progress.

Companies of all sizes have received ATP grants. Increasingly, the program has aimed at expanding cooperation among the business community, academia, and government. This may take the form of strategic alliances among developers and potential users.

The ATP has received recent attention. The NIST reports that while larger technology firms know about its program, the ATP is less well known among smaller firms, so an outreach effort has been conducted to get these entrepreneurial firms involved. The intent is to spread funding to promising new technology enterprises.

Since it began, the ATP has contributed to 177 technology initiatives. The areas funded are quite varied, including materials, manufacturing, electronics and information systems, biotechnology, chemical processing, and DNA analysis. Thus, the ATP has covered most of the technologies we have defined as high technology. It represents a prime funding source for new and established high technology companies. A future direction for the ATP is to make it truly national scale, which it is not now. The pilot was slated for a $200 million appropriation. The

Clinton administration sought to increase support in certain, well-defined programs, as well as maintain open competition for all types of technology innovation. A $750 million 1997 appropriation was submitted for this program.

In our review of state high technology programs, the importance of ATP for many states was in evidence. Now it is available for urbanized and rural states alike, and has become something of a fixture as a funding source.

In all, the ATP is structured in its purposes and decision criteria, and is most suitable for firms that can clearly demonstrate the potential value of their innovations. Because of the partial funding element, the ATP cannot be considered a deep-pocket funding source, even if the capital requirements are very low. It is a well for the intrepid, not the faint-hearted.

The *Small Business Innovation Research* (SBIR) program is certainly one of the most original to be developed. It is more than a one-shot investment, and it is directed to business development rather than just research and development support. It also has a successful track record.

Small Business Innovation Research is a federal program that has actually built small technology businesses, not just innovations. The program is aimed at creating more opportunities for small businesses to participate in research done by the federal government. As with the ATP, the congressional intent was to use the program to increase U.S. competitiveness in the world market. It has the purposes of improving domestic innovation, making use of small business to conduct federal research and development programs, and improving opportunities for the disadvantaged in the technology area, as well as increasing private-sector innovation commercialization.

There are 11 federal agencies that provide SBIR grants. These are the larger federal agencies, which assign a small percent (around 1.5 percent) of their noninternal research and development budgets to the program. The Small Business Administration (SBA) administers the overall program, and each participating agency administers the details of the program. This causes different types of grants to be funded, which is a consequence of different participating agency priorities.

There are three phases to the SBIR programs and small businesses must compete for grants in each phase. The first phase is a feasibility study, the second is the core research and development work, and the third is commercialization of the product or service.

As an example of a successful SBIR project, NASA contracted with a very small technology business, Cybernet Systems Corporation of Ann Arbor. The firm developed a teleoperation device with a force feedback (a hand controller) that would let astronauts position the remote robot arm. Since the original project, the company grew to 25 employees and it has worked on 16 SBIR projects.

Aside from these three programs, there are a number of bodies that have been put together by the U.S. government to solve technology problems with a social basis. A particularly interesting form of support has been through industrial consortia. These groupings of businesses are intended to ensure that government-supported projects are driven by market needs. Among these is the Clean Car Initiative, a partnership among the Big Three automakers and the federal government. Some federal basic research laboratories cooperate with Big Three's applied engineering laboratories in the design of vehicles that are three times more energy efficient than current vehicles.

Another federally supported consortium is the U.S. Display Consortia, a ten firm group aimed at developing and manufacturing high definition displays. The Sematech project to develop improved semiconductor manufacturing technology is another consortium. And there are many smaller high technology businesses that form ad hoc consortia to share resources for a mutually beneficial development. Providing that true competition emerges, the consortia can pass muster on the usual restraint-of-trade criteria. Their actions are not antitrust magnets, providing collusion does not continue if products are commercialized.

These federal support programs for high technology businesses are both formal and informal. The formal SBIR and ATP programs are quite prescriptive about what will be supported, yet there are many informal means of support at the federal level, including congressional staff assistance for new technology businesses and plain old-fashioned publicity for venture launches, as mentioned earlier.

A major federal power—fiscal and monetary policy—has implications for the cost of capital for small entrepreneurial firms, yet the effects of these policies are felt by all businesses. The effect on high technology businesses is not readily determined, though it certainly exists.

Protecting High Technology Ideas and Information Gathering

Having surveyed the executive and legislative branches, an appropriate shift is to the judicial branch, where consideration turns to a great concern for high technology—protection of innovation.

International high technology business is not solely a matter of what individual nations do to support their own businesses. There are other issues, and they stem from doing any kind of international business, but they are especially applicable to high technology. They may also be points at which international high technology business cooperation will fail.

The first of these is *technology transfer*, a longstanding sore point in our relations with other industrial countries. There are American engineers who could see, rivet by rivet, the Soviet duplication of the Douglas DC-3 aircraft. Complaints about the Japanese copying of patented American electronics was a constant event in the immediate postwar period. Those complaints have been displaced by claims of patent infringement by other Pacific rim nations.

The transfer of intellectual work, as technology transfer is, can happen if the conditions of transfer satisfy both parties. A protected technology has been licensed to firms outside the innovator's country. This has worked, but it is the quiet side of the patent protection issue. The challenges have and will come in those instances when patent-holding firms have already lost their protection. These are especially awkward if there is a history of stolen ideas and an apparently legitimate effort is made to obtain a new protected invention.

The second point of international friction has to do with patent law. Patent law has not been shown in this book to be a major obstacle for domestic ventures, but it can be on an international level. Each country establishes it own patent laws, and there are differences in the level of protection accorded the originator. It is safe to at least suppose that patent protection will be a barrier to operating high technology businesses overseas if U.S.-patented procedures are exposed to unlicensed replication and competition. Microsoft has been particularly outspoken about the failure of other governments to protect its technology. The same might also be said for technologies devised in other countries and left unprotected in the United States market.

The need for greater protection of technology ideas has been recognized by international agreement. The Uruguay round of trade agreements provides for considerably stricter international agreement on protection of patents, copyrights, trade names, and other intellectual property. Since U.S. technology firms lose billions of dollars in royalties and product sales to worldwide intellectual pirates, they have welcomed the results.

A particular lament expressed by the Clinton administration has been that there is simply no systematic gathering of data on the technology advances and applications of other countries. As this is corrected by the creation of a federal international technology monitoring function, the potential is at hand for discovering piracy. Suspicious innovations can be reviewed as duplications of U.S.-protected intellectual property, so the single point database can be a first step toward discovering a patent violation by an unscrupulous technology thief. Thus there is identification of another federal role, that of monitoring invention internationally. Together with other federal roles, this emerges as a picture of what could be a varied and active federal influence on technology development.

How Far Should Government Go?

The federal role thus far has been that of federal direct aid, especially in the defense conversion process. But some have argued that this is a form of corporate welfare. According to Ralph Nader, these entitlements are not deserved and will not produce the kind of critical mass that will lead to well-disbursed economic benefits. This really leads to the question of what the costs and benefits are of public investment in private technology operations—a question that has only been nudged one way or another with very modest federal programs, an essentially separate public and private strategic planning process, free-market advocacy, and trade protectionism, all producing little resolution. But that may be the most powerful feature of our particular fashion of pluralism. There is a role for the federal side to both encourage high technology and limit its excesses, and the balancing between the two is so constant and case-specific that being dogmatic is both impractical and short-sighted.

In the final chapter, we will draw together our research on ventures and established businesses, along with our review of federal and state governmental programs. Then some overall conclusions will be offered on the state of high technology business in the United States and its prospects for the future.

Note

1. Karl Vesper has written on this topic in *Entrepreneurship and National Policy* (Pittsburgh: Heller Institute for Small Business Policy Papers, Carnegie-Mellon University, 1983). Also recommended in this book are initial seed and early-stage equity financing, debt financing of entrepreneurial firms, more contracting with entrepreneurial companies, greater patent quality and patent protection, the teaching of entrepreneurship, federal recognition of the discipline of entrepreneurship, more knowledge sharing of entrepreneurship, and prizes and recognition for entrepreneurial success stories.

19

CONCLUSIONS ON HIGH TECHNOLOGY BUSINESS

In this chapter, summary comments are made on high technology business in the United States. Those lead the way to additional comments about the future for high technology business, representing a future for future firm speculation.

The basis for this summary is the findings from our research conducted for this book and research done by others. We also have an opportunity in this closing chapter to draw on the insights of some theorists, the purpose being to integrate the parts and phases of high technology business that until now have been view separately.

The Idea of High Technology

At the outset, we laid out a conceptual basis for looking at the activity commonly known as high technology. This started with a definition of high technology and led to a look at the classification systems for high technology businesses. As science and engineering-based businesses, these enterprises occupy part of the business spectrum, not all of it. The standard classification systems have identified high technology businesses as those with a higher percentage of technical workers and a greater percentage of expenditures on research and development than other businesses. For many reasons, this is not a satisfactory way of deciding who is in the business of high technology and who is not. These are only two defining criteria of the many available for classifying businesses. They have arbitrary cutoff points for separating high technology businesses from their traditional brethren. Lastly, using simple percentages does not really get to the heart of the business.

Those in high technology see themselves as either producers or users of high technology. There are companies like Intel that build the microprocessors and biotechnology companies like Agracetus that use technology for discovery. There are more technology user firms than producer firms; this is a consideration to remember in debating what high technology business is.

Popular usage has complicated the definition problem. As used by the media, the term *high technology*, or *high tech*, has come to mean all kinds of things in architecture, design, clothing, furniture, communications, and manufacturing. Such usage tempts trivialization, but it is a fact of our time. The business media are so enamored of the concept that most leading national publications have high technology sections, reports, columns, or special reporters.

The financial community, likewise, has contributed to popularization by grouping stocks as high technology stocks. When Hewlett-Packard had its initial public offering in 1947, it was identified as an electronics stock, not a high technology stock. Yet about a decade after the term emerged in the mid-1960s, Wall Street started grouping the computer, electronic, and other industries into a high technology classification.[1]

Our approach to defining high technology businesses was not to be prescriptive by using formulas, but to call science and engineering-based businesses high technology businesses. Existing definitions based on workforce type and research and development intensity did help us select some business classifications for our research, however.

The Nature of High Technology

Although technology has always been a part of human organization, there was an acceleration of scientific discovery in the twentieth century. The roots of contemporary high technology are found in World War II, when national governments and private industry joined resources to intensively develop the fields of aerospace, nuclear energy, electronics, medicine, and communications. These areas fulfilled their promise for both good and bad. They also changed the way technology was developed, from small private laboratories to large team, private-public research and development exercises.

When we visited several present-day high technology companies, we saw many examples of the modus operandi of team orientation, project management, and assiduous use of the scientific method in research and development work. Scientists and engineers lead projects in high technology work, but the business disciplines set the objectives for the scientists and engineers. There is a formula for conflict here because the sciences seek truth whatever the cost, and business occupations seek asset productivity with minimal cost. The successful high technology companies recognize the conflict, but are still able to align the organization to meet agreed-upon purposes.

Another aspect that was quite apparent from our observations of day-to-day technology work and from our interviews with founders was that technology development is quite deliberate and methodical. Francis Bacon described inventions as coming from either science—the discovery of scientific principles, a knowledge base for inventiveness—or from happenstance—essentially lucky inventions. The days of happenstance are gone in technology businesses. Even the most basic science operations—the biotechnology companies—are directed toward specific discoveries and are evaluated for progress all along the line by the founder, founder team, or parent firm. These technology firms aimed at applying science rather than discovering it. In the research and development laboratories, the cry was not "Eureka" but "Whew" as laboratory budgets were approved and work could go on.

Contemporary high technology companies are not as insular as they once were. Good technology and technological insight spread quite quickly. Some of this has to do with the networking done by technological workers, such as the women engineer support groups. There are many professional associations and frequent conferences where scientists and engineers learn about each other's work. The technology companies themselves urge to have their technical staffs publish applications. Finally, there is quick dissemination of ideas through the numerous journals and the Internet.

In the past, the main ways technology was spread was through imitation and by adaptation. Much more recently, these methods have become embodied in the concept of technology transfer. The federal government itself has stimulated technology transfer through defense-industry conversion programs so the spread of information is well sponsored. It is now a major facet of the high technology business.

In this information age, the process of doing technology has become immensely socialized. Try to think of a technical development that emerged from a single person. Everyone in the organizations we observed constantly referred to the cooperative nature of technology work. These are team-oriented businesses, in both horizontal and vertical directions. Multidisciplinary teams produce innovation, and these teams take or pass on the product or project to other teams.

Modern, nonaccidental technology businesses require a fusion of science, engineering, economics, and management. These skills are not within the scope of a single individual. That's part of the reason why an entrepreneurship, if successful, always becomes a business organization with numerous members. You would be hard pressed to find another activity that combines so much from so many different disciplines to be successful, yet that is the case with high technology companies.

To arrive at any summary of technology companies as good or bad is to arrive nowhere, because they are both, having produced a means to end humankind through nuclear war and a way to prolong it until the limits of biological destiny are reached. What can be said is that the impact of high technology is enduring and the future is promising.

Venture Capitalists, Founders, and Entrepreneurships

The promise of high technology companies does not rest on the their technology alone. It depends on outside resources as well. Though many ventures are founder funded, those that have made the major leaps to prominence have drawn financial support from others. Venture capitalists are sources for startup and, much more frequently, expansion capital. The venture capitalists we interviewed indicated that they looked for business acumen over the product or service. This demonstrates that successful technology companies must have business and technical skills to attract outside investment.

Many times the venture capitalists form an extended team for the internal teams, offering advice and personal support. They can also be the cold hand of death for the struggling venture, passively declining requests for help and then standing aside as the venture collapses.

The founders of technology firms have themselves been found to be as varied as individuals can be, with a few important common threads. They are persistent and energetic. The founders in our survey group were college educated and had prior business experience. Although there was no provable connection between the business performance of the venture and the educational background and business experience of the founder, education and business experience are valued in the high technology field. Business experience especially helps in getting aid from venture capitalists. Educational achievement of founders builds credibility with employees of technology companies.

Team founding of technology ventures is now quite common, with and the team members each bringing some special expertise to the venture. Team founding was not linked to better or worse business performance, however.

The origins of technology ventures was also reviewed. High technology businesses in the survey group have some origin relatedness, but not a high degree of it. They frequently locate in the same area as the origin organization. The products they produce are often similar to those of the incubator organizations. The entrepreneurs seem to have a fondness for their former companies and organizations, which in very many cases were large organizations. The relative degree of relatedness, either high or low, did not have performance consequences, however.

High technology business ventures write business plans, but they do far less written planning in marketing and human resource areas. The existence of written plans does not have associations with how well ventures do.

Strategy matters for high technology ventures. The strategic decision of where to locate the business is important. Closer proximity to customers and suppliers has business performance consequences. But marketing strategies did not produce performance differences, nor did the degree of cooperation between marketing and research and development functions.

As far as the environment for technology ventures was concerned, this too has an impact on technology ventures. The fewer capital and access to technical skills barriers there were, the better the performance.

The three main venture subgroups studied—computers, biotechnology, and electronics—had significant differences among themselves in virtually all aspects of their operations, leading to a dismissal of the idea that high technology ventures are a singular and coherent business grouping.

Technology ventures differ considerably from established technology companies in such fundamental areas as marketing, products, customers, pricing, promotion, production, financial and planning operations, personnel, and leadership of the organization. So drastic are these differences that any single technology company evolves to a point where it no longer resembles itself. This adds importance to considerations of business maturation and industry life cycle in understanding how and why high technology firms act as they do.

Established Businesses

Established high technology companies—those in business for five or more years—were another source of research. Established high technology companies in our respondent group provided information about their employees. About half of each firm's workforce is technical workers and well over half have college or postgraduate degrees. They are paid above average salaries, but they are not unionized. The group has seen increasing employee productivity in recent years. Within these broad strokes, there are some telling characteristics of these workers. They work together as teams on projects that lead to products. Technology workers seem to stay on the job if there are equity sharing arrangements and opportunities for recognition. There is also a wanderlust factor, especially in the software business. If employee expectations are not met or if firms falter, employees have no reservations about moving on to other companies.

As with the ventures, the CEOs of established companies were experienced in business and were well educated. They preferred to use a team-oriented management style. The consequences of any of these characteristics was not demonstrated in business performance results, however.

For the established high technology company, it is important that the strategy fit the environment of the business. Companies that had this fit did better.

In considering environmental and organizational factors, the firms in our group had flat structures and were organized by customer or product line. They were very aware of competitors and believe they react faster than their competitors. They also believe they are growing faster than their industries. The firms feel a strong need to adapt to technological change. These companies tend to use many suppliers. Joint business ventures are not a mainstay business mode for these technology firms. There is also a very high degree of research and development spending. There was no linkage found between environmental or organizational factors and performance, though.

Regarding marketing operations, our responding firms use few distribution channels. They have several hundred general customers and a handful of more important customers. The firms have national-scope markets and product management is an important aspect of marketing. Sales and promotion practices are similar to those of conventional firms. If there is a most consequential area of marketing, it is pricing. Practices in pricing are extremely important, with firms that price by customer demand doing better than those that use other guidelines. Firms that customize products also do better than those that don't.

Although there are high technology businesses all across the United States, there are about 20 areas of concentration where there are broad bases of university and community resources for high technology business development. All states have some kinds of economic programs that high technology firms can utilize, and a few states have very comprehensive programs specifically for high technology businesses. The federal government has an array of grant and technology transfer programs as well.

Some Summary Views

Let's move to some general observations about the businesses we have been collectively labeling high technology.

Time Is of the Essence. This is a business grouping that has as its main threat, not competition, not escalating costs but time itself. Time is an enemy because innovation has never relented and is now on so many parallel courses that technology consumers and users have constant technology improvement as an expectation for all goods and services.

So speed is the milieu of success. Speed to market is not just a slogan but a mandate for technology businesses. That is obvious in the close watch on competitors and the speed with which new products are brought to market. As Alexis de Tocqueville said of the inhabitants of America, "The lands of the New World belong to the first occupant; they are the natural reward of the swiftest occupant." It is not always so— "The race does not always go to the swiftest . . . but that's how you bet" might be the Runyonesque corollary.

A Business of Fashion. There appear to be waves of popularity for the more specific components of high technology. In the 1950s, it was an age of the atom and, perhaps to another degree, an age of aviation as aircraft doubled their speeds and range. From the 1960s through the 1970s, it was evidently the space age, where imaginations leapt and resources were cast into the skies. From the 1980s through today, it is the age of the computer. These are well known depictions of time periods, but they serve to illustrate that no single form of technology has dominated the postwar period. Some forms have become popular and then fallen, such as the aviation sector, but they are replaced by other forms. Looming ahead could be another replacement in popularity, perhaps an integrated information age or a biotechnology revolution. Whatever the form, it will bob on the rising tide of technological development in all fields.

The Dualities. Not everything we have encountered about the technology businesses gleams. There is tarnish on the goblet. We refer to the labor base for these businesses, especially in the software business. Although we have not closely looked at this in this study, we have noted that there is, among some major software companies, a tendency to exploit the energies and intellects of young, new programmers. Among a few companies, and these are well known within the industry, programmers are overworked to the point of burnout. Burning the candle at both ends and being satisfied by the temporary informality of bucolic settings, these programmers work on the big projects constantly, far past the 40 hours per week that workers a generation ago considered a fair exchange for their wages. Exhaustion creeps up on them and takes full possession within a couple of years.

We've shown how little unionization is a part of high technology, and perhaps this explains why people are free to extinguish themselves with overwork. But probably it has more to do with the culture of the business itself and who it attracts—those who want intellectual challenges and who consider themselves professionals. All the CEOs we met wanted dedicated people, and none of them wanted burned-out people, yet they did not see that they had created the circumstances that transform dedication into exhaustion.

Another edge of the business that takes the luster off high technology is the exportation of software development work overseas. This has happened already among a few larger software companies. If it increases, many of the expected benefits of high technology, such as an improvement in U.S. competitiveness in the world, will greatly diminish.

All of what are call high technology businesses rest on the discoveries of science. These discoveries have resulted in the possibility of very suddenly propelling ourselves to betterment or oblivion. The fact that businesses organize themselves around these advances in science is no guarantee of either outcome. The business organization is simply the tool of the human tool-making activity. This is not to relieve those who run business organizations of responsibility. It is, in contrast, a way of reinforcing it. Technology businesses could end T. S. Eliot's vision of a world with a bang or a whimper. That, most fundamentally, is reason to direct the technology business, because this activity can produce instant catastrophe or slow environmental poisoning.

On a less Olympian scale, the promise of technology business may not even be realized for those whose work depends on it. "Technological advances in production have different effects upon employment in different industries" is the finding of Schecter (1993, p. 72). This investigator conducted two hundred case studies on the effects of technological innovation and process improvement, and found they did not always lead to an economic upgrade for all. The innovations and improvements often eliminated high-skill jobs and led to more low-skill jobs. On the other hand, the shift to PCs and word processing in offices had the opposite effect.

Historical reference to some of the negative effects of technology business can be found in the arms race, where defense workers in essential aerospace industries were paid over prevailing market wages yet did not produce the goods for private consumption that could have slowed inflation. Their wages crept upward for decades and inflation rose as well. And since so many technology businesses were also heavy energy users, the oil crisis of the mid-1970s exacerbated productivity problems with this sector.

Celebrants of high technology point to its growth and wealth. They use high technology's consumption of science discoveries and the importance of entrepreneurship as links in young technology firms. As an example, Keely and Tabrizi (1995) have argued in favor of the point. They note that technology ventures have been seen as contributors to the decline of American competitiveness. Critics say new technology companies fragment industries yet fail to become major contributors with the net effect of weakening industries. Looking at 49 ventures in nine industries, their study found few cases in which the venture could be considered to compete with or damage a former employer. In many cases, both the venture and former employer benefitted.

Costs of Building Technology Companies

Views about the costs and benefits of technology companies have been quite varied. An article by John S. McClenahen appearing in the September 16, 1996, issue of *Industry Week,* reported that Tenneco CEO Dana Mead believes the U.S. needs to grow 50 percent faster (about 3 percent a year instead of 1.5 percent) to pay for the investment needed in high technology. The costs would be in training and retraining workers, improving education, and paying for career-changing trauma. In this view, high technology businesses do not emerge from nothing but instead from indirect investments that should be repaid as the firm goes into business operations.

The investment costs of high technology might be recouped by gains by the business sector itself. Writing in the July-August 1996 issue of the *Harvard Business Review*, W. Brian Arthur argued that technology companies are evidence that the classical economic theory about diminishing returns no longer applies. He held that firms could get only so far ahead of competitors, only so efficient before falling back to the production efficiencies of the industry itself. The technology revolution has supported a new theory of increasing returns through which technology advantage leads to even greater advantages, letting leading companies maintain or increase their lead. This view about the special nature of technology—a nature that makes it different from other businesses—is not widely held. Business strategy researchers such as Michael Porter (1996) caution against considering technology businesses as exceptions to the rules of the business game.

The productivity gain from newer technologies can benefit U.S competitiveness, but it can also close the productivity gap with other nations. Other national firms will be able to become more productive quicker by using U.S. technology. That's the point of an article in *The Economist*, March 18, 1995, "Of Strategies, Subsidies and Spillovers." Productivity gains of other countries using imported technologies from the United States may eradicate much of the comparative advantage of U.S. technology firms.

The benefits and costs of high technology businesses as they operate and as they reach their destiny, producing either social betterment or doom, are quite clearly still in debate. This effort has aimed at producing information about how they do this but not if they should operate. Being normative about this business activity is the domain of public policy making.

Coming to Terms with the Term

Singling out businesses under the term *high technology* out of other businesses has the apparent power of separating all businesses into two groups. But the cleaver is dull and slices only gelatin. The distinction only separates a portion of the mixture, since most businesses are not high technology. A mass remains where the businesses stay apart only as long as R&D expenditures and employee base keep them apart. Without these factors, organization structure, marketing, and production practices remain and the high technology businesses are like other businesses.

High technology firms are an amalgam of organized activities. Picking out only two measures—research and development and technical workforce—obscures many other activities. There are thousands of other dimensions a definition could be based on, but for one reason or another they are not. Such activities and measures as economic value added, payroll cost in relation to sales, patent production, product innovation, and manual vs. automated production could be used. The R&D and technical employment criteria are insensitive to company yet we found that situations change in technology firms of different sizes.

An associated problem with the definition of high technology is that of tagging the firms themselves—pinpointing exactly which firms among the multitude is SIC-code dependent. As we saw with the larger technology companies, the major firms span many SIC classifications, high technology and traditional alike. Even smaller companies can have mixed high technology–general business operations and end up being misassigned to one category or another.

Commonalities

Although there are many ways that high technology businesses are different, what can be said about similarities? The individual businesses and ventures we examined have similarities. Though these are not as given to measurement as the research and development and the technical employee share, they are significant. All the businesses *add substantial value*. The top end of this is the semiconductor manufacturing business, which crafts very inexpensive raw material into very expensive microprocessors.

These companies are the *first users of scientific discovery*. They are the commercializers of discovery and are the link between science and business. They are the big idea developers. In every research field where there are commercial possibilities, high technology business is the delivery vehicle of science to the consumer.

Another commonality is *change*. No matter what the high technology field—electronics, aviation, or test instruments—the businesses are constantly replacing the current with the new. They depend on new knowledge and are the first adapters of it. It could be said they do the science on science by setting up the marketing laboratory for scientific discovery. They are also very responsive to competition and are thus environmentally focused—as much as any other business grouping. Like sharks, high technology companies must keeping moving to survive, and must move through turbulent waters, a situation of instability and complexity. They must be highly adaptive to their environment, rapidly mutating—a real contrast to the natural world, where life forms have evolved slowly. Sometimes the only thing that doesn't change is nothing. Products are dropped. New contract workers come and go. Owners and executives are replaced because they are highly marketable or they become restless.

By and large, these companies tend to be more *service oriented* than product focused. This is not because we encountered so many software consulting businesses. We found the larger firms tend to deemphasize the iron and promote the support functions, especially as TQM principles and a marketing orientation began to assert themselves, even in the staid electronic components and semiconductor manufacturing businesses.

Though high technology companies can be fixated on profits, the firms we saw were not exclusively so. They view themselves as classically *technology organizations*, providing tools to help people master their environments. The articulated purposes of the technology business leaders resound with this direction. Put another way, these were technology businesses not businesses with technology.

This is also a business area that is essentially *not bounded by size*. Take the extremes—individual software consultants and Boeing Aircraft. They are both in the technology business. This is not a business activity that starts at a certain threshold, as a utility business does because there is a need for an economic scale of operation, and ends at another threshold. It is as inclusive as any business can be. It also needs to be underneath a big tent, since growth is so commonplace among these firms.

The young firms that start in high technology tend to *remain in high technology*. There are exceptions such as G.E., but most remain true to their initial cause. This is another point of commonality.

The Nature of the Business

Definitions of high technology are unsatisfactory because they are not really definitions of high technology. They are simply ways of separating some forms of business from others. A definition of any business needs to do much more.

Our view is that the definition must give a sense of what the business actually does. Our suggestion is to say that *high technology businesses are science-based businesses engaged in the business of commercialization of innovation or the utilization of innovation*. This captures firms that either use technology to produce goods, such as simple liquid pumps as Isthmus Engineering and Manufacturing does, or produce technology even if the production process is simple, as in the case of biotechnology companies where most of the work is cerebral.

This definition allows us to become free of the relativist ratios whereby a company could be high technology one year and not another. It also avoids the problem of a company that is high technology in the employee ratio but not in the research and development ratio.

It must be acknowledged that this definition does not have the precision to send one descending marble one way and another a different way, but it does fill a need. There is no accepted definition now, and that has been a problem because public conceptions of high technology are loose, especially if you are a state economic development director and you want to boast about your business environment. We have been told by some state economic directors that marginally high technology businesses—for example, printing—are high technology businesses.

Deciding what companies qualify as high technology from the myriad existing businesses misses the essential nature of these businesses. They are future businesses. They represent the way that work will be done.

Indeed the present businesses were formerly high technology businesses in almost every instance. When it might be argued that coal mining or agriculture are not high technology businesses, remember that coal mining uses high technology equipment to map, extract, weigh, and load coal. Agriculture has been changed fundamentally by technology in every aspect, from genetically altered seed to fertilizers, equipment, and storage. All businesses have been affected by the products of high technology because these products keep changing the businesses themselves.

What high technology companies have most in common is their *substitution effects* on other businesses. By either replacing the businesses themselves or by altering the way they do business, the science and engineering-based technology companies are a force for change in all industries.

Successful high technology firms have *rapid growth* as a common reference. Phillips (1991) has examined just how much faster high technology firms grow. During the last decade with complete data (1976–1986) high technology businesses grew faster than all industries combined. In 1976, the businesses employed 7.98 percent of workers and in 1986 it was 8.44 percent. About 7.7 million workers are employed in high technology. High technology firms expanded employment by nearly 40 percent while overall employment grew by 32.2 percent in that decade.

Phillips also demonstrated that smaller high technology firms are responsible for most of the growth. Companies with fewer than 500 employees created nearly 40 percent of the high technology jobs during the period. About a quarter of the jobs in high technology are in firms with 500 or fewer workers. As described by Phillips, business services, computer manufacturing, instrument manufacturing, and nonelectrical machinery are responsible for most of the growth. This again emphasizes the importance of small firms as feeders for the fast-growth high technology area.

Yet growth in this business area produces real differences in the way the businesses operate. They become more different. A new embryonic electronics company has very little in common with General Electric; it may have more in common with a biotechnology venture. These high technology firms are quasars of the business universe, growing fast, then scattering everywhere.

Differences

Different types of technology firms are sufficiently different to say that considering them all together does little to help us understand this diverse group of businesses. Other than the general commonalities we listed, high technology businesses are really a mixed lot. Some, like biotechnology companies, hover close to doing pure science. Others, like electronic

Table 19.1   Classification and Evolution of Technology Firms

Time	Product Firms	Process and Service Firms
Single business	Entrepreneurial firms (single commercialized inventions)—e.g., Reining Intl.	Entrepreneurial firms (examples are small software consulting firms found in local business directories)
Multi-business	Established product firms (Hewlett-Packard)	Established process firms (Amoco)

component manufacturers, are really just factories with no science and all production. There are processing industries, such as petroleum refineries and product businesses, such as computer manufacturers. To obfuscate the distinction between process and product—something useful for organizational theorists—is to accede to the idea that product and process do not matter. That is a conclusion that cannot be reached in this look at high technology businesses.

Even as a single stream of technology, there are many consequential differences in the nature of these firms. If the fairly narrow scope of biotechnology research is considered as an example, there are firms involved in human biotechnology and others in plant biotechnology. Some biotechnology companies do pure research on genetic engineering possibilities and others produce laboratory equipment for molecular biologists.

As a way of depicting what high technology companies do and how they evolve, Table 19.1 illustrates these ideas and has some example firms. The depiction reflects the reality that high technology business that produce products tend to evolve along that line and that process and service businesses tend to evolve along a separate line, although crossovers are seen in the very large technology businesses.

Final Observations

When Alvin Toffler wrote *The Third Wave*, he observed that the third wave started in the 1950s, along with the emergence of Route 128 and Silicon Valley. The third wave is a wide-ranging social, political, and economic transformation that will lead to more democratization and individual freedom. Toffler (1980) is a positivist, viewing this wave as an advance over the second wave (industrialization), which was itself an advance over the first wave (the agricultural revolution). While showing how high technology in the third wave can improve daily life, Toffler also prominently features its counterpoint—those catastrophes of the environment and nuclear weaponry that have engendered a citizenry of opposition. He calls this citizenry the "techno-rebels," a broad-based social consciousness that tempers technology. These are a force within the third wave.

Our first step in this book has been to acknowledge and define the high technology business. The second was to measure it, and the third was to evaluate it. The first step has been taken. The second step was begun with this book and other research. The third step has not been taken, other than by ethical philosophers. Evaluation is a matter left open, as indicated earlier. Science has not been done on science-based businesses. We do not know what the costs of high technology businesses are, but we know these are future businesses.

What we have come to is the revelation that high technology businesses are fragmented, not unified. They are not distinctive as a set of businesses, yet they are nevertheless special in qualitative measures. They appeal to us. They are enterprises of the mind. They are clean and attractive immigrants. They grow fast. They produce things that help other businesses grow. Their employees are highly educated. All these are attractions of this loose collection of science and engineering based businesses.

High technology is a term that is becoming far less useful as popularizers such as economic development agencies push the definitional border to the extremes. What remains is the core of what is done by scientists and engineers who work in the business. That is the commercialization of tools to better humankind.

Prospects for High Technology

The prospects for high technology businesses can be looked at in both the short and long term. In the short term, there are a few perspectives on investment value and corporate dominance. Most analysts believe high technology stocks are good investments as corporate spending on computers, communications, and related equipment is expected to grow through the next decade. Vendors most likely to gain from this are proven global sellers and those with histories of effective partnerships. Some trends likely are; a shakeout in the PC industry. Other trends are: home computer gadgets; the Intranet, a market for connectivity equipment; Internet companies; online services using the Web; domination by Microsoft and Intel; digital video disks; wireless data services; and leading-edge semiconductor manufacturers.

Further predictions about the breaching and centralizing tendencies of high technology are presented in Jerome Glenn's November–December 1993 *Business Horizons* article, "The Post Information Age: New Horizons for Business and Education." Glenn's essential point is that the division between humans and technology will become more fluid in the future. People will be transformed into cyborgs with the ability to communicate and share information through worldwide networks.

Recognition of the limitations of technology in business transactions is a subject of consequence now and in the future. Author Faye Rice, writing an article entitled, "The New Rules of Superlative Service" in the Autumn-Winter 1993 issue of *Fortune*, notes that many companies are using high technology to better serve their customers. According to Rice, the technology should not overwhelm the customer. The successful technology and traditional companies will produce customized products and services and get feedback from customers.

While predicting continued worldwide domination by U.S. computer, semiconductor, and other high technology products with technology spending expected to increase by 70 percent through 1999, Simmons, Fischer, and Whitelaw (1995) raise the issue of greater unemployment resulting from restructuring. U.S. firms are looking to foreign markets for continued growth.

These views all present visions of high technology business as continuing to grow in the near future. That is a well-accepted idea now. For the long-term, though, the vector of growth is not as clear. For the first time in history there are many rapidly developing high technology areas—biotechnology, computer memory, automation, and the integration of technologies. Despite the known and expected progress, there could be something new afoot that a business will quickly form around to deliver to a market. This business used to be dominated by singular invention but it now produces a mechanism for invention collections—groups of related innovations that emerge from disciplined teams of scientists and engineers in the research and development laboratories of established companies and in the companies of their own minds as they push out on their own with the opportunity and privileges of technical entrepreneurship.

This has been a tale of two cities, the emerging firms and the established firms. Technology itself has produced the best and worst of times with the constant and imposing promise of much more of either.

Note

1. The high technology business sector has become so much a part of investing that mutual funds invest in the group. Some investors are Alliance Technology A, Fidelity Select Technology, Invesco Strategic Technology, John Hancock Global Tech A, and T. Rowe Price Science and Technology. These funds specialize in high technology investment and diversify the high technology risk. The funds have performed well with five year total returns between 17.4 and 29.3 percent. An August 28, 1995, *Business Week* article stated that technology stocks outperformed the Standard & Poor's 500 stock index. But high technology stocks collapse as well. The mid-1980s boom in biotechnology stocks was short-lived, as the companies could not commercialize their patents. Perhaps as a sign of enduring faith in a high technology future, when Netscape made a public offering, the company was valued at $2.7 billion, even though it had not generated a profit.

References

Arthur, Brian, W. "Increasing Returns and the New World of Business." *Harvard Business Review* (July–Aug. 1996), pp. 100–109.

Glenn, J. (1993, Nov–Dec.). "The Post Information Age: New Horizons for Business and Education." *Business Horizons*, pp. 44–49.

Keely, R. H., and B. Tabrizi (1995). "High-Tech Entrepreneurs: Serious Competitors or Troublemakers?" *Journal of High Technology Management Research* 6: 1, pp. 127–143. As reported by G. Dale Meyer.

McClenahen, John S. "Grow 50% Faster." *Industry Week* (September 16, 1996), p. 20.

Phillips, B. (1991). "The Increasing Role of Small Firms in the High Technology Sector: Evidence from the 1980s." *Business Economics*, January, pp. 40–46.

Porter, M. (1996 Nov–Dec). "What is Strategy." Harvard Business Review, pp. 61–78.

Schecter, H. (1993). *The Global Mismatch: High Technology and Low Pay.* Westport, CO: Praeger.

Simons, J., D., Fischer, and K. Whitelaw (1995). "High-Tech Karma: U.S. Dominance in the Digital Age Should Keep Paying Big Dividends." *U.S. News & World Report*, August 21.

"Of Strategies, Subsidies, and Spillover." *The Economist* (March 18, 1995), p. 78.

Toffler A. (1980). *The Third Wave.* New York: William Morrow and Company.

APPENDIXES

A

VENTURE COMPANIES
Notes on Research Design, Methodology, and Sample Selection

This note establishes the research design for our work on high technology ventures. The central purpose was to develop a research method that evaluates the ideas concerning technology startups. The second section of the appendix discusses the research design for established companies.

This venture model emerged from the basic research questions we posed in a survey about operations and performance along with a review of past research and theory and an analysis of the venture capitalist survey discussed in Chapter 4. The research was aimed at testing a revised high technology venture performance model as described in Chapters 5–8, instead of several existing models, since existing models are incomplete explainers of venture performance. The new model proposed that venture economic performance is related to prior experience of the founder, relatedness of the venture to the incubator organization, industry characteristics, industry life cycle, entry barriers, customer-supplier proximity, entry level-business strategy, organizational adaptation to environment, and functional integration of business units.

Research Objectives

Our research objectives were to learn more about the nature of high technology business ventures and to discover how more successful ventures are distinguished from less successful ventures. These broad objectives are captured in three research questions:

1. Are there any factors or variables that significantly influence high technology venture performance? If so, what are they?
2. Which of these factors or variables have the greatest impact on performance and which have a lesser impact?
3. Are the factors or variables that have been shown to have an impact on performance comparatively the same or different across high technology business groups?

These specific questions are addressed by the research effort, which is detailed in subsequent sections of this appendix.

Research Design

The research for this evaluation of high technology venture performance followed a nonexperimental design. It was based on observations of high technology venture operations as reported by the founders or managers of high technology ventures.

The approach used was a survey research study, in which sample ventures were studied and observations were made on a one-time basis after venture founding. Statistical associations among the dependent and independent performance variables were derived through various data analysis techniques.

Survey research was employed on a sample of high technology ventures drawn from a commercial database of high technology companies. The unit of analysis was the high technology venture.

Selection of High Technology Venture Business Subgroups

Not every high technology business is represented in this study. Since there are at least 50 potential three-and four-digit SIC high technology business classifications, it would have been difficult to adequately investigate all business classifications, so a more narrow but practical alternative was pursued. Three classes of high technology businesses were selected. They are biotechnology, electronics, and computer and computer-related ventures.

One reason for this selection was that these are separate and distinctive groupings. Biotechnology businesses use different technologies, employ different professionals, and market different products than electronics and computer-related businesses. While there is technology overlap between computer and electronics firms, there is a large technology gap between these two types of firms and biotechnology firms, which justifies their consideration as distinctive businesses.

Another reason for selecting these categories was that they have received more public attention than lesser known high technology activities, such as new materials fabrication or chemicals. The highly visible high technology sectors have stimulated public-sector economic development issues. Implications for public policy are addressed at the conclusion of this study, so it was appropriate to use this as a selection attribute.

The businesses were selected because they constitute a sample of different aspects of high technology. Computer businesses exhibit the fastest growth in the three subgroups. Electronics businesses have a large workforce. Compared to these two groups, biotechnology businesses are not nearly as production intensive and rely more on advanced technical instrumentation and processes.

These business categories include 3 of the 28 Bureau of Labor Statistics–defined high technology industries. Other major industries in this group are aircraft, communications equipment, plastics, and computing and accounting equipment. For reasons mentioned, these were not included in the study.

This selection of different high technology industries, and the analysis of economic performance, origin, and environment information, helped establish if heterogeneous high technology industries have homogeneous performance and environmental characteristics. That, in turn, helped determine if high technology ventures themselves are a set of related or unrelated businesses. Put another way, it was a test of the worth of a separate classification for high technology business apart from other business ventures.

Sample Selection

A simple random sample was drawn from a single database, the entire population in the Venture Economics Inc. database. This database is the list of Venture Economics clients that reside in the company's computer files.

Venture Economics was chosen to supply the sample for two reasons. First, it is a centralized source for the names and other data on high technology ventures. Second, the organization is an authority on sources and financing for new businesses and leveraged buyouts. The organization also publishes *Venture Capital Journal*, a 28-year-old trade journal for the venture capital industry. It maintains a number of databases that contain information on over 1,000 financing sources and over 8,000 companies that have received startup, development, or buyout funds. The organization also publishes and distributes *Pratt's Guide to Venture Capital Sources*. Used by entrepreneurs, venture capitalists, investment bankers, accountants, institutional investors, corporations, financial advisors, and lawyers, this is a directory of approximately 790 active U.S. and Canadian venture capital firms and provides a primary link between those who seek venture capital and those who can provide it.

Receipt of venture capital was not a requirement for inclusion in the survey; however, the ventures included are likely to be interested in venture capital. The names of the ventures, their chief contacts, and their addresses were supplied by the research office of Venture Economics.

The list of ventures was derived from three Venture Economics industry codes. The first group was code 2000, Computer-related, which includes computers, peripherals, computer services, and software. The second group was code 3000, which includes electronic components, batteries, power supplies, electronics-related equipment, and analytical and scientific instrumentation, the third group was code 4000, Biotechnology, which includes human medical biotechnology, medical biotechnology, agricultural and animal biotechnology, and industrial biotechnology.

A special advantage of the Venture Economics data source was that venture subgroups were already placed in the appropriate venture classifications of interest. This relieved us of the task of assigning subgroup classifications.

In the course of acquiring information about these venture subgroups, we directly contacted the director of the Small Business Administration to identify sources of high technology venture information that could be used to supplement the Venture Economics database. The director suggested an additional source of high technology data, the *Corp Tech Directory of Technology Companies*. (For our survey of established technology firms, described in Chapters 10–16, the *Corp Tech Directory of Technology Companies* was a crucial database.) The *Corp Tech Directory of Technology Companies* was useful in providing key personnel information, sales histories, and profiles of ventures supplied by the Venture Economics database. That was the reason it was used for the technology venture study.

In order for us to obtain participation from those surveyed, the venture respondents were told that they would retain their anonymity in any reporting of information. The intent was to overcome the presumed reluctance of some participants concerned with dissemination of information. We believe this made respondents more comfortable in providing answers.

Sample Size and Response Rate

There are generally recognized advantages to larger rather than smaller sample sizes. Increasing the sample size reduces the possibility of Type I and Type II errors. There are practical disadvantages to larger samples, though. Cost can be a constraint, as can time. Given the empirical imperative for a larger sample size and the practical constraints favoring smaller samples, it is obvious that a compromise sample number is the solution.

Our initial sample of candidate ventures was 2,194. That was the number of firms provided by Venture Economics, based on random selection in the three aforementioned categories and the requirement that the firms be formed before 1985 (to ensure at least a five-year perfor-

Table A.1 Distribution and Responses of Venture Companies in Survey

	Initial Sample	Responses
Computer and computer-related	1,454	59
Electronics	576	25
Biotechnology	164	17
TOTAL	2,194	101 returned of 1,503 mailed

mance history). The distribution and responses of the ventures were as shown in Table A.1.

The ventures were contacted by mail and provided the questionnaire with a cover letter explaining the purpose of the research. Each cover letter was personalized. Before the cover letter was finally drafted, it was reviewed by Mr. Kirk Stantis, a direct-response marketing manager. The last step taken with respect to the cover letter was to obtain permission from Nova University (where author Eric Bolland was writing a dissertation) to use its letterhead. This step was accomplished with the assistance of Dr. Ronald Needleman.

There was further screening of the venture lists, with firms incorporated before 1960 eliminated in an effort to reduce the effects of different historic economic conditions. The CorpTech Directory of Technology Companies was used for this. A mailing of questionnaires went to 1,503 firms, after factoring in a percentage for firms that would not respond because of address changes. Most of these companies were probably out of business because attempts to reach them by phone produced not-in-service phone messages.

The initial response rate was 6 percent, so in order to reduce the nonresponse bias, random phone interviews were undertaken. The eventual overall response was 101 replies, for a 6.7 percent response rate. Because of the low response, nonresponse bias is still a factor in this study, although attempts were made to minimize that. Response rates for subgroup responses per mailed questionnaire was not tracked; however, the response rate for replies per ventures provided by Venture Economics was as shown in Table A.2.

Data Sources

To obtain descriptive information about the financial performance, environment, and origin characteristics in the three high technology sectors, we devised a written questionnaire. This was the main source of data for the study, along with results of our venture capitalist survey (see Chapter 4). Secondary source materials such as business and trade publications were also used on occasion—for no more than 10 percent of the ventures.

Table A.2   Response Rate by Industry

Computer and computer-related	4.1%
Electronics	4.3%
Biotechnology	10.4%

These secondary materials consisted of annual reports, IRS filings, Dun & Bradstreet reports, product information sheets, catalogs, and media information kits. The secondary material was used mainly to confirm information, but in a few cases also to obtain sales information.

Data Collection Instrument

The data collection instrument was a crucial piece of research. In survey research, items that are vague or not identified as important lead to nonresponse or unreliable responses. Consequently, the data collection instrument must adhere to standards that maximize survey participation and authentically represent information offered by the participants.

The initial questionnaire was reviewed by three experienced survey research designers so as to identify and correct any vague or misleading questions. Suggestions offered by these designers were incorporated into the questionnaire. The questionnaire was also reviewed by two academic-based researchers who regularly assist with postgraduate university-based research. The questionnaire was developed over several months and evolved from numerous reviews and discussions.

The initial questionnaire was pretested on ten Madison, Wisconsin, area high technology venture firms. In one case, the questionnaire was orally reviewed. The pretest uncovered no problems in answering the questionnaire and the respondents' minor organizational suggestions were accommodated in the final version.

The final questionnaire collected information on the economic performance of each venture, how it was organized, who founded it, the background, and the founder's environmental conditions that presumably affected the venture. In order to improve the response rate, sensitive information about economic performance was asked in the last part of the questionnaire instead of the first part.

There were 41 questions asked, which fell into the general categories as shown in Table A.3.

Each of the general categories was addressed by no fewer than four questions. The intent was to collect information on several aspects of each category using multiple questions.

Respondents selected mainly one answer from several response categories. There were scaled response questions in the strategy category, as well as an open-ended performance and organization question. Since

Table A.3   Questions by Subject

Category	Number of Questions
Entrepreneur	6
Origin	6
Venture environment	12
Strategy	7
Organization	4
Venture performance	6
Total	41

there were no suitable earlier questionnaires on this subject, the survey was constructed for this particular research process.

Questionnaire Distribution and Screening

A first-class mailing was assumed to be more effective because it was believed that recipients would not categorize it unsolicited fourth-class mail. Furthermore, undeliverable first-class mail was returned for address correction and remailing. To improve the chances of a reply, a postage-paid business return envelope was included.

The questionnaire was printed on inkjet letter-quality plain white paper. To avoid the look of a mass-mailed piece, the questionnaire was prepared with a Courier "typewriter" font.

Using the *Corporate Technology Directory,* which showed if firms were divisions of larger companies and thus excluded, we sorted the returned questionnaires by subgroup and checked for completeness. (Corporate-sponsored ventures were excluded because they do not incorporate the important variable of individual, or group, venture founder, thus the entrepreneur factor is absent. They also embark on business operations with distinct advantages over independent ventures, making a controlled investigation much more difficult.)

There were several possible approaches to research design for this study. Both the potential and the limitations of the design must be considered, along with the freedom and constraints in approaching the population under study. In this study, the primary population constraint was in identifying and locating less successful ventures. The original intent was to select ventures that had failed and compare them to ventures that had succeeded. This was not feasible, however. There were no data sources for failed high technology ventures, which made it difficult to locate the founding entrepreneur, along with the venture records. It was also true that, even when failed ventures were located by trade publication sources or other sources, there was reluctance by the founder to share data about the failed venture. The founders often preferred to let their mistakes remain buried. Coaxing the founder was often necessary to obtain performance information.

Because of this limitation, there would not be sufficient data on failed ventures. It was therefore decided to not make any a priori selections of ventures based on performance, but instead to establish degrees of performance after the venture data were collected. This necessitated a shift from an extended case-study analysis to a survey-research project involving larger numbers of ventures.

The research instrument also has a longitudinal component in that it included the effects of time on the ventures studied. The time period was five years from the start of business operations. This is certainly a period of great dynamics for a venture, often with high business mortality.

The research environment was a field study rather than a laboratory study. Laboratory study was impractical for business policy research on high technology research. A field study was necessary to collect data and observe venture performance within its operating environment. That complexity cannot be replicated by laboratory conditions.

Research Environment and Data Analysis Considerations

This study was formalized, in that it was aimed at the testing of hypotheses. It is a descriptive study of the characteristics of high technology ventures and the hypotheses were descriptive in nature. A descriptive study facilitates the development of a profile of the venture population. There were estimates of the proportions of venture populations with given characteristics. There was measurement of the degree of association among variables, but there was no intent to establish definitive causal relationships among the independent variables. Posed this way, the study was primarily descriptive, but included use of inferential statistics.

Response Processing

The returned questionnaires were reviewed and a coding sheet was developed on which numerical scores were recorded for all responses and for incomplete responses. The SAS statistical program was utilized to analyze the responses.

Profiles of Respondents

There were 17 biotechnology respondents. These firms were mainly involved in applied biotechnology research; several were in immunology, and others were in genetic research. No single classification of a type of work applies to all in the sample. Average sales for these ventures ranged from zero to over $6 million per year. Five of these ventures achieved no sales in the first five years. It is entirely possible that some or all of these firms may have pursued patent development goals instead of sales goals.

The staff growth for the biotechnology firms ranged from 4 to 190 people.

Compared to the biotechnology ventures, the electronics firms were greater in number, with 25 responding firms. Some of these firms produced electronic devices; others developed electronic subassemblies, and one built robotic equipment. Average sales were from a low of zero (in one case) to over $1 billion. Staff growth ranged from only 1 person to 694.

Fifty-nine ventures fell into the computer and computer-related category. There was considerable diversity, including an industry-leader computer manufacturer and many smaller software development companies. It is not possible to state beyond this what the typical computer and computer-related firm was. Sales averages were from zero to $20 million and employee growth was from none to approximately 500.

Data Analysis

Hypothesis testing using nonparametric statistics was the data analysis method, along with the venture capitalist survey. An importat criterion for this type of hypothesis testing is the independence of the samples. It is assumed that the samples are independent of each other as freely operating high technology firms. This assumption was supported, based on selection of ventures that, on the basis of information from the *Corp Tech Directory of Technology Companies*, exhibited separate organizations and management.

The process of selecting a usable data analysis technique often involves considering a variety of methods and selecting one that is both feasible to accomplish and rigorous enough to reveal important relationships. This study was no exception, and the first step was to consider the several customary analytical tools.

Regression analysis could show relationships between the dependent and independent variables, but the sample size is too small to undertake regression analysis. In some cases, the number of observations of independent variables was ten or fewer. It is possible to generate a regression line in such cases, but the line is not likely to depict the true relationship between the independent and dependent variables. The lack of ratio data also inhibited use of regression analysis.

Smaller sample sizes lead to higher standard errors. A higher standard error reduces the capacity of the regression equation to predict values and serve as a useful analysis technique.

Another useful technique involves parametric tests, but it, too, had to be rejected for this study. The study uses only nominal and ordinal scales; it is not possible to do t and f tests and so nonparametric tests are appropriate.

This leads to the consideration of nonparametric methods of two independent populations. Among the advantages of nonparametric methods is that they do not make as many assumptions about population distribution, and the assumptions that are made are less constraining than with parametric methods. Moreover, they are generally only slightly less efficient than their parametric counterparts. This leads to greater applicability and more general, broadly based conclusions. Finally, nonparametric techniques are useful when it is not supposed that the sampled population is normally distributed. They are useful with small samples, a feature that should be helpful if there are only small samples.

The application of nonparametric techniques is especially justified in this case because of the relatively small samples sizes and the reluctance to make an assumption that the sample population was normally distributed in this study.

While there are advantages to the use of nonparametric analytical techniques, there are drawbacks as well. A primary disadvantage is the inability to assess the strength of statistical relationships. The technique measures direction of such relationships, but does not quantify the extent. This makes it difficult to evaluate the relative importance of two different variables. Since the high technology venture performance model here proposes that numerous variables affect economic performance, that limitation is of concern, especially as it regards the second research question.

Statistical Tests

There are several possible statistical tests that could be used. For nominal data, the descriptive statistics may be percentage and mode, and the inferential statistics could be the chi-square or binomial test. The sign test could also be used. It is an easy and versatile test of significance.

For ordinal data, the descriptive statistics are typically the median, and the inferential statistics, the Mann-Whitney U, Friedman two way ANOVA, and rank-order correlation.

In this study, all these tests were considered for employment in data analysis. The exact tests are related to each of the variables and the nature of the data. This information is presented in the various findings chapters. The primary tests used were Spearman test and Kruskal-Wallis chi-square approximation.

These two particular tests were deemed appropriate because of the nature of the data obtained. Spearman's rank-order correlation coefficient is one of the best known coefficients of association for rank-order data, according to former Nielsen Research Center director Gilbert Churchill. It is employed for data that were rank-ordered. For example, one hypothesis speculated on a relationship between venture performance and experience of the founder. Founder experience was rank-ordered from the most experienced to least experienced. Correlation coefficients were then derived, and there was acceptance or rejection of hypotheses based on the coefficients.

The Kruskal-Wallis test is the nonparametric analog for analysis of variance. It is especially useful when several results can occur, as in the case with several venture categories. Kruskal-Wallis has also been well established as a powerful nonparametric test, so it has been utilized.

The sign test was also used when warranted. This is a well-known and generally accepted test, though it has more limited capacity to measure the strength of associations among variables than the other tests mentioned.

Statement of Hypotheses and Statistical Significance

In all the tests, the hypotheses of interest have been treated as the alternative hypothesis. Prior theory or informed speculation about high technology performance is contained in the alternative hypothesis. The null hypothesis states that there is no difference or no association between the venture's performance in the different classifications. Rejection of the null hypothesis is taken as some evidence in favor of the alternative hypothesis.

As an example, if average sales was the dependent variable and founder experience the independent variable, the null hypothesis would be that there was no correlation between venture performance and founder experience. The alternative hypothesis would be that the venture's performance and founder experience are positively related.

In most cases, one-sided rather than two-sided tests are done, since most alternative hypotheses are of the following form: ventures with a higher level of X characteristic tend to have a higher performance. Also, ventures that belong to a certain subgroup tend to have higher performance than those in another subgroup.

Two-sided tests were performed when the hypothesis of interest was of the following general form: there is no difference between ventures across different classifications. This became the null hypothesis. The alternative hypothesis is that there is a difference. An example can be constructed with the same dependent and independent variables as before. The null hypothesis would be that there is no difference in the performance of ventures related to founder experience. The alternative hypothesis would be that there is a difference in the performance of ventures as related to founder experience.

Summary

This note established the research methods used to test a revised theory of high technology venture performance. A discussion of research design, data sources, and selection of candidate ventures was also presented. A review of statistical methods of data analysis was conducted, and nonparametric methodology was adopted because of the nature of the population under study and the inaccessibility of a sample, which resulted in a small sample size.

B

RESEARCH NOTES ON SURVEY OF ESTABLISHED COMPANIES

This appendix describes how the survey research was done for established companies. In the first part of the book, our database consisted of 101 high technology ventures. In the second part, where the research results are reported on high technology companies that have passed their venture period (a demarcation point we have designated as being five years after the business launch), another separate set of companies was analyzed. This set comprised 233 high technology companies. This database was made up of businesses that had been in operation for five or more years.

The research objectives for established companies were to develop the business characteristics of the firms and to discover if there were any relationships between business performance and these characteristics. The research design was again nonexperimental. Performance data was collected for a three-year period from 1993–1995. The independent variables were considered to be the business characteristics and the dependent variables were performance measures. The association between the variables was evaluated to draw conclusions about business characteristics and performance.

Sample Selection

The database for established companies was compiled from firms listed in the Directory of Technology Companies detailed in the resource guide at the end of the book in Appendix D. This source is frequently cited as the leading directory of high technology companies. Although there were other sources of technology company information, this source was the largest and most comprehensive.

A sample of 10,000 firms was taken from the Directory. The sample contained high technology firms as defined in Chapter 1. This was a relatively large sample but it was deemed necessary because of the need to have a wide range of different types of companies in the responding group. There was also a need to have a sufficient number of respondent companies to conduct the analysis. As mentioned though, the sample itself is not representative of the entire population of high technology companies in the United States.

The 10,000 firms was a random sample of firms from the Directory's complete database of technology firms. Mailing labels for each of these firms was provided to the authors. A control set of labels was also provided.

Data Collection

A written questionnaire was prepared, along with a cover letter explaining the purpose of the research. The questionnaire was pretested with five technology entrepreneurs to help correct wording and interpretation problems. The pretest revealed no wording, length, or interpretation problems with the questionnaire.

The questionnaire was done in booklet form with a prepaid first-class panel on the back of the booklet. It was entitled "1996 Technology Company Survey." The questionnaire and letter went into an outer envelope with an address window. The package was sent first-class mail to the 10,000 firms. The package was individually addressed to an executive at each one of the firms. Instructions for completing the questionnaire were a part of the booklet.

After we accounted for nondelivered, returned mail, our net outgoing mailing was 9,800. There were 233 usable replies for a response rate of 2.4 percent, which was low and introduces the consideration of a nonresponse bias. This is a caution for the reader about making generalizations about all high technology firms based on this data.

There were 65 questions on the mainly closed-end, self-administered survey. The responses were categorical in certain ranges. Nominal, ordinal, interval, and ratio data were collected. The questions themselves fell into the broad categories of background information, human resources, management, market characteristics, distribution channels, product characteristics, pricing, new product development, and market strategies. These categories cover the span of mature technology company activities.

As an example of questions within the categories, the "background information" category asked if the company was privately or publicly held, whether it was a unit of another company or an independent company, where the company was headquartered, the primary products, and

SIC codes. When SIC code data was missing, it was completed using the Directory of Technology Companies.

There were a few open-ended questions that were used to capture possible responses where the researchers were not confident all the likely response choices were adequate (e.g., What primary strategy the company followed in the last three years?).

Performance data that was collected was annual sales, number of innovations per year (defined as a commercially feasible breakthrough technology), patents held and new patents per year, total return on assets to stockholders per year, number of employees per year, as well as annual market share. Sales data and employee numbers tended to be reported the most frequently and could be verified so these two measures were the predominant performance indicators used in the analysis.

The returned surveys were counted and checked for completeness. When possible, missing data on sales performance was completed using the Directory. Any other missing data was coded as such.

Completed questionnaires were entered onto an electronic spreadsheet and analyzed with the use of SPSS. The same person who did data entry also did the statistical tests. Responses were checked against the technology venture database to ensure no duplication between the earlier venture group and this established company group. Sales data reported from a random sample of twenty companies was checked against the sales data reported in the Directory to help establish if sales information was being consistently reported.

The SPSS program was used to develop all the descriptive and inferential statistics.

Data Analysis

The analysis of the data was done in way similar to that done for the technology ventures. (See Appendix A.) The principal test for statistical significance was the chi-square. The strength of association between the performance (dependent) and the characteristics (independent) variables was mainly Pearson's chi square. For hypothesis tests, the procedure used in the venture group was applied. Null hypotheses were generally rejected at the .05 level. During the hypothesis testing, the number of cases in each cell was noted to avoid small cell number problems and the incorrect interpretation of statistical significance. The results of analysis of variance were displayed in contingency tables for the interval data.

Sample Composition and Characteristics

Some characteristics of the sample as a whole are described here. The largest number of firms in the database, about a fifth of the total, were business services. Typically these were software firms. The other businesses well represented were industrial and commercial machinery and computer equipment; chemicals; fabricated metal products, as well as the business category of measuring, analyzing, and controlling instruments; photographic, medical, and optical goods; and watches and clocks. Remaining companies were widely dispersed among the other business classifications.

Other details about the methods and results are more fully detailed in Chapters 9–16. In these chapters, the results of the analysis is presented along with interpretation of results.

Other Research Sources

The survey research of established high technology companies was not the only source of information used to discover the characteristics of high technology companies. Because of the complexity of organizations, especially high technology organizations, survey research alone cannot capture many aspects of organizational life. As a remedy, other information was used. An additional source of information, often mentioned in the chapters on established companies were sixteen high technology companies that the authors had direct personal access to. This provided a much less formalized but equally useful way of doing first-hand observation. In most of these instances, the CEOs or managers were interviewed on site. This helped enrich much of the survey research information. The on-site visits occurred between 1987 and 1996. They involved biotechnology, electronics, custom engineering, automation, and computer software firms. The mode of collecting information was most often done by asking open-ended questions and by touring facilities. Included in this number are the four technology company profiles in Chapter 3 as well as two high technology companies where author Eric Bolland worked. This additional research can best be described as qualitative research used to supplement the quantitative research used to supplement the quantitative data collection through survey research.

As a result of the survey data, personal experience and other on-site visits and business associations, the research on established high technology companies is a research effort of multiple dimensions.

C

HIGH TECHNOLOGY
SUPPORT IN
OTHER COUNTRIES

Worldwide High Technology

High technology business is now a worldwide phenomenon. Owing to the very rapid transfer of technology breakthroughs—itself a function of high technology activities such as telecommunications—and to the internationalization of higher education, many nations now have many productive high technology sectors.

A review of worldwide high technology demonstrates, if nothing else, that national support is quite varied. Some industrialized countries have organizations that have advanced particular forms of technology above all other countries while other countries have led the way with breakthrough technology-transfer machines or well-coordinated government-industry projects. Exploring what is occurring in other countries also reveals the great variety of technology organizations and their relationships with governments.

As we know, much of what develops into high technology emerges from ventures. Our contact with academic colleagues such as Robert Brockhaus at St. Louis University, who has taught entrepreneurship around the world, verifies the view that entrepreneurship is a very similar exercise no matter what country it happens in. The problems of having a viable concept, a business plan, and adequate capitalization are shared throughout the world. High technology, in its entrepreneurial form, is similar throughout the world as well. Much of the education occurs in leading research universities in the United States. New advances are quickly communicated through the global media.

High technology long ago left the domain of the individual genius. It has also long since left the province of a handful of postindustrial nations. The somber apex of high technology—the nuclear club—has doubled membership since the postwar United States, England, France, and Soviet Union. Advanced technologies, especially factory-floor automation, are in place in Japan. And China, Pakistan, and India have joined the nuclear club.

Differences among the countries in high technology investment can be partly indicated by relative research and development spending. Research and development expenditures for the top 200 foreign companies recorded by Standard and Poor's Compustat Services is one usable indicator. Although not an exact measurement of high technology intensity because it excludes resources devoted to new company formations, the indicator is nonetheless revealing of the technology predispositions of leading companies, if nothing else. There are considerable differences among countries. Japan has led the way with about 41 percent of spending directed to research and development. West Germany followed at 18 percent, with Britain at 9 percent France at 8 percent, and the Netherlands at 6 percent. Expenditures drop off considerably among the ten other industrialized countries listed. U.S. companies were not included in this listing, but when expressed as a percent of pretax profits for all companies, U.S. firms reported 3.4 percent of research and development as a percentage of sales.

Following is a glimpse at high technology organizations in three countries—Russia, the United Kingdom, and Japan. These countries have very different ways of dealing with the development of high technology. Much of that originates from their very different historical experiences: Russian defense conversion, the economic stagnation of the United Kingdom, and the reemergence of Japan. Just how these circumstances shaped their current status is reviewed in each section.

High Technology Support in Russia

It is quite apparent that high technology is seen as having a substantive role Russia's economic revitalization. As with the United States, this is the result of defense reductions.

One small but telling indication of this change is the publication of the *Russian Defense Business Directory*, intended to help U.S. businesses identify facilities they may want to explore for possible investment. Ninety-one such enterprises are identified in the directory.

The greater evidence of the importance of high technology in Russia is the Conversion Law. In March 1992, this law took effect, with Boris Yeltsin remarking that it "states [that] the main principle of defense conversion is to use the high technology capability of the defense industry to manufacture products capable of competing on the world market." Yeltsin's view reinforces the linkage of high technology with the defense sector. In his world, the most advanced part of that industry will create opportunities for global competition, a theme of U.S. presidents as well.

The Conversion Law, representing the cleaving off of the Russian defense sector to facilitate conversion, is a profound change in the state's economic orientation. It means the transfer of people and resources from the defense industry to the nondefense industry, either by serving industrial or consumer markets.

The conversion of former Soviet defense enterprises to industrial and consumer purposes is different from conversion in the United States. Defense plants in the Soviet Union were generally larger than their U.S. counterparts, they had few subcontractors, and they had a greater social infrastructure. These differences are significant for American firms interested in joint development of high technology. American firms, particularly defense firms, make extensive use of subcontractors. They also have not created a social infrastructure, or "company town," the way the Communist regime had. Even more important, though, is the necessity of agreeing upon mutual business goals. One apparent obstacle has been whether Russian high technology will be used to develop products for internal or external consumption. Russian leadership opinion, as evident in Yeltsin's comment, stresses global market opportunities; but American views, at least as far as business executives are concerned, favor domestic markets. Clearly, some resolution is needed to ensure Russian–U.S. cooperation.

Russia still has an industrialized infrastructure and academic institutions capable of training the needed scientists and engineers. However, a strong high technology sector has not emerged from the postcommunist state for either domestic or international markets. Should the nation complete the transition to a more Westernized economy, promising high technology areas in Russia could be aircraft airframes and powerplants, petroleum refining, telecommunications, and the chemical industry. If this happens, there will a need for more than a climate for entrepreneurship, there will need to be policies aimed not solely at defense conversion but also at innovation stimulation. The new Russian technology companies won't be old defense establishments, but completely different, home-grown science-based firms.

United Kingdom

The British experience with support for high technology support originated in the mid-1970s, as the Labour government of Prime Minister Harold Wilson directed government resources to Cambridge University laboratories. There were a number of spinoff companies formed out of these labs. Applied Research was an example of a CAD company, and Cambridge Scanning made scanning electron microscopes. Another wave of startups followed, and these were led by computer companies such as Acorn and Sinclair Computers.

The last great wave of Cambridge-centered technology growth happened in the early 1980s. This was fueled by the Business Expansion Scheme, which linked venture capital to technology starts. High technology firms were founded in electronics, medical instrumentation, computers, and basic research. Numbering as many as 400, these technology companies proved to be high in technology but low in business acumen.

Still, the Cambridge-based firms tended to be different from their American counterparts. Although the British at Cambridge open very similar types of companies, the British establishments are more technology focused and less broadly based as businesses, at least judging from advertising materials. These companies are also somewhat more insular, not having a large cadre of smaller supplier firms as in the United States.

In other areas of the British Isles, Oakey (see note 9, Chapter 7), found two types of innovating companies in Scotland and the southeastern part of England. They were subcontracting firms and product-based firms. This categorization could apply to U.S. companies as well, with contract software houses being subcontracting firms and semiconductor manufacturers being product-based firms.

There is something akin to Silicon Valley in the United Kingdom, and that is the M4 corridor from London to southern Wales. Most of the jobs are located in the cities. Here, there are many different high technology businesses, not solely software businesses. The mix of businesses is similar to what is found in the United States.

The success record for U.K.'s high technology businesses has been mixed. While there have been substantial employment gains in electronic consumer goods, there have been major job losses in aerospace and telecommunications.

Governmental support for high technology development has not been a deliberate coordination of resources. Instead, high technology firms have benefited from road, airport, telephone, and utility improvements directed to business and industry as a whole. Where governmental support did matter, it was in the area of defense spending. As in the United States, many of the M4 business launches trace themselves back to initial government subcontracting from the 1950s through 1980s.

Japan

The Japanese experience with high technology businesses is very different from the other countries reviewed here. Much of that has to do with the joint business and public-sector planning done by MITI, the central strategic planning body in Japan. The agency has helped private businesses develop technologies for the market that take advantage of the special Japanese competencies. As a result, there is a less broad base in technology and more concentration in what Japan considers critical fields. The machine industry is the example. Japan developed this industry quite quickly, and now it has a considerable share of the world market in robotics and automated machinery.

A similarity of Japan to Western nations is in its concentration of high technology businesses. Japan has a broad belt of high technology from Tokyo southeast to Yokohama. In Japan, much high technology work is done in "science cities" such as Tsukuba or the "technopolis concept" as advocated by MITI, which has pushed the development of high technology for national growth in the wake of the oil crisis. Tsukuba has a concentration of research and development facilities, both public and private—It was developed as a national project for that specific purpose. Almost all of the national research laboratories are in Tokyo, but high technology firms have settled all around the country, demonstrating the effectiveness of technology transfers and infrastructure improvements.

High technology in Japan has been marked by a concerted national policy to build and support these businesses. As such, it is a unique climate for technology development—one that has succeeded. The "science city" concept has also emerged in Singapore, China, and India. It could well be that the science city—where the academy, quality housing, and private research and development all come together—will be the eastern variant on our high technology spatial location.

D

RESOURCES FOR
TECHNOLOGY
BUSINESSES

This guide to resources for technology driven businesses briefly describes some services available to those connected to high technology, be they incipient entrepreneurs, managers and technical professionals in existing companies, investors, writers, and others with an interest in the field.

There is no single compendium of resources for high technology companies. There are, however, scattered services and products aimed at the technology business. There are other sources of assistance not directly targeted for high technology businesses, but are close enough to being useful for technology firms that they are included in this guide.

Readers are encouraged to follow up on the listings here for additional information.

There are quite a few public resources. In Chapters 16–18 we have already described the many federal, state, and local resources available for high technology companies. All states have economic development agencies, usually in the state departments of commerce, business, or development. These agencies are excellent sources of information about high technology businesses and business opportunities.

CorpTech Directories and Services

CorpTech is a comprehensive source of information on over 45,000 manufacturers in 18 industries—computer hardware, advanced materials, energy, computer software, biotechnology, factory automation, telecommunications, chemicals, manufacturing equipment, subassemblies and components, pharmaceutical, transportation, test and measurement, medical, defense, photonics, environmental, and holding companies.

Company Information. CorpTech's company profiles are extremely detailed, including company name, alternate and former names, address, phone number, 800 number, fax, e-mail, Internet home page, ownership (including the names of parent companies) year formed, minority or female ownership, high-tech operating units in the United States, annual sales, percentage of international business, employee number, employment growth rate. Also key executives with names, titles, and job responsibilities; company description including product activity, primary industry focus, and markets served; government contractor status; and product details using a 3,000-plus-product category key.

Information Collection. Information for the company profiles is gathered by professional staff researchers. A company senior executive is initially interviewed, and the results are put into an internal database. After editing, the profile is verified by the executive. Every year the information is updated.

CorpTech Directory of Technology Companies

A 6,000-plus-page, four-volume directory is the primary standard reference work on United States high technology businesses. It is found in most public libraries with major business collections. Profiles are listed alphabetically with indexes by company name, city, and 3,000 product categories. Also included is a detailed 26-page High-Tech Market Analysis with over 100 tables that overview the industry and rank fastest growing companies within the industry.

Regional Technology Guide

These directories are available for 13 regions. They are published annually in soft-cover format. The directories detail the local technology manufacturers in a format useful for job hunters and salespeople. Profiles are organized by 250 major product categories and indexes are provided by company name and city. Also included is a detailed 26-page High-Tech Market Analysis with over 100 tables that overview the industry and rank the fastest growing companies within each of the industries tracked.

CorpTech EXPLORE Database

This is a research tool with direct access to the Internet. The entire CorpTech file of 45,000+ companies can be searched by over 30 different criteria, such as company name, location, products, URL, ownership, size, and growth; or text search through product descriptions.

There is a direct link, using a Netscape or Explorer browser, to a firm's home page or e-mail, or you can access CorpTech's Website at www.corptech.com. Summary or full profile reports can be viewed and printed. Quarterly updates and a print copy directory are included.

CorpTech EXPLORE Database GOLD

In addition to the features of the EXPLORE database, this database includes mailing label capability and exporting files in comma-delimited format. The profiles can be customized to include only the data elements needed. Sorting and data analysis reports can also be done. This version also includes extended profiles providing a four-year history of changes, including ownership, executives, employees, sales, location, and products.

Comma-delimited Data

CorpTech company information is available on ASCII diskette in a comma-delimited format that can readily be imported into most popular contract managers and databases. Formats include mailing label, telemarketing, and summary and full profiles. Standard selections include states, regions, and industries. Custom selections are available.

Mailing Labels

These can be provided to target specific executives by job responsibility, in a variety of print formats, for custom and standard selections. In ASCII diskettes with comma-delimited data files.

Custom Reports

Users requiring access to the CorpTech database on a project basis can call 800-333-8036, request a custom selection and report, and have it shipped within 24 hours.

Monthly Job Growth Newsletter

Called the "Technology Industry Growth Forecaster," the newsletter overviews the anticipated job growth in each of the 18 industries covered and provides a detailed view of a specific industry each month. It is based on thousands of interviews conducted each month with senior high technology executives.

Typical Uses of Information. The array of Corptech services can be used in the following ways.

Job Search and Recruiting

Job seekers can use the directory to target a business sector that may be both growing yet difficult to identify Use this CorpTech information for:

- Strategy—selecting the type of company that offers greatest job growth and security, identifying the companies that anticipate the greatest job growth.
- Company identification—locating companies that most closely match the job seekers expertise.
- Initial contact—discovering who to contact, finding out what the company does so the resume and cover letter can show what the potential employee can contribute.
- Interview preparation— learning about the company and its competitors.

In addition, recruiters can use the company information to fill positions from search assignments and identify possible new employees.

Sales Prospecting and Analysis

Listed technology companies are often customers. They are often small enough to be approachable and often have not formed rigid supplier rules and relationships. As they grow, so do their suppliers. The CorpTech data allow sales and marketing staff to identify potential customers, create mailing lists, and plan for sales presentations. Doing market segmentation through CorpTech information is also possible.

Other Uses

For those interested in the emerging technology manufacturing sector, CorpTech offers a comprehensive source of company data. CorpTech also offers the venture capital business a comprehensive list for competitive analysis from tiny startups, emerging private companies, and hidden operating units of foreign companies to the *Fortune* 500 giants. This aids in evaluating investments.

This source also can be used for technology transfer operations. The companies with specific technologies are noted as part of CorpTech's product classification system.

To receive information:

Phone: 781-932-3100 or 800-333-8036
Mail: 12 Alfred Street, Suite 200, Woburn, Massachusetts 01801-
   1915
Fax: 781-932-6335
E-mail: sales@corptech.com
Internet: http://www.corptech.com

Public libraries with good business reference sections have the *CorpTech Directory of Technology Companies.*

Business Classification System

The types of business operations that distinguish one industry from another and one high technology business from another is found in the *Standard Industrial Classification Manual.* Produced by the Executive Office of the President, Office of Management and Budget, the 1987 publication was published by National Technical Information Service, 5285 Port Royal Road Springfield VA, 22161.

Assistance from Utility Corporations

Many utility companies, through their economic development departments, keep track of available sites for locating headquarters. Profiles of sites and buildings are available, along with assistance on utility costs, access, and energy savings programs. Some utilities publish a directory of high technology companies located in the utility's territory.

Private Commercial Developers

Commercial property developers have taken notice of the growth in high technology companies and some offer pooled office facilities especially designed for high technology ventures. These include shared meeting rooms and laboratory areas. Contact the major area commercial developers to establish if such locations are available.

Financial Resources

Directory of Computer and High Technology Grants

By Richard Eckstein, published by Research Grant Guides, P.O. Box 1214, Loxahatchee, Florida 33470. A state-by-state listing of funding organizations and their addresses and phone numbers. Also describes the types of projects that the organizations consider for funding. There is a section on federal programs, with descriptions of objectives for invested funds and eligibility criteria.

Venture Capitalist Sourcebook

By A. David Silver, published by Probus Publishing, Chicago, Illinois. Contains the names of venture capitalists, their preferred industries, and their profile.

Small Business Administration

The United States Small Business Administration is a source for business startup loans, as described in Chapter 18. There are SBA offices in each state.

Marketing and Sales Help

Certainly one of the most difficult tasks is to discover if there is a market and how sizable the potential market is. This task is one of the new firm's marketing research needs.

One source for marketing research is the major marketing research agencies themselves, listed in *Marketing News*, but there are other sources. *Marketing News* is published bi-weekly (except at Christmas) by the Publishing Group of the American Marketing Association Suite 200 250. S. Wacker Dr. Chicago, Illinois 60606-5819. Also there are advertising agencies in almost all cities over 50,000 people and some have marketing research capabilities. Advertising agencies also offer services such as the creating and placing of print and electronic advertisements.

Education and Training

Prospective technology entrepreneurs can take classes on a variety of business subjects offered by state small business development centers. The centers, usually located on or near college campuses, offer courses in business planning, financial planning, marketing, sales, international operations, and entrepreneurship.

INDEX